FUNDAMENTALS OF ELECTROANALYTICAL CHEMISTRY

Analytical Techniques in the Sciences (AnTS)

Series Editor: David J. Ando, Consultant, Dartford, Kent, UK

A series of open learning/distance learning books which covers all of the major analytical techniques and their application in the most important areas of physical, life and materials sciences.

Titles Available in the Series

Analytical Instrumentation: Performance Characteristics and Quality
Graham Currell, University of the West of England, Bristol, UK

Fundamentals of Electroanalytical Chemistry
Paul M. S. Monk, Manchester Metropolitan University, Manchester, UK

Forthcoming Titles

Polymer Analysis
Barbara H. Stuart, University of Technology, Sydney, Australia

Environmental Analysis
Roger N. Reeve, University of Sunderland, UK

FUNDAMENTALS OF ELECTRO-ANALYTICAL CHEMISTRY

Paul Monk
Manchester Metropolitan University, Manchester, UK

JOHN WILEY & SONS LTD
Chichester • New York • Weinheim • Brisbane • Toronto • Singapore

Baffins Lane, Chichester,
West Sussex PO19 1UD, England

National 01243 779777
International (+44) 1243 779777
e-mail (for orders and customer service enquiries): cs-books@wiley.co.uk
Visit our Home Page on http://www.wiley.co.uk

Other Wiley Editorial Offices

John Wiley & Sons, Inc., 605 Third Avenue,
New York, NY 10158-0012, USA

WILEY-VCH Verlag GmbH, Pappelallee 3,
D-69469 Weinheim, Germany

Jacaranda Wiley Ltd, 33 Park Road, Milton,
Queensland 4064, Australia

John Wiley & Sons (Asia) Pte Ltd, 2 Clementi Loop #02-01,
Jin Xing Distripark, Singapore 129809

John Wiley & Sons (Canada) Ltd, 22 Worcester Road,
Rexdale, Ontario M9W 1L1, Canada

Library of Congress Cataloging-in-Publication Data

Monk, Paul M. S.
Electroanalytical chemistry / Paul Monk.
p. cm. – (Analytical techniques in the sciences)
Includes bibliographical references and index.
ISBN 0-471-88036-1 (alk. paper) – ISBN 0-471-88140-6 (alk. paper)
1. Electrochemical analysis. I. Title. II. Series.

QD115.M58 2001
543′.0871–dc21 00-043577

British Library Cataloguing in Publication Data

A catalogue record for this book is available from the British Library

ISBN 0 471 88036 1 (HB) 0 471 88140 6 (PB)

Typeset in 10/12pt Times by Laser Words, (India) Ltd
Printed and bound in Great Britain by Antony Rowe Ltd, Chippenham, Wiltshire
This book is printed on acid-free paper responsibly manufactured from sustainable forestry, in which at least two tress are planted for each one used for paper production.

Contents

Series Preface

There has been a rapid expansion in the provision of further education in recent years, which has brought with it the need to provide more flexible methods of teaching in order to satisfy the requirements of an increasingly more diverse type of student. In this respect, the *open learning* approach has proved to be a valuable and effective teaching method, in particular for those students who for a variety of reasons cannot pursue full-time traditional courses. As a result, John Wiley & Sons Ltd first published the Analytical Chemistry by Open Learning (ACOL) series of textbooks in the late 1980s. This series, which covers all of the major analytical techniques, rapidly established itself as a valuable teaching resource, providing a convenient and flexible means of studying for those people who, on account of their individual circumstances, were not able to take advantage of more conventional methods of education in this particular subject area.

Following upon the success of the ACOL series, which by its very name is predominately concerned with Analytical *Chemistry*, the *Analytical Techniques in the Sciences* (AnTS) series of open learning texts has now been introduced with the aim of providing a broader coverage of the many areas of science in which analytical techniques and methods are now increasingly applied. With this in mind, the AnTS series of open learning texts seeks to provide a range of books which will cover not only the actual techniques themselves, but *also* those scientific disciplines which have a necessary requirement for analytical characterization methods.

Analytical instrumentation continues to increase in sophistication, and as a consequence, the range of materials that can now be almost routinely analysed has increased accordingly. Books in this series which are concerned with the *techniques* themselves will reflect such advances in analytical instrumentation, while at the same time providing full and detailed discussions of the fundamental concepts and theories of the particular analytical method being considered. Such books will cover a variety of techniques, including general instrumental analysis, spectroscopy, chromatography, electrophoresis, tandem techniques, electroanalytical methods, X-ray analysis and other significant topics. In addition, books in the series will include the *application* of analytical techniques in areas such as environmental science, the life sciences, clinical analysis, food science, forensic analysis, pharmaceutical science, conservation and archaeology, polymer science and general solid-state materials science.

Written by experts in their own particular fields, the books are presented in an easy-to-read, user-friendly style, with each chapter including both learning objectives and summaries of the subject matter being covered. The progress of the reader can be assessed by the use of frequent self-assessment questions (SAQs) and discussion questions (DQs), along with their corresponding reinforcing or remedial responses, which appear regularly throughout the texts. The books are thus eminently suitable both for self-study applications and for forming the basis of industrial company in-house training schemes. Each text also contains a large amount of supplementary material, including bibliographies, lists of acronyms and abbreviations, and tables of SI Units and important physical constants, plus where appropriate, glossaries and references to original literature sources.

It is therefore hoped that this present series of textbooks will prove to be a useful and valuable source of teaching material, both for individual students and for teachers of science courses.

Dave Ando
Dartford, UK

Preface

This present book is no more than an *introduction* to electroanalytical chemistry. It is not a textbook, but is intended for those wanting to learn at a distance, or in the absence of a suitable tutor. Accordingly, the approach taken is that of a series of tutorial questions and worked examples, interspersed with questions for students to attempt in their own time. In no way is this meant to be a definitive text: students who have mastered these topics are recommended to consult the books and articles listed in the Bibliography at the end.

Electroanalysis is a relatively simple topic in concept, so the first few chapters are intended to be extremely straightforward. Some aspects of the later chapters are more challenging in scope but, as students build on the earlier sections, these latter parts should also appear relatively painless.

A word about errors. I have used the phrase 'treatment of errors' fairly liberally. A few of my colleagues take this phrase to mean the statistical manipulation of data once the latter have been obtained. I have followed a different tack, and mean here those errors and faults which can creep into an actual experimental measurement. Indeed, this book is not long enough to describe the actual manipulation of data.

Perhaps I should mention a few of these colleagues. I am delighted to work with such professionals as Dr Brian Wardle and Dr David Johnson of my own Department, namely Chemistry and Materials, at the Manchester Metropolitan University (MMU). They have both read this book in manuscript form from end to end. Additionally, my friends Dr Séamus Higson of the Materials Science Centre, University of Manchester Institute of Science and Technology (UMIST) and Dr Roger Mortimer of the Department of Chemistry, Loughborough University, have also read the entire manuscript. The kind encouragement of these four, together with their perceptive and shrewd comments, have made the preparation of this book much more enjoyable. I also wish to thank Professor Arnold Fogg of Loughborough University and Dr Alan Bottom for their comments within the context of a stimulating correspondence. I have incorporated just about all of the comments received from these wise men, and I extend my heart-felt thanks to all of them. Nevertheless, all errors remaining are entirely my own.

I also wish to thank Dr Lou Coury and Dr Adrian Bott of Bioanalytical Systems, Inc. for their enthusiasm, and permission to reproduce Figures 6.16, 6.18, 6.19, 10.1 and 10.3. I gladly thank Dr Manfred Rudolph for his description of the DigiSim^{TM} program, Dr Mike Dawson of E G & G for his help concerning the Condecon program, and Dr Keith Dawes of Windsor Scientific for his help, and the permission to reproduce Figure 10.2.

Further thanks are also in order, namely to Professor Derek Pletcher of Southampton University for permission to reproduce Figures 6.12, 6.23 and 6.25 and the first two of the three computer programs presented in Chapter 10, to Elsevier Science for permission to reproduce Figures 7.14, 8.1, 8.5 and 8.16, to The Royal Society of Chemistry for permission to reproduce Figures 7.11, 8.3, 8.6 and 8.7, to Wiley-VCH for permission to reproduce Figure 4.3, and to John Wiley & Sons, Inc. for permission to reproduce Figures 3.12, 4.9, 4.10, 6.7, 6.8 and 6.28, plus the third computer program given in Chapter 10. In addition, I wish to acknowledge the following organizations for permission to reproduce further material used in the text, namely The Electrochemical Society, Inc. for Figure 8.14, Oxford University Press for Figure 7.8, the American Chemical Society for Figure 6.27, and the International Society for Optical Engineering (SPIE) for Figure 6.14.

Finally, I would like to thank John Wiley & Sons Ltd and Dave Ando (Managing Editor of the AnTS Series) for commissioning this book, the second title in this series of texts, my friends who have not seen very much of me over the past year, and not least, my precious wife Jo.

Paul Monk
Manchester Metropolitan University

Acronyms, Abbreviations and Symbols

General

A	ampere (amp)
AC	alternating current
ASV	anodic stripping voltammetry; adsorptive stripping voltammetry
AU	absorbance unit
C	coulomb
CE	counter electrode
CME	chemically modified electrode
CSV	cathodic stripping voltammetry
CT	charge transfer
CV	cyclic voltammogram
DC	direct current
DME	dropping-mercury electrode
EIS	electrochemical impedance spectroscopy
EPR	electron paramagnetic resonance
ESR	electron spin resonance
F	farad (unit of capacitance)
G	gauss
HMDE	hanging mercury-drop electrode
IR	infrared
ISE	ion-selective electrode
IUPAC	International Union of Pure and Applied Chemistry
J	joule
LSV	linear-sweep voltammetry
m	metre

MFE	mercury-film electrode
min	minute
NMR	nuclear magnetic resonance
OTE	optically transparent electrode
Pa	pascal
RDE	rotated disc electrode
RE	reference electrode
RRDE	rotated ring-disc electrode
s	second
SCE	saturated calomel electrode
SHE	standard hydrogen electrode
SI (units)	Système International (d'Unités) (International System of Units)
STP	standard temperature and pressure
TIR	total internal reflection
TISAB	total ionic strength adjustment buffer
UV	ultraviolet
UV–vis	ultraviolet and visible
V	volt
WE	working electrode

a	activity
A	area
Abs	absorbance
$\boldsymbol{B}$	magnetic field strength (flux density)
c	concentration of analyte
$c^{\ominus}$	standard concentration under standard conditions, i.e. 1 mol per unit volume
C	capacitance
D	diffusion coefficient
e	electronic charge (charge on an electron)
E	potential; electrode potential; energy
$E^{0'}$	formal electrode potential
$E^{\ominus}$	standard electrode potential
$E^{\ominus}_{\mathrm{O,R}}$	electrode potential for the O,R couple under standard conditions
E_{b}	baseline potential in pulse voltammetry and polarography
E_{f}	final potential in a voltammogram
E_{i}	initial potential in a voltammogram
E_{in}	electrode potential of a redox indicator
$E_{\mathrm{O,R}}$	electrode potential for the O,R couple
E_{λ}	switch potential in cyclic voltammetry
$E_{1/2}$	polarographic half-wave potential
emf	electromotive force
f	linear rotation rate

F	Faraday constant
G	Gibbs function (Gibbs energy)
G'	Gibbs function (Gibbs energy) at frustrated equilibrium
$G^{\ominus}$	standard Gibbs function (Gibbs energy)
H	enthalpy
H'	enthalpy at frustrated equilibrium
i	current density
I	current; ionic strength; intensity of an EPR absorption
I_{circuit}	current through a circuit
I_{d}	diffusion current in polarography
I_{D}	current at a disc electrode
I_{K}	kinetic current
I_{p}	peak current in a voltammogram
I_{R}	current at a ring electrode
j	flux
k	general constant, e.g. calibration constant
k_n	rate constant of a chemical reaction (the subscript, where indicated, gives the order of reaction)
k'	rate constant of a pseudo-order reaction
k_{et}	rate constant of electron transfer
K	equilibrium constant
K_{S}	solubility constant (solubility product)
l	length; optical path length
L	inductance; Avogadro constant
m	mass
n	number of electrons transferred
N	collection efficiency at an RRDE
N_0	RRDE collection efficiency when no homogeneous reaction occurs
N_{κ}	kinetic collection efficiency at an RRDE
p	pressure
$p^{\ominus}$	SI standard pressure
q	electric charge density
Q	electric charge
r	radial distance from the centre of an electrode's surface
R	resistance; molar gas constant
R_{cell}	resistance of a cell
s	selectivity coefficient of an ion-selective electrode
S	entropy; intensity of an EPR signal
S'	entropy at frustrated equilibrium
t	time; transport number
T	thermodynamic temperature; optical transmission (transmittance)
v	velocity
V	volume; applied potential

V_f	flow rate of solution
X	width coefficient of a channel electrode
z	charge on a particle
Z	impedance
Z'	real component of Z^*
Z''	imaginary component of Z^*
Z^*	overall impedance
Z_W	Warburg impedance
α	transfer coefficient; irreversibility coefficient
$\gamma_{\pm}$	mean ionic activity coefficient
δ	thickness of Nerst depletion region (layer)
ε	extinction coefficient
η	overpotential
η_S	viscosity
θ	phase angle between current and potential in impedance
λ	wavelength
Λ	ionic conductivity
ν	scan rate ('sweep rate') in polarography and voltammetry; stoichiometric number
ξ	extent of reaction
ρ	density
σ	electronic conductivity
τ	time-scale of observation; length of a cycle, e.g. lifetime of a dropping-mercury electrode drop
υ	kinematic viscosity
ω	angular rotation rate
Ω	ohm

Standard Electroanalytical Prefixes and Suffixes

a	anodic
c	cathodic
D	disc
et	electron transfer
f	flow
j	junction
lim	limiting
p	peak
R	ring

Chemical Species

bipm	4,4′-bipyridine
bipy	2,2′-bipyridine
cp	cyclopentadiene
det	diethylenetriamine
DMF	*N*,*N*-dimethylformamide
DPB	diphenyl benzidine
e^-	electron
EDTA	ethylenediaminetetraacetic acid
EtOH	ethanol
GC	glassy carbon
HV	heptyl viologen (1,1′-diheptyl-4,4′-bipyridilium)
ITO	indium–tin oxide
M	general metal species
MB	Methylene Blue
MV	methyl viologen (1,1′-dimethyl-4,4′-bipyridilium)
PC	propylene carbonate
X	general anion species

About the Author

Paul Monk

The author was brought up in Hastings, on England's south coast, where he attended a local comprehensive school. Despite this education, he achieved entrance to the University of Exeter to read Chemistry. Having obtained a B.Sc. degree and then a doctorate (in 1989) on the electrochemistry of the viologens, he was awarded a fellowship at the University of Aberdeen to study the electrochromism of thin films of tungsten trioxide.

He joined the staff of the Department of Chemistry and Materials, Manchester Metropolitan University in 1991 as a Lecturer in Physical Chemistry. He was promoted to Senior Lecturer in 1998.

He enjoys writing, and is also the author of the books *Electrochromism: Fundamentals and Applications* and *The Viologens*, both of these published by John Wiley & Sons.

Paul Monk is married, in which state he is instructed to profess great happiness, and is also a Methodist Local Preacher.

Chapter 1
Explanatory Foreword

Learning Objectives

- To appreciate that electroanalysis is an analytical tool in which electrochemistry provides the analytical methodology.
- To understand the fundamental differences between potentiometric and amperometric electroanalytical measurements, namely potentiometric measurements are those of the potential made at zero current (i.e. at equilibrium), while amperometric measurements are of the current in response to imposing a perturbing potential (dynamic, i.e. a non-equilibrium measurement).
- To learn the standard nomenclature of electroanalysis.
- To appreciate that while the majority of electroanalytical variables follow the IUPAC system of units, a majority of the common electrochemical equations, if containing variables of length (or units derived from length), will still use the unit of centimetre.
- To appreciate that the fine-detail of electroanalytical nomenclature is important, for example, the way an electrode potential or concentration is written has important implications.
- To notice that the way in which a complicated electrochemical word or term can be split up into its component parts will aid the understanding of its meaning.

1.1 Electroanalysis

Analysts always ask questions such as 'what is it?', 'how much of it is present?', and sometimes, 'how fast does it change?'. Electrochemistry is an ideal analytical tool for answering each of these questions – sometimes simultaneously. Here,

Table 1.1 Comparisons between potentiometry and amperometry

Feature	Potentiometry	Amperometry
Property measured	Potential of electrode (at zero current)	Current through an electrode
Analytical methodology	Quantitative and monitoring	Qualitative and quantitative
Concentration range	10^{-7} to 1 mol dm^{-3}	Generally 10^{-8} to 10^{-4} mol dm^{-3}, but can be as low as 10^{-15}
Relative precision	0.1 to 5%	2 to 3%, and lower detection limits
Particular advantages	Useful for titrating coloured or turbid solutions (and also for direct measurements in these solutions)	Can readily follow kinetics of fast reactions. More versatile than potentiometry for the elucidation of reaction mechanisms
Major disadvantages	The measured quantity is activity and not concentration	Apparatus can be expensive (but much cheaper than many non-electrochemical alternatives). Measurements sensitive to contaminants
	Measurements are slow (unless automated)	Measurements sensitive to dissolved oxygen (although this is not true for *square-wave* voltammetry)

we will use the word 'electroanalysis' to mean the use of electrochemistry in an analytical context.

In this present book, we will look at the analytical use of two fundamentally different types of electrochemical technique, namely potentiometry and amperometry. The distinctions between the two are outlined in some detail in Chapter 2. For now, we will anticipate and say that a **potentiometric** technique determines the **potential** of electrochemical cells – usually at zero current. The potential of the electrode of interest responds (with respect to a standard reference electrode) to changes in the concentration[†] of the species under study. The most common potentiometric methods used by the analyst employ voltmeters, potentiometers or pH meters. Such measurements are generally relatively cheap to perform, but can be slow and tedious unless automated.

An **amperometric** technique relies on the **current** passing through a polarizable electrode. The magnitude of the current is in direct proportion to the concentration of the electroanalyte, with the most common amperometric techniques being polarography and voltammetry. The apparatus needed for amperometric measurement tends to be more expensive than those used for potentiometric measurements alone. It should also be noted that amperometric measurements can be overly sensitive to impurities such as gaseous oxygen dissolved in the solution, and to capacitance effects at the electrode. Nevertheless, amperometry is a much more versatile tool than potentiometry.

The differences between potentiometry and amperometry are summarized in Table 1.1. It will be seen that amperometric measurements are generally more precise and more versatile than those made by using potentiometry, so the majority of this book will therefore be concerned with amperometric measurements.

1.2 Nomenclature and Terminology (IUPAC and Non-IUPAC)

The experimental practice of electrochemistry has a long history. For example, more than 200 years have passed since Volta first looked at the twitching of animal tissues in response to the application of an electric impulse. The literature of electrochemistry was huge even before the International Union of Pure and Applied Chemistry (IUPAC) first deliberated in a systematic code of electrochemical symbols in 1953. Accordingly, many of the IUPAC recommendations will not be followed here.

We will now look at each of the major variables in turn.

Redox couple. Two redox states of the same material are termed a 'redox couple', e.g.

[†] Strictly speaking, it responds to changes in **activity**, as defined in Chapter 3.

$$AgCl_{(s)} + e^- = Ag_{(s)} + Cl^-{}_{(aq)}$$

The electron in this equation will come from (or go to) an electrode if current flows.

Electrode potential. Potentiometric experiments determine potentials. The two components of the redox couple are only together at *equilibrium* at a single energy specific to the system under study and at the conditions employed. This energy, when expressed as a voltage,[†] is termed the **electrode potential, E**. The 'electrode potential' is also known as the 'redox potential' or 'reduction potential'. Some texts for physicists call E the 'electrode energy'.

Unless stated otherwise, we normally assume that the solution is aqueous.

E is normally written with a subscript to indicate the two redox states involved. $E_{Fe^{2+},Fe}$, for example, is the electrode potential for the ferrous ion–iron metal system. Note that we expect a different electrode potential if different redox states are involved, so $E_{Fe^{2+},Fe} \neq E_{Fe^{3+},Fe}$. It is the usual practice to write the oxidized form of the couple first.

SAQ 1.1

Write symbols for the electrode potential for the following couples:

(a) bromine and bromide;
(b) silver and silver cation;
(c) ferrocene, $Fe(cp)_2$, and the ferrocene radical cation, Fe $(cp)_2^{+\bullet}$.

There are several different electrode potentials we shall need to use, for example, E, which is the potential of a **half-cell** reaction. It is not usually described with any form of superscript, but will have subscripts, as shown above.

$E^{\ominus}$ is the **standard electrode potential**, and represents a value of E measured (or calculated) when all activities are 1, when the applied pressure[‡] p is 1 atmosphere and with all redox materials participating in their standard states. As for E, $E^{\ominus}$ should be cited with subscripts to describe the precise composition of the redox couple indicated. Note that $E^{\ominus}$ is often written as E^0, thus explaining why standard electrode potentials are commonly called 'E nought'. The symbol '$\ominus$' implies standard conditions i.e. 298 K, $p^{\ominus}$ and unit activities throughout.

[†] This exposition has been greatly simplified. At equilibrium, the sums of the electrochemical potentials, $\overline{\mu}$, within each of the two half cells comprising the overall cell are the same, and $\overline{\mu}$ is related to the chemical potential μ a by the relationship $\overline{\mu} = \mu + nF\phi$. The occurrence of a potential E at the electrode is a manifestation of the difference in electric field, $\Delta\phi$ between the electrodes and their respective couples in solution, as a function of their separation distances.

[‡] The SI unit of pressure is the pascal, Pa. The SI standard pressure is 1 bar (10^5 Pa) and is denoted by the symbol $p^{\ominus}$. For historical reasons, the electrochemical standard pressure is taken as being 1 atmosphere of pressure; $p^{\ominus}$ for the electroanalyst is therefore 101 325 Pa, a 1% difference from the SI value, which causes negligible differences in $E^{\ominus}$.

Related to $E^{\ominus}$ is the **formal electrode potential**, $E^{0\prime}$ (as discussed in Chapter 6), which can be called 'the standard electrode potential at 298 K, $p^{\ominus}$ and unit *concentrations* throughout'. The differences between $E^{\ominus}$ and $E^{0\prime}$ are discussed in Chapter 6.

The electrode potential obtained with linear-sweep polarography, for example, at a dropping-mercury electrode (DME), is different again and is called the **half-wave potential**, $E_{1/2}$, which is also discussed in Chapter 6.

The phrase 'electrode potential' implies a single electrode, but such potentials are in fact determined or calculated from measurements of **cells** comprising two or more electrodes. This procedure is necessary since it is not possible to measure the energy of a single redox couple at equilibrium:† in practice, we have to measure the *difference* or *separation* in energy between two (or more) electrodes. This separation is termed the ***emf***, following from the somewhat archaic expression 'electromotive force'.‡ In other texts, the alternative name E_{cell} is sometimes given to the *emf*; some texts (rather confusingly) call it just E.

The magnitude of E does not depend on the size of the electrodes – nor does it depend on the material with which the electrode is made, or on the method of measurement. It is therefore an intrinsic quantity.

Current. Amperometric experiments measure current. The **current** I is the rate at which charge is passed, while the **current density** is symbolized as i. Current density is defined as the current per unit electrode area A, so we can write the following:

$$i = \frac{I}{A} \tag{1.1}$$

where current has the unit of the ampere (or 'amp', for short).

Note Both area and ampere have the symbol A, but A for ampere is shown in upright script, while A for area is shown as italic – because it's a variable.

SAQ 1.2

What is the current density if an electrode of area 0.35 cm^2 is passing a current of 12 mA?

---

† While this potential cannot be determined for a single electrode, a potential *can* be derived if the potential of the other electrode in a cell is defined, i.e. the potential of the standard hydrogen electrode (SHE) is arbitrarily taken as 0.0000 V. In this way, a potential scale can then be devised for single electrode potentials – see Section 3.2.

‡ The abbreviation '*emf*', in upright script, is often used in other textbooks as a 'direct', i.e. non-variable, acronym for the electromotive force. Note, however, that in this present text it is used to represent a variable (cell potential) and is therefore shown in italic script.

DQ 1.1

Why use current density at all?

Answer

Current is not measured at equilibrium. Each electroanalytical laboratory will own its own set of electrodes, some large and some small. The current actually measured will be a simple function of the electrode area since charge is passed when electrolyte impinges on a electrode (if it is sufficiently polarized). We see that current is an extrinsic *quantity because its value depends on how much 'something' we employ during the measurement. In this case, the 'something' i.e. the current, relates to the electrode area.*

In contrast, current density is intrinsic *and does not depend on the electrode area, since, by its definition, the current measured has been adjusted to compensate for differences in area.*

In a similar manner to current density, we next distinguish between charge and charge density. The charge that flows is Q, while the charge density, i.e. the charge per unit area, Q/A, is symbolized by q.

Length. It is when we come to units of length that the problems begin. The SI unit of length is the metre, m. Accordingly, the SI unit of concentration is mol m^{-3}. Interconversion between concentration in mol m^{-3} and concentration expressed in the more familiar[†] units of mol dm^{-3} is simple, i.e.

$$\text{concentration in mol m}^{-3} = 10^3 \times \text{concentration in mol dm}^{-3} \qquad (1.2)$$

because there are 1000 cm^3 in 1 dm^3, and thereby 1000 litres in one cubic metre. We see that concentrations in SI units appear as larger numbers.

SAQ 1.3

11g of solid KCl are dissolved in 250 cm^3 of water. What is the concentration when expressed with the units of mol m^{-3}?

It is rare for electrochemists to use SI units in this way, so, like most analysts, they will usually talk in terms of the concentration units that are most convenient.

Unfortunately, many of the standard equations encountered in electrochemistry require the concentration unit of mol cm^{-3} (moles per cubic centimetre). The conversion between mol cm^{-3} and the familiar mol dm^{-3} is as follows:

$$\text{concentration in mol cm}^{-3} = 10^{-3} \times \text{concentration in mol dm}^{-3} \qquad (1.3)$$

[†] Many texts use the symbol 'M' for mol dm^{-3}. We will not use 'M' here in order to emphasize the requirement for interconversion.

because there are 1000 cubic centimetres in one cubic litre. In this case, we see that concentrations expressed in electrochemistry units appear as smaller numbers than when expressed in SI units.

SAQ 1.4

The concentration of a solution of copper nitrate is 0.7 mol dm^{-3}. Re-express this concentration in units of mol m^{-3} and mol cm^{-3}.

Area. When describing the charge density or current density, the usual electroanalytical convention is to cite an area in units of square centimetres. For this reason, most current densities have units of A cm^{-2}. Interconversion between current density in A m^{-2} and current density, expressed in the preferred units of A cm^{-2}, is simple, i.e.

$$\text{current density in A m}^{-2} = 10^4 \times \text{current density in A cm}^{-2} \qquad (1.4)$$

because there are 10 000 cm^2 in 1 square metre (1 m^2).

SAQ 1.5

A current I of 23 mA is passed through an electrode with an area A of 1.07cm^2. What is the charge density, i, as expressed in both SI units and electrochemistry units?

The units of area appear everywhere in electroanalysis. For example, the diffusion coefficient D has the SI unit of m^2 s^{-1}, but such usage is so rare that a good generalization suggests that they are *never* used. In practice, electrochemists cite D with the units of cm^2 s^{-1}.

Nomenclature. The form of words we employ in electroanalysis will tell us much about the parameters under study, as shown by Table 1.2. Being able to take a word apart, bit by bit, will tell us what the overall electroanalytical term means. We have already mentioned potentiometry and amperometry as being techniques for following potential and current, respectively. As another example, there is a commonly used *joint* term, i.e. 'voltam-', which implies measurement of current in response to potential variation. An example of this is a voltammogram, which is a trace ('-gram') of current ('ammo-') as a function of potential ('volt-').

Note that a voltmeter is an exception to the general rules outlined in Table 1.2, since a voltmeter measures potential alone. The use of the name 'voltmeter' has arisen largely on account of manufacturers' advertisements. The old-fashioned name of **potentiometer** is preferred, but won't be used here.

Worked Example 1.1. What does the 'chronopotentiometry' experiment follow?

Table 1.2 Some of the more common word roots encountered in electroanalytical usage

Word root	Meaning	Example
Am-, amp-	Current	An *am*meter is a device for determining the magnitude of a current
Chrono-	Time	*Chrono*amperometry is the following of current (as 'amp-') with time
Coul-	Charge	A *coul*ometer measures charge passed
-gram	Trace or graph	A voltammo*gram* is a graph of current ('amm-') against potential ('volt-')
Poten-	Potential[a]	*Poten*tiometer – a high-resistance device for measuring potential A *poten*tiometric titration is a titration in which the end point is indicated by following the potential
Voltam-	Potential[b] and current	A *voltammo*gram is a trace of current measured as a function of applied potential

[a]The prefix 'poten-' is generally used for measurements of potential determined at zero current, so the potential is essentially constant, i.e. measured in an *equilibrium* experiment.

[b]The prefix 'volt-' is used for measurements involving potential, but generally measurements in which current is determined as a function of potential, i.e. in a *dynamic* experiment. A voltmeter (a device for determining the magnitude of a potential) is the exception to this general rule.

The two roots in this 'mystery' word are *chrono* (meaning 'time', or 'as a function of time') and *poten* (meaning 'potential'), so chronopotentiometry is a study of the way in which potential varies with time.

Cells. The rules for constructing a cell schematic are given in Appendix 2.

Number of electrons (n). There is one final divergence from standard IUPAC usage that may cause confusion. In normal thermodynamics, the symbol n is used for 'amount of substance'. An older convention is followed in electroanalytical work, and electrochemistry in general, such that n means simply the number of electrons involved in a redox reaction. 'Normal' IUPAC representation would use ν for this latter parameter since the number of electrons is a **stoichiometric quantity**. The opposition from electrochemists has been so concerted that IUPAC now allows the use of n as a permissible deviation from its standard practice.

Modified electrodes. Where relevant, we have followed the recent IUPAC directive on the recommended list of terms for 'chemically modified electrodes' (CMEs) [1]. A CME is thus 'an electrode made up of a conducting or semiconducting material that is coated with a selected monomolecular, multimolecular, ionic or polymeric film of a chemical modifier and that, by means of faradaic $\cdots$ reactions or interfacial potential differences $\cdots$ exhibits chemical, electrochemical and/or optical properties of the film'.

Additional deviations from the IUPAC conventions will become apparent in the following text.

1.3 This Book

This book is a part of the series, Analytical Techniques in the Sciences (AnTS). Please assume from the outset that this is neither a *textbook* of analytical chemistry or electrochemistry, nor is it a textbook of their hybrid, 'electroanalysis'. There are many extremely good texts on these topics already available (e.g. see the Bibliography for a list). There is simply no need for a new textbook; if a new one was required, then this would not be it.

No – this book is designed for **open learning**, i.e. for students and practitioners who are new to electroanalysis and require *tutorial support* in the form of a book. Students in such a position are advised to work through the examples at their own pace, starting at the beginning and progressing in a *linear* fashion. There may be a few readers with some prior experience of electroanalysis who prefer to 'pick' at the text, and to use it as a form of 'refresher course'. It is hoped that they will also benefit from the approach used here.

There is a substantial index given at the end of the book, and cross-referencing has been regularly employed throughout. We hope that this 'user-friendly' approach will prove to be successful. This book is not at all exhaustive, though, so students may wish to use it as just an initial component within a larger study programme on electroanalysis.

It has been assumed that students have a working knowledge of the basics of analytical chemistry and its methodology. Accordingly, no attempt has been made to cover the fundamentals of analysis or of the statistical manipulation of analytical data. We will use the minimum of mathematics, and no derivations will be included. The rudiments of algebra will be sufficient.

In order to avoid duplication, there will not be much mention of topics such as sensors and modified electrodes, since they will be covered comprehensively in other AnTS titles in the series.

Finally, it will be seen that many footnotes have been included for those wanting further clarification or additional information. A student following this book for the first time could be advised to ignore some, if not all, of these: while they might add completeness to the text, they are in no way essential.

Summary

Electroanalysis is the science of carrying out analytical chemistry by the use of electrochemistry. At the heart, the two electroanalytical observables are the potential E (also called the voltage) and the current I (or its integral, with respect to time, of the charge Q). The potential is determined at zero current, while the current is determined as a function of careful voltage manipulation. The current density i and charge density q are also defined (e.g. see equation (1.1)).

The rudiments of the IUPAC convention are outlined, insofar as they impinge on electroanalysis, and common exceptions are discussed. The conversion factors between the IUPAC and non-IUPAC usages are given in equations (1.2) and (1.3).

It is seen that although electroanalysis is fraught with rather gruesome-looking words, the learning of a few common roots allows such terms to be decoded by taking the word apart, piece by piece.

Finally, having introduced the basic terms and concepts, this book is itself introduced, and the recommended ways of reading and usage are described.

Reference

1. Durst, R. A., Baumner, A. J., Murray, R. W., Buck, R. P. and Andrieux, C. P., *Pure Appl. Chem.*, **69**, 1317–1323 (1997).

Chapter 2

Introductory Overview and Discussion of Experimental Methodology

Learning Objectives

- To appreciate that a cell *emf* should only be determined at equilibrium.
- To know that the value of the *emf* can only be accurate if equilibrium conditions apply.
- To appreciate that the best means of ensuring equilibrium is to work at zero current, as can be achieved by using a voltmeter of infinite resistance.
- To learn how to calculate the ohmic current induced by current flow (i.e. from Ohm's law, equation (2.3)).
- To learn that there are two basic types of electrode reaction, namely oxidation (loss of electrons) and reduction (gain of electrons). These reactions are the sum total of redox chemistry.
- To learn that if reduction occurs at one electrode, then an equal amount of oxidation occurs at the other.
- To appreciate that the occurrence of reduction or oxidation implies that material is consumed or generated at the electrode, so concentration changes always occur in accompaniment with redox chemistry.
- To appreciate that current is the derivative of charge, so large currents accompany large rates of electrochemical change during analysis.
- To learn that analytical measurements involving concentration change (in response to redox chemistry) are termed dynamic.
- To appreciate the fundamental differences between equilibrium and dynamic measurements.

- To realize that the rate-limiting process during dynamic measurement is usually mass transport of analyte to the electrode | solution interface.
- To know that the overall process of mass transport occurs via three mechanisms, namely convection, migration and diffusion. Convection is the physical movement of solution, migration is the movement of charged analyte in response to Coulomb's law and diffusion is an entropy-driven process. In terms of mass transport, the order of effectiveness is as follows: convection $\gg$ migration $>$ diffusion.
- To know how to alter the experimental conditions in order for convection or diffusion to be the dominant form of mass transport.

2.1 Overview of the Differences between Equilibrium and Dynamic Measurements

For the purposes of this book, the term **electroanalysis** will be taken to mean the analysis of an analyte by using electrochemical methods. The analyte will be termed the **electroactive** species (or **material**). Sometimes, however, we will also call it the **electroanalyte**. At heart, analysts working with electrochemical techniques monitor the behaviour of an electroactive species by performing two types of experiment: they will either measure a potential (or occasionally variations in potential), or they will measure the charges, Q, and changes in the charge, ΔQ.

Measurement of charge. In practice, the most common way of measuring a charge is to determine a current I, which is defined as the rate of change of charge passage:

$$I = \frac{\mathrm{d}Q}{\mathrm{d}t} \tag{2.1}$$

We need to be aware that whenever we talk about a measured charge, we might find that the charge was *calculated* from the measured current. Charge can be obtained from current as its first integral, e.g. as the area under a graph of current against time. Alternatively, charge can be obtained directly from a **coulometer**.

If the current is constant, then equation (2.1) simplifies to give the following:

$$Q = I \times t \tag{2.2}$$

SAQ 2.1

A current of 1.42 mA is passed for 2 min and 44 s. What is the charge passed?

Measurement of potential. Analysts employing electrochemistry usually measure potentials with a **voltmeter**, or any device capable of replicating the behaviour of an accurate voltmeter. Typically, the potential is measured under conditions of **equilibrium**. While there are many definitions of 'equilibrium' in the broad topic of electrochemistry, the simplest, when measuring a potential, is to say that the measurement was performed at zero current.

DQ 2.1

Is it technically possible to perform a measurement at zero current?

Answer

It is impossible to make an electrochemical measurement without consuming some energy. Accordingly, all *voltmeters consume a small amount of energy during measurement.*

When measuring a potential, the equation we will consider to work out the magnitude of the energy consumed is Ohm's law, as follows:

$$V = IR \tag{2.3}$$

where V is the potential, I is the current and R is the resistance. A simpler 'words-only' form of Ohm's law says that if a current is passed through a resistor, then a potential is induced. Equation (2.3) is merely a quantitative form of this 'words-only' definition.

For the purposes of answering DQ 2.1, we will take the resistance term R as representing the resistance of the circuit constructed to measure, say, a cell *emf*. The resistance comes from the leads connecting the voltmeter and all components within the cell; the resistance of the cell itself (R_{cell}) will comprise resistance contributions from both the electrodes and the solution. We will take I to be the current passed during the measurement. The potential V is therefore the voltage induced by current flow through the voltmeter.

We are now in a position to see why measurement at *zero current* is important: the *emf* that is measured for a cell will be incorrect if V from equation (2.3) is other than zero. In fact, we can say:

$$emf_{\text{(measured)}} = emf_{\text{(real)}} + V_{\text{(induced at voltmeter)}} \tag{2.4}$$

where V (from the voltmeter) $= IR$, from Ohm's law. We soon realize that V should be minimized by restricting the current flowing to as small a value as possible. The resistance of the leads is usually negligible (of magnitude 0.01 Ω), so the resistance of the solution is a larger cause of additional potential. The error in *emf*, caused by the resistance of the solution, etc., is called the ***IR* drop** (see Section 3.6.4).

Generally, the best way of minimizing the current is to measure the *emf* with a voltmeter that itself has a huge *internal* resistance. For this reason, it is usual to see references given to 'high-resistance' voltmeters.†

Worked Example 2.1. If the internal resistance of a voltmeter is 10^{12} ohms, and the *emf* to be measured is 1.1 V, what is the current passed during measurement?

From Ohm's law:

$$V = IR, \text{ so } I = \frac{V}{R} = \frac{1.1 \text{ V}}{10^{12}\Omega}$$

We see that $I = 1.1 \times 10^{-12}$ A. Because this current is induced according to Ohm's law, we call it an **ohmic current**.

This current is so close to zero that we can with confidence call it a 'zero current'. In this way, we can then calculate the magnitude of the current passing through the circuit (i.e. the leads and cell). We will call such a current the $I_{circuit}$.

If we know the resistance of the leads and cell, and we also know the current passing through them, we can then use Ohm's law (equation (2.2)) to determine the *IR* drop (i.e. the additional potential induced by current flow during measurement).

SAQ 2.2

An analyst finds that the only voltmeter available is of particularly poor quality since its internal resistance is a mere 10^3 ohms. What is the current passed during measurement if the *emf* to be measured is 3.02 V? (Take the resistance of the leads and solution to be 120 Ω.)

SAQ 2.3

For the same system as that described in SAQ 2.2, what would be the additional potential added to the measured *emf*?

Occasionally, we may wish to measure a potential while it changes with time, for example, chronopotentiometric measurements are often a useful means of studying the processes occurring during corrosion.

Most modern voltmeters have such a large internal resistance that we can indeed say that a cell potential will actually be measured at zero current.

† It is common to see these also referred to as high-*impedance* voltmeters, which in fact means the same thing.

SAQ 2.4

If the *emf* is 1.4 V and the maximum current allowed is 10^{-8} A, what is the minimum value of R we can allow our voltmeter to have?

Measurement of current. In order to measure a current, we must use an ammeter, or any device capable of acting as an ammeter. (Remember: the root 'amm-' will always mean current here.)

For now, we will note that electrons are acquired or given during a redox reaction. In an **oxidation** reaction, electrons are given away by the electroactive species:

$$R \longrightarrow O + ne^- \tag{2.5}$$

By IUPAC conventions, we say that an oxidative current is negative.

The electron formed as a product of equation (2.5) will usually be received (or 'collected') by an **electrode**. It is quite common to see the electrode described as a 'sink' of electrons. We need to note, though, that there are two classes of electron-transfer reaction we could have considered. We say that a reaction is **heterogeneous** when the electroactive material is in solution and is electro-modified at an electrode which exists as a separate phase (it is usually a solid). Conversely, if the electron-transfer reaction occurs between two species, *both* of which are in solution, as occurs during a potentiometric titration (see Chapter 4), then we say that the electron-transfer reaction is **homogeneous**. It is not possible to measure the current during a homogeneous reaction since no electrode is involved.[†] The vast majority of examples studied here will, by necessity, involve a heterogeneous electron transfer, usually at a *solid* electrode.

The other type of redox reaction that can occur is the reverse of oxidation, in which an electroactive species receives ('acquires', 'takes up' or 'gets') an electron in a **reduction** reaction:

$$O + ne^- \longrightarrow R \tag{2.6}$$

where, by IUPAC convention, we say that a reductive current is positive. If the electron comes from an electrode, we say the electrode is the source of the electron.

An easy way to remember which reaction is oxidation and which is reduction is to note that reduction involves the oxidation number being reduced[‡] by '1' per electron received. Alternatively, remember the acronym 'O I L R I G', which spells out the initial letters of 'oxidation is loss, reduction is gain' of electrons.

[†] It *is* possible, however, to measure the potential *at* an electrode immersed in a homogeneous system, for example measuring an *emf* during a potentiometric titration – see Chapter 4.

[‡] In order to avoid all possible ambiguity, we will not use the word 'reduce' again in this text to mean either 'decrease' or 'get smaller', but will only use it to mean that an electroactive material has acquired electrons.

SAQ 2.5

Decide which of the following reactions are oxidation and which are reduction (all electrons have been omitted):

(a) $Br_2 \longrightarrow 2Br^-$;
(b) $Sn^{2+} \longrightarrow Sn^{4+}$;
(c) $[Fe(CN)_6]^{4-} \longrightarrow [Fe(CN)_6]^{3-}$.

Equations (2.5) and (2.6) are the two types of **redox** reaction we will follow. (The word 'redox' comes from the joining and then shortening of the two words 'reduction' and 'oxidation'.)

Redox changes will always accompany current passage because each electron is used to perform an electrochemical reaction. We will occasionally talk in terms of **electromodifying** the electroactive species.[†] We will see later in Section 5.1 that the amount of material electrochemically oxidized or reduced is in direct proportion to the number of electrons passed: a large number of electrons consumed or given out implies that a large amount of material has been electrochemically modified during the current passage, while a small current implies that only a small amount has been modified.

Each electron that moves to and from an electrode represents a charge transferred, either during oxidation or reduction. We see now why equation (2.1) is so important, i.e. current is the rate at which electrons flow. If we know the current I, then we also know the number of electrons which have passed, and hence we can find out the extent and *rate* of the chemical changes occurring. A large current means that much material is changed per unit time, whereas a smaller current implies a smaller amount of change. In summary, the passage of current tells us about the changes in composition inside an electrochemical cell.

We are now in a position to see *the* major difference between measurement of potential and measurement of current. We recall that we want a *zero* current when measuring the potential. We see from the argument above that a zero current implies that *no* compositional changes occur inside an electrochemical cell. Conversely, compositional changes *do* occur during measurement of current, precisely because charge is transferred.

The electrochemist will say that a measurement that accompanies compositional changes is **dynamic** in nature, while a measurement that is performed without compositional changes occurring is **static** or **at equilibrium**. We will use this latter term in this present text.

[†] The extent of such modification derives ultimataly from Faraday's laws. We will look at these laws in a little depth later (Section 5.1).

2.2 The Magnitude of the Current: Rates of Electron Transfer, Mass Transport, and their Implications

DQ 2.2

Current flows during an electrochemical reaction at an electrode. If this current represents a *rate* of charge flow (from equation (2.1)), then what dictates the magnitude of this current I?

Answer

Straightaway, because we are talking about the magnitude of a current, we are assuming that we do not wish to measure a potential. If we did want to measure a potential, we would have employed a high-resistance voltmeter to intentionally stop *any current flow. We are also assuming that the reaction is a heterogeneous electron-transfer process since we are talking about electron transfer at an electrode.*

Equation (2.1) defines current as the rate of charge movement. An electroanalyst could have re-expressed equation (2.1) with, in words, 'the magnitude of an electrochemical current represents the number of electrons consumed or collected per second'. Each electron consumed or collected represents a part of a heterogeneous redox reaction at an electrode (equations (2.3) or (2.4)), so the magnitude of the current also tells us about the amounts of material consumed or formed at the electrode *surface per unit time.*

The rate of a redox reaction and the resultant current are seen to be interconnected and inseparable. We shall see later how this simple statement underlies the whole subject of dynamic electroanalytical measurement.

If current and the rate of electrochemical reaction are two sides of the same coin, then the question, 'What dictates the magnitude of I*?' can be simplified, since we are, in effect, asking the related question, 'What dictates the rate of the electrochemical reaction occurring at an electrode?'.*

DQ 2.3

In a dynamic electroanalytical experiment, if the magnitude of the current is dictated by the rate of the electrochemical reaction, then what is the rate-limiting process during the electrode reaction?

Answer

The phrase 'rate-limiting process' tells us straightaway that several kinetic processes are present (as discussed below), of which one is slower that the others. It is this slow step which we call ***rate determining****.*

Before a heterogeneous electron-transfer reaction can take place, be it oxidation or reduction, we must appreciate that the redox reaction occurs at the **interface** that separates the electrode and the solution containing the electroanalyte. Some electrochemists call this interface a **phase boundary** since either side of the interface is a different phase (i.e. solid, liquid or gas). An electrochemist would usually indicate such a phase boundary with a vertical line, '|'. Accordingly, the interface could have been written as 'solution | electrode'.

Electrons are transferred across the solution | electrode interface during the electrode reaction.

During reduction, electrons travel *from* the power pack, through the electrode, transfer across the electrode–solution interface and enter *into* the electroactive species in solution. Conversely, during oxidation, electrons move in the opposite direction, and are conducted away from the electroactive material in solution and across the electrode–solution interface as soon as the electron-transfer reaction occurs. (Incidentally, these different directions of electron movement explains why an oxidative current has the opposite sign to a reductive current, cf. Section 1.2.)

We will need to consider three rate processes when deciding what dictates the overall size of the current during redox chemistry, i.e. the amount of material that undergoes electrochemical reaction per unit time, as follows:

(i) movement of electrons through the electrode, either to or from the interface;

(ii) movement through the solution of the electroactive material to be electromodified at the same interface;

(iii) movement of electrons across the interface, i.e. the electrode reaction itself.

Electron movement through the electrode. The movement of electrons through an electrode will usually be extremely fast since the material from which the vast majority of electrodes are made will have been chosen by the analyst precisely because of its superior electronic conductivity. Electrodes made of liquid mercury and of solid metals such as platinum, gold, silver or stainless steel, are all used for this reason. Accordingly, it is extremely unlikely that the rate-limiting process during a redox reaction will be movement of the electrons *through* the electrode.

SAQ 2.6

Look up, for example, in Reference [1], the conductivities of metallic platinum, gold, silver, mercury and tungsten, and hence explain why graphite is not the best choice of electrode material.

If the movement of an electron is this fast, we see that the overall rate of charge movement will depend either on the rate at which the electroactive material reaches the electrode or on the rate at which the electrons transfer across the electrode | solution interface once the material has reached it.

Electron movement across the electrode | solution interface. The rate of electron transfer across the electrode | solution interface is sometimes called k_{et}. This parameter can be thought of as a rate constant, although here it represents the rate of a *heterogeneous* reaction. Like a rate constant, its value is constant until variables are altered. The rate constants of chemical reactions, for example, increase exponentially with an increasing temperature T according to the Arrhenius equation. While the rate constant of electron transfer, k_{et}, is also temperature-dependent, we usually perform the electrode reactions with the cell immersed in a thermostatted water bath. It is more important to appreciate that k_{et} depends on the *potential* of the electrode, as follows:

$$k_{et} \propto \exp\,(V_{electrode}) \tag{2.7}$$

The resemblance between this equation and the Arrhenius equation should automatically alert us to the fact that k_{et} is an **activated** quantity. Variation of k_{et} with $V_{electrode}$ will allow us (ultimately) to obtain the activation energy for electron transfer.

We will need to look at the $k_{et}-V$ relationship in some depth later (in Section 7.5), but until then we will merely make a mental note that the more extreme the potential of an electrode, then the faster will be the transfer of electrons across the electrode | solution interface.

We need to consider two possibilities, i.e. both oxidation and reduction. During oxidation, if the potential of the electrode is *less* positive than $E^{\ominus}$, then k_{et} will be fairly small; if the potential of the electrode is *more* positive than $E^{\ominus}$ during oxidation, then k_{et} will be very fast (meaning that I will be large). Similarly, during reduction, if the potential of the electrode is less negative than $E^{\ominus}$, then k_{et} can be fairly small, and if the potential of the electrode is more negative than $E^{\ominus}$ during reduction, then k_{et} will be very fast (and $I_{reduction}$) will be large. The relationship between I and E is illustrated schematically in Figure 2.1.

If k_{et} is made very fast (i.e. by the choice of an extreme V), then we can safely say that the transfer of the electron across the electrode | solution interface will not be rate-limiting. If k_{et} is slow, though, then we must be aware that electron transfer may or may not be rate-limiting.

Electroanalyte movement through solution. If electron conduction through the electrode and electron transfer across the interface are both fast, then the rate that limits the overall rate of charge flow will be that at which the electroactive material moves from the solution to approach close enough to the electrode for electron transfer to occur.

Such movement is termed **mass transport**, and proceeds via three separate mechanisms, namely **migration, convection** and **diffusion**. The overall extent of

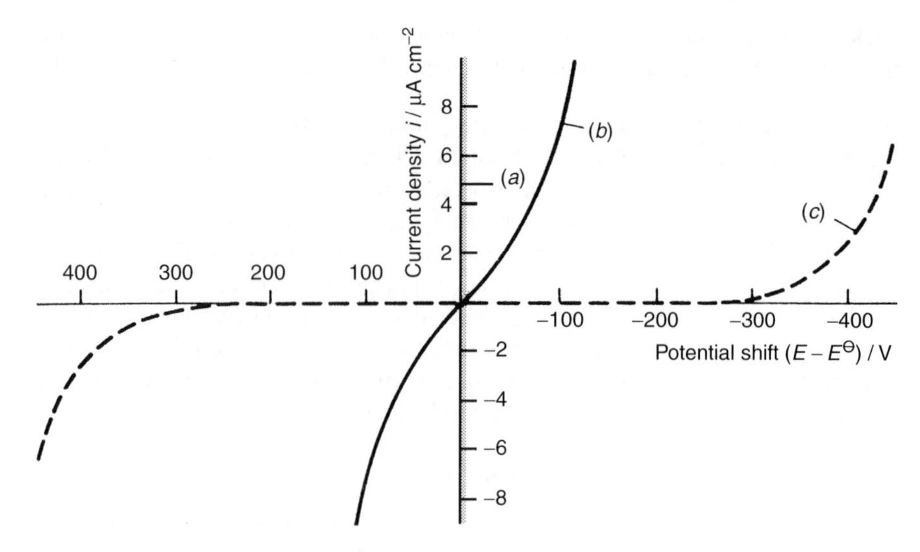

Figure 2.1 Simplified schematic plots showing the exponential relationship between the current density i and the potential of the electrode, E. (The latter is represented here as being relative to $E^{\ominus}$, the standard electrode potential of the couple undergoing electromodification; for now, the abscissa $(E - E^{\ominus})$ can be thought of as deviation from equilibrium.) Three examples of electron-transfer rate (k_{et}) are shown: (a) (coincident with the y-axis) representing a very fast rate of electron transfer of 10^{-3} A cm^{-2}; (b) representing an 'average' rate of electron transfer of 10^{-6} A cm^{-2}; (c) representing a slow rate of electron transfer of 10^{-9} A cm^{-2}. For each trace, $T = 298$ K and the reaction was 'symmetrical', i.e. $\alpha = 0.5$, as defined later in Section 7.5.

mass transport is formally defined as the **flux**, j_i, of the electroactive species i to an electrode, as follows:

$$j_{\mathrm{i}} = \text{function of} \begin{pmatrix} \text{component due to migration} \\ \text{component due to diffusion} \\ \text{component due to convection} \end{pmatrix} \quad (2.8)$$

Each one of these three **modes** of mass transport, i.e. migration, diffusion and convection, will now be considered in turn.

Equation (2.8) looks rather fearsome, but we will not need it to refer to it again except to note the way that flux is a function of the three modes of mass transport, with each acting in an *additive* way.

Migration. This is the movement of charged species in response to an electric field. The charged species in solution are called **ions**: the positively charged ionic species are called **cations**, while the negatively charged species are called **anions**. Cations are attracted to negatively charged electrodes, which in this context are called **cathodes**. Anions are attracted to positive electrodes, which we shall call **anodes**.

The rate of migration depends on the charge on the ion, the size of the ion (including its full solvation spheres), and on the strength of the interaction between the ion and the field. A stronger field will form at an electrode bearing a larger potential, that is, being more negatively or more positively charged. This definition is one way of summarizing **Coulomb's law**.

Following on from Coulomb's law, we see that migration will not occur if the electrode has no potential (and for this reason does not exert a field). Such a situation is not useful, though, since no potential implies no current, and hence no measurement. More importantly, though, a neutral atom or an uncharged molecule does not and cannot undergo migration.

DQ 2.4

Can migration be minimized, even if the electroactive species is ionic?

Answer

The effects of migration may indeed by neglected if a liquid electrolyte, e.g. water, is employed which contains an excess of unreactive ionic salt (often termed a ***swamping electrolyte****). By 'excess', we usually mean that the concentration of such an electrolyte is about 100 times greater than the concentration of the electroanalyte (if at all possible).*

Having an excess of unreactive electrolyte induces the phenomenon of **electrode polarization**, with a buildup of oppositely charged electrolyte ions at the electrode | solution interface. Such polarization decreases and hence suppresses all migration effects.† The most commonly used salts for the purpose of suppressing migration in water are KCl and KNO_3.

SAQ 2.7

Look up why KCl or KNO_3 are the most common choices of swamping electrolytes. (Hint: think in terms of transport numbers, t, and conductivities, Λ.)

If adding an inert electrolyte is undesirable, or if no swamping electrolyte is sufficiently inert or soluble, then the effects of migration can be lessened somewhat by performing the analysis at low field, for instance with an electrode having a very small potential. Such a practice is seldom useful, though, from considerations of mass transport (see below).

† Note that migration *is* important for ionic movement within solid polymer electrolytes or solid-state electrochromic layers since the transport numbers of the electroactive species or of the mobile counterions become appreciable. These aspects are unlikely to trouble us further in this present text.

Convection. This is the physical movement of the solution in which the electroactive material is dissolved. In practice, convection arises from two causes, i.e. from deliberate movement of the solution, e.g. by mechanical stirring (sometimes called **hydrodynamic control**, see Chapter 7) or, alternatively, convection is induced when the amount of charge passed through an electrode causes localized heating of the solution in contact with it. The convective 'stirring' in such instances occurs since the density ρ of most solvents depends on their temperature: typically, ρ increases as the temperature decreases.

DQ 2.5

For the analyst, which is superior – convection as caused by stirring or convection from electrode heating?

Answer

Imagine a cold winter's day: the radiator is hot and the rays of the wintry sun shine through the window and cast shadows on the wall opposite. We will note how the shadow is in a state of continual movement. The patterns are caused by eddy currents around the heater as the air warms and then rises. After just a quick glance, it's clear that the movement of the warmed air is essentially random. By extension, we see that, as an electroanalytical tool, electrode heating is not *a good form of convection, because of this randomness. Conversely, a hydrodynamic electrode gives a more precisely controlled flow of solution. In consequence, the rate of mass transport is both reproducible and predictable.*

The extent of convection can be minimized by not stirring the solution of electroanalyte, and by careful experimental technique, i.e. not banging the bench, or other things that will cause movement of the solution. A solution under such ideal conditions is said to be **still, quiet**, or (occasionally) **quiescent**.

Diffusion. Often, the most important mode of mass transport is diffusion. The rate of diffusion can be defined in terms of **Fick's laws**. These two laws are framed in terms of **flux**, that is, the amount of material impinging on the electrode's surface per unit time. Fick's first law states that the flux of electroactive material is in direct proportion to the change in concentration c of species 'i' as a function of the distance x away from the electrode surface. Fick's first law therefore equates the flux of electroanalyte with the steepness of the **concentration gradient** of electroanalyte around the electrode. Such a concentration gradient will always form in any electrochemical process having a non-zero current: it forms because some of the electroactive species is consumed and product is formed at the same time as current flow.

Fick's first law gives the concentration profile, i.e. a distance-dependence. His second law gives the time-dependence and quantifies the rate at which a concentration gradient dissipates.

SAQ 2.8

What will the concentration profile look like?

Note that because diffusion is entropy-driven, it is one of the means whereby the system attempts to minimize the magnitude of the concentration gradient, i.e. to maximize the overall disorder that is present. Accordingly, diffusion will always be present even in the absence of both convection and migration.

In summary, if a solution is 'quiet' and contains a swamping electrolyte, then diffusion is the sole means for an electroactive species to approach the electrode.

2.3 The Implications of Using Diffusive or Convective Control

DQ 2.6

If the overall mass control comprises three components – migration, convection and diffusion – which form of mass transport is the most effective?

Answer

We can answer this question in part by looking at a cup of hot tea: after adding a spoonful of sugar, we know that we should stir the tea because the sugar will not mix (except after a long time) by just 'standing'. An analyst making a cup of tea (and using electrochemical terminology) would say that convective control – the stirring – is a more efficient mode of mass transportation than is diffusion – mere standing.

Migration in the absence of a swamping electrolyte is somewhat more effective than is diffusion, but migration can be ignored if a swamping electrolyte is added to the solution. Diffusion still occurs even if the solution is stirred, but convection is so much more efficient that we can ignore the diffusion completely.

Summary

Potential differences can only be measured accurately at zero current. Accordingly, when measuring a potential at an electrode, it is best to use a voltmeter with as high an internal resistance as possible. It is also advisable for the resistance of the leads and other experimental paraphernalia to be as small as possible, thereby minimizing the *IR* drop.

The passage of current through an electrochemical cell implies that the internal composition of the cell changes. The magnitude of the current tells us how much change has occurred, while the sign of the current tells us whether the reaction at the electrode of interest was oxidation or reduction.

In summary, then, the biggest difference between measurements of potential and current (potentiometry and amperometry, respectively) is that potentials are measured with a zero current wherever possible, implying that *no* compositional changes occur inside the cell during measurement, whereas compositional changes *do* occur during the measurement of current.

Analysts say that measurements accompanying compositional changes are 'dynamic', whereas measurements performed without compositional changes are 'at equilibrium'.

Dynamic electroanalytical measurements at a solid electrode involve heterogeneous electron transfer. Electrons are transferred across the 'solution | electrode' interface during the electrode reaction. In fact, the term 'electrode reaction' implies that such an electron-transfer process occurs.

The observed current is proportional to the *rate* of such a transfer of electronic charge. This rate of the electrode reaction can depend on the rate of electron conduction through the electrode, mass transport of electroanalyte through the solution, or electron transfer across the electrode–solution interface. At low voltages, the observed rate may be due to either mass transport or the rate of electron transfer. However, at more extreme potentials, the observed current is proportional to the rate of mass transport, which is easier to treat.

Mass transport comprises three different 'modes', i.e. convection, migration and diffusion. Convection (stirring) is the most efficient form of mass transport. Migration can be minimized by adding a swamping electrolyte to the solution. Diffusion occurs even in the absence of migration and convection.

Reference

1. Aylwood, G. and Findlay, T., *SI Chemical Data*, 3rd Edn, Wiley, Brisbane, 1994, pp. 5–13.

Chapter 3

Equilibrium Measurements: 'Frustrated' Equilibrium with No Net Electron Transfer

Learning Objectives

- To learn that accurate potentiometric measurements must be obtained at equilibrium, and that a true 'equilibrium' measurement can only be obtained at zero current.
- To appreciate that the majority of equilibrium cells comprise two half cells that are physically separated, so transfer of energy and material between such half cells is prevented – this state of enforced equilibrium is called 'frustrated' or 'quasi'-equilibrium.
- To learn what a reference electrode is, and understand why a current cannot be allowed to pass through it.
- To learn that the primary reference electrode is the standard hydrogen electrode (SHE), and that the potential of all other reference electrodes (so-called secondary references) are determined with respect to the SHE.
- To understand the fundamental relationships between *emf* and the thermodynamic parameters $\Delta G'$, $\Delta H'$ and $\Delta S'$ (where the prime indicates a frustrated equilibrium).
- To recall the basic concepts of the thermodynamics of cell operation, such as the electrode potential E, the standard electrode potential $E^{\ominus}$ and the electromotive force (*emf*).
- To learn that E and $E^{\ominus}$ are related by the Nernst equation (equation 3.8)), which describes them as a function of activity.
- To appreciate that the Nernst equation is not written in terms of concentration, but of activity.

- To learn that activity a and concentration c are related by the expression $a = c \times \gamma_{\pm}$ (equation (3.12)) where $\gamma_{\pm}$ is the mean ionic activity coefficient, itself a function of the ionic strength I.
- To learn that an approximate value of $\gamma_{\pm}$ can be calculated for solution-phase analytes by using the Debye–Hückel equations (equations (3.14) and (3.15)).
- To learn that the change of $\gamma_{\pm}$ with ionic strength is a major cause of error in electroanalytical measurements, and so it is advisable to 'buffer' the ionic strength (preferably at a high value), e.g. with a total ionic strength adjustment buffer (TISAB).
- To learn how a pH electrode can be constructed and why the potential developed across a thin layer of glass can be related to pH by a variant on the Nernst equation.
- To understand the basics of ion-selective electrodes (ISEs), and relate the activities of analyte ions to the *emf* from an ISE by using the Nernst equation.
- To learn how the fluoride ISE works, and to understand how it should be used.
- To appreciate that an ISE must be selective if its measurements are to be useful, and how the Nernst equation can be adapted to take account of the ISE selectivity by using the concept of a selectivity coefficient (or 'ratio').
- To learn how the problems of selectivity can be overcome by using a 'standard-addition' method, such as drawing a Gran plot – one of the more useful of such methods.
- To look at an ISE which operates with a liquid membrane, and appreciate how and why it works.
- To learn how the solubility constant, K_s, can be measured potentiometrically.
- To learn about how a liquid junction potential, E_j, arises, and appreciate how it can lead to significant errors in a calculation which uses potentiometric data.
- To appreciate how an IR drop can affect a potentiometric measurement, and learn how to overcome it.
- To appreciate how potentiometric measurements are prone to errors as caused by current passage and the mathematical formulation of the Nernst equation.

3.1 Introduction: What is 'Equilibrium'? Concepts of 'Frustrated' Equilibrium

In the previous chapter, we saw that a simple definition of an 'equilibrium' electroanalytical experiment is simply one in which the analyses are performed at zero current.

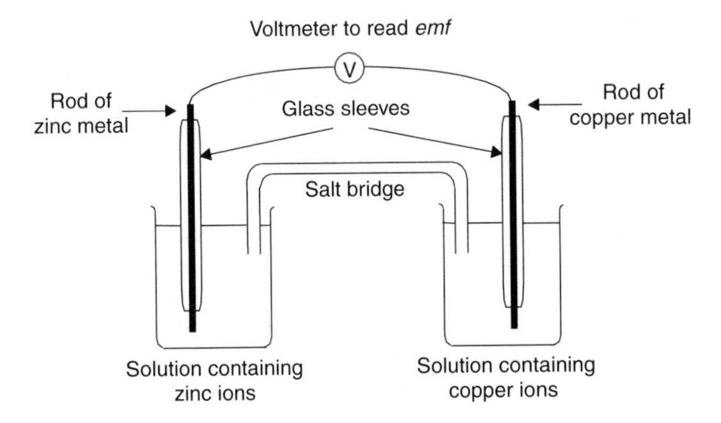

Figure 3.1 Schematic representation of a non-sophisticated cell for equilibrium electrochemical measurements. The example shown is a Daniell cell comprising Cu^{2+}, Cu and Zn^{2+}, Zn half cells. The need for the glass sleeves is discussed in Chapter 9.

A typical electrochemical **cell** is shown in Figure 3.1. The cell comprises two **half cells**, with each comprising a redox couple. The electrode from each half cell is connected to a voltmeter to enable the cell *emf* to be determined. Finally, a **salt bridge** is added to enable ionic charge to transfer between the two half cells.

DQ 3.1

What would happen if the *emf* was measured on a voltmeter not having a zero resistance?

Answer

The simplest way for the cell to discharge would be for the leads to the two electrodes to touch i.e. 'shorting' the cell. Using a voltmeter of low resistance has the same effect, allowing charge from one electrode to conduct to the other. If the cell was allowed to discharge, then reduction would occur in the half cell that is more positive, while oxidation would occur in the half cell that is more negative.

We will illustrate the above point with the following example. Consider the cell Zn | $ZnSO_{4(aq)}$ | | $CuSO_{4(aq)}$ | Cu, which is commonly called the Daniell cell.† The actual process of 'cell discharge' involves an electrochemical reaction at both electrodes. Since the zinc is the more negative of the two half cells, oxidation would occur on the zinc side of the cell, as follows:

$$Zn^0_{(s)} \longrightarrow Zn^{2+}{}_{(aq)} + 2e^- \tag{3.1}$$

† The rules for 'translating' a cell schematic are given in Appendix 2.

An ammeter placed in the electrical circuit between the two electrodes would register the electrons from equation (3.1) as **current**.

Additionally, reduction would occur in the copper half cell in accompaniment with the zinc oxidation:

$$Cu^{2+}{}_{(aq)} + 2e^- \longrightarrow Cu^0_{(s)} \qquad (3.2)$$

The current through the copper half cell will be equal and opposite to the current through the zinc half cell. In other words, the number of moles of copper metal formed in equation (3.2) will be the same as the number of moles of zinc metal consumed by dissolution of the zinc electrode according to equation (3.1). The two half cells have to be **complementary** in this respect.

If the potential of the Daniell cell is to be determined accurately, we have already seen that the measurement has to be made at zero current. In order to ensure zero current, the internal resistance of the voltmeter shown in Figure 3.1 must be vast, as discussed in the previous chapter. The voltmeter operates in much the same way as a switch or 'circuit breaker' does: reaction (in this case, cell discharge) would occur but for the incorporation of the voltmeter in the circuit.

DQ 3.2

Why do we call any data obtained from a Daniell cell 'equilibrium', if the two half cells are physically separated?

Answer

In everyday chemical usage, the word 'equilibrium' means that a reaction has stopped, e.g. because it has reached its position of minimum chemical potential or because one reactant has been consumed completely. In this electroanalytical context, however, we say that we are making a measurement of potential 'at equilibrium', yet the system has clearly not reached a 'true' equilibrium because as soon as the voltmeter is replaced with a connection having zero resistance, a cell reaction could commence. What then do we mean by equilibrium in this electroanalytical context?

When electroanalysts talk of equilibrium in a cell of the type described here, they do not mean a normal thermodynamic *equilibrium. Furthermore, in a 'normal' sense, the reactants would be allowed to mix, whereas the complementary oxidations and reductions in a cell occur within two containers that are physically separated.*

We say that the system experiences ***frustrated***[†] **equilibrium**. *In short, the starting materials are 'forbidden' from mixing with each other, and*

[†] The term *quasi-* equilibrium is preferred to *frustrated* equilibrium in many texts. The descriptor 'quasi' can sometimes imply that equilibrium will eventually be attained but only after kinetic limitations have been reached. In contrast, the word 'frustrated' is employed here to emphasize that a true equilibrium is not even possible owing to the limitations of the cell design, with the two half cells being physically separated.

hence from reacting, because they are physically separated. Despite the apparent artificiality of this situation, the laws of thermodynamics may still be applied to such a cell as that discussed below.

We have laboured this point in order to emphasize how it is that a system can be apparently at equilibrium and yet at the same time could react spontaneously if allowed.[†]

3.2 Revision: *emf*, Electrode Potentials and the SHE

3.2.1 Terminology and Symbolisms

The fundamental basics of electrochemistry are fairly well known, so we will not spend too much time on their revision here.

Cells. An electrochemical cell comprises two (or more) redox couples, with the energy of each being monitored by an electrode. (As we have seen already, the electrode may itself be one part of the redox couple.) By convention, we say that the more positive electrode is the **right-hand** electrode, while the **left-hand** electrode is the more negative. The difference in potential between the right-hand and left-hand electrodes is called the cell ***emf***:

$$emf = E_{\text{right-hand side}} - E_{\text{left-hand side}} \tag{3.3}$$

The acronym *emf* derives from the somewhat archaic term 'electromotive force'.[‡] Physicists tend to employ the term **potential difference** (and abbreviate it to 'pd'). Another term that is used is E_{cell} (cell potential).

We must note that the way equation (3.3) has been written means that the *emf* of a cell is defined as positive since $E_{\text{right-hand side}} > E_{\text{left-hand side}}$ by definition.

The cell *emf* can yield much thermodynamic information. If we assumed that, for once, the cell reaction was allowed to occur, i.e. was not frustrated, the Gibbs energy of cell reactions would be given by the following:

$$\Delta G'_{\text{cell}} = -nF\,emf \tag{3.4}$$

where n is the number of electrons transferred in the cell reaction (if it was allowed to occur) and F is the Faraday constant, which represents the charge on 1 mole of electrons (see Chapter 5). The prime is added to indicate that ΔG is not a true Gibbs energy since the reaction is frustrated. The term $\Delta G'_{\text{cell}}$ is always negative because *emf* is always positive. This is again a definition. The sign of $\Delta G'_{\text{cell}}$ implies that the cell reaction would occur spontaneously if the two

[†] We will not use concepts such as reaction quotients here: the equilibrium is 'frustrated' and therefore not the same as a *true* equilibrium.

[‡] Recall from Chapter 1 that the acronym '*emf*' is used in this present text to represent a variable (cell potential) and is therefore shown in italic script.

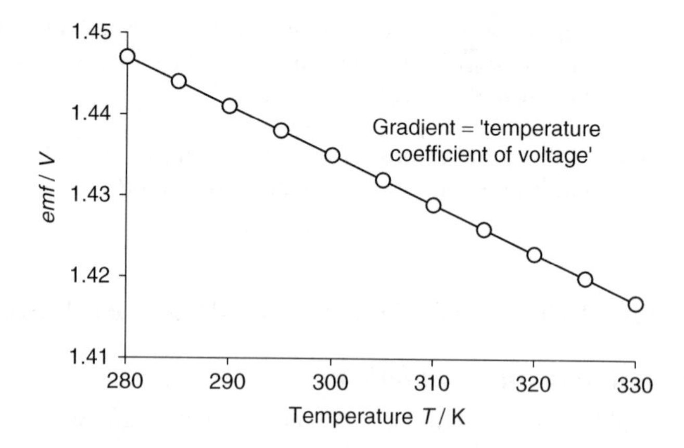

Figure 3.2 Determination of the temperature coefficient of voltage, from which $\Delta S'_{\text{cell}}$ may be calculated. Data relate to the Clark cell, i.e. Zn | $ZnSO_{4(\text{saturated})}$, $HgSO_{4(s)}$ | Hg.

halves of the cell were brought into contact – or at least if the two electrodes were allowed to touch, i.e. shorting the cell.

The entropy change associated with the cell reaction is given by the following relationship:

$$\Delta S'_{\text{cell}} = nF\left(\frac{\partial emf}{\partial T}\right)_p \tag{3.5}$$

at constant pressure, p, and hence the addition of the subscript to the bracket. Again, notice the prime which reminds us yet again of the frustrated equilibrium. The term in brackets has a whole list of names, namely temperature coefficient, temperature coefficient of voltage, voltage temperature coefficient, etc. These names all mean the same, i.e. the *gradient* of a graph of the cell *emf* against the temperature T (see Figure 3.2). The gradient can be either positive or negative, meaning that $\Delta S'_{\text{cell}}$ can have either sign, i.e. depending on the half cells involved.

From straightforward second-law thermodynamics, $\Delta H'_{\text{cell}} = \Delta G'_{\text{cell}} + T\Delta S'_{\text{cell}}$, so (from the equations above) we can write the following:

$$\Delta H'_{\text{cell}} = -nF\,emf + TnF\left(\frac{\partial emf}{\partial T}\right) \tag{3.6}$$

so the enthalpy of the cell reaction may also be obtained via the temperature dependence of the cell *emf*.

Worked Example 3.1. Consider the Clark cell, Zn | Zn^{2+}, Hg_2SO_4 | Hg. The *emf* of this cell is 1.423 V at 298 K, and its temperature coefficient of voltage is -2.3×10^{-4} V K^{-1}. What are the values of $\Delta G'_{\text{cell}}$, $\Delta H'_{\text{cell}}$ and $\Delta S'_{\text{cell}}$?

On discharge, the spontaneous cell reaction would be as follows:

$$Zn + Hg_2SO_4 \longrightarrow Zn^{2+} + 2Hg^{+} + SO_4{}^{2-}$$

which is a two-electron reaction. Accordingly, from equation (3.4) we obtain:

$$\Delta G'_{cell} = -nF\,emf = -2 \times 96\,485 \text{ C mol}^{-1} \times 1.423 \text{ V}$$

so:

$$\Delta G'_{cell} = -275 \times 10^3 \text{ C V mol}^{-1} = -275 \text{ kJ mol}^{-1}$$

The unit 'C' is the coulomb, the unit of charge, while J is the joule (note the unit manipulation, i.e. 1 C × 1 V = 1 J). Next, the entropy of the cell reaction is given by equation (3.5), as follows:

$$\Delta S'_{cell} = nF\left(\frac{\partial emf}{\partial T}\right)_p = 2 \times 96\,485 \text{ C mol}^{-1} \times -2.3 \times 10^{-4} \text{ V K}^{-1}$$

$$= -44.4 \text{ J K}^{-1}\text{mol}^{-1}$$

Finally, $\Delta H'_{cell}$ is given by equation (3.6) as $\Delta G'_{cell} + T\Delta S'_{cell}$:

$$\Delta H'_{cell} = (-275 \text{ kJ mol}^{-1}) + (298 \text{ K} \times -44.4 \text{ J K}^{-1}\text{mol}^{-1})$$

so:

$$\Delta H'_{cell} = -288 \text{ kJ mol}^{-1}$$

SAQ 3.1

A cell has an *emf* of +1.100 V at 298 K, and the temperature coefficient of voltage for the cell is +5 mV K^{-1}. Calculate $\Delta G'$, $\Delta H'$ and $\Delta S'$ for the cell at 298 K, taking $n = 2$.

Notice how the *emf* has a '+' sign in SAQ 3.1. This sign is often inserted but, from the definition of *emf*, the sign is superfluous since *emf* is positive *by definition*.

Although electrochemical methods are experimentally easy, the practical difficulties of obtaining accurate data are severe. Unless daunting precautions are enacted, values of $\Delta G'_{cell}$, $\Delta H'_{cell}$ and $\Delta S'_{cell}$ are best regarded as only approximate. The two most common errors are (i) allowing current passage to occur, and (ii) not performing the measurement **reversibly**. The most common fault which causes problem (ii) is changing the temperature of the cell too fast, such that the temperature T inside the cell is not the same as the temperature of, e.g. the water bath *outside* of the cell, where the thermometer monitoring T is placed.

Half cells. Although we can calculate the thermodynamic parameters for a complete cell, we cannot do the same for a half cell because we cannot measure the electrode potential for the O, R couple ($E_{O,R}$): in fact, we already know that its value cannot actually be known for a cell. All that we can measure, e.g. with a potentiometer or voltmeter, is the separation between two electrode potentials.

The reason why we are able to cite values of $E_{O,R}$ arises because we *arbitrarily* define the value of the $H^+ \mid H_2$ couple as 0.0000 V under standard conditions. By setting the value of this one couple to zero (0 V), we can obtain the values of all other couples in relation to it.

The SHE. The $H^+ \mid H_2$ couple is the basis of the primary standard around which the whole edifice of electrode potentials rests. We call the $H^+ \mid H_2$ couple, under standard conditions, the **standard hydrogen electrode (SHE)**. More precisely, we say that hydrogen gas at standard pressure,† in equilibrium with an aqueous solution of the proton at unity activity‡ at 298 K has a defined value of $E^\ominus$ of 0 *at all temperatures*. Note that all other standard electrode potentials are temperature-dependent. The SHE is shown schematically in Figure 3.3, while values of $E^\ominus_{O,R}$ are tabulated in Appendix 3.

The SHE also utilizes the catalytic properties of platinum, so the electrode monitoring the energy of the $H^+ \mid H_2$ couple is made of platinum coated with **platinum black**. 'Platinum black' is a layer of finely divided Pt metal, which catalytically speeds up the dissociation and association of H–H molecules. The catalysis here aids the speed at which equilibration is attained so that the measurement is more likely to be performed reversibly.

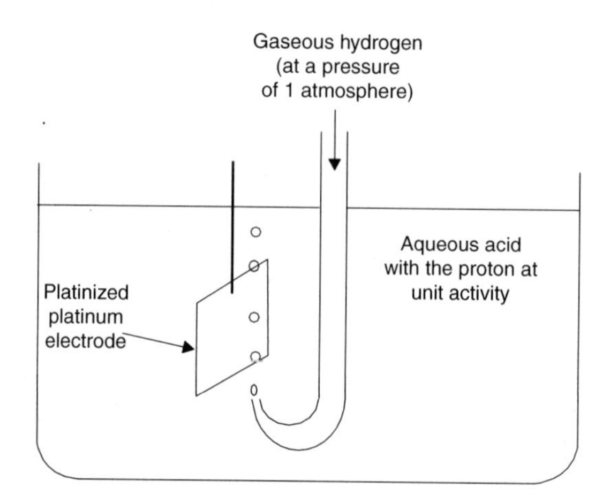

Figure 3.3 Schematic representation of the primary reference, the standard hydrogen electrode (SHE).

† While the standard pressure $p^\ominus$ is 10^5 Pa, most electrochemists use a $p^\ominus$ value of 1 atm since the original $E^\ominus$ scale was formulated in this way. The differences between using $p^\ominus = 10^5$ Pa and $p^\ominus = 1$ atm are negligible.

‡ We will define activity in the following sections. For now, we will assume that activity and concentration are the same thing.

Since Pt is a solid, the schematic for the SHE half cell is 'Pt | $H_{2(gas)}$ | $H^+_{(aq)}$'.

3.2.2 Reference Electrodes

We should remind ourselves that in a potentiometric experiment, we cannot measure individual electrode potentials: we can only measure the *emf* of a cell. The *emf* comprises two half cells (see equation (3.3)). The value of one electrode potential will be known, while the other will be unknown: the value of E for the known half cell is that of the electrode potential for the reference electrode. Accordingly, the value of E for the half cell that contains analyte is only as good as the value of E for the reference electrode.

The SHE is chosen as the ultimate reference electrode since its value is *defined*. By simply making a cell in which one half cell is the SHE, then straightaway we also know the potential of the second half cell. For this reason, we say that the SHE is a **reference electrode**. Since all potentials are ultimately cited with respect to the SHE, the latter is the reference electrode from which all other electrode potentials are derived: we say that the SHE is the **primary standard**. It is also called the **primary reference electrode**.

A **reference** electrode is defined as 'a constant-potential device'. In other words, if we make a cell in which one half cell is the SHE and then measure the *emf*, we then simply employ the equation $emf = E_{\text{right-hand side}} - E_{\text{left-hand side}}$ (equation (3.3)) to determine the other half-cell potential.

Worked Example 3.2. The cell 'SHE | $Cu^{2+}_{(aq)}$ | $Cu_{(s)}$' is constructed with all species in their standard states and the *emf* is determined as 0.34 V. The copper is the positive electrode. What is $E^{\ominus}_{Cu^{2+},Cu}$?

From equation (3.3):

$$emf = E_{\text{right-hand side}} - E_{\text{left-hand side}}$$

so by inserting the appropriate values:

$$0.34\ \text{V} = E^{\ominus}_{Cu^{2+},Cu} - 0$$

and hence:

$$E^{\ominus}_{Cu^{2+},Cu} = 0.34\ \text{V}$$

SAQ 3.2

The cell 'SHE | $Zn^{2+}_{(aq)}$ | $Zn_{(s)}$' is constructed with all species in their standard states, and the *emf* is determined as 0.76 V. The zinc electrode is the *negative* electrode. Deduce the value of $E^{\ominus}_{Zn^{2+},Zn}$.

We see that calculations with the SHE are so easy as to be almost trivial. In practice, however, the SHE is so difficult to operate experimentally, and not particularly safe because it involves elemental hydrogen gas, that we avoid the SHE if it is at all possible. Instead, we employ a **secondary reference electrode**, where the word *secondary* implies a reference electrode other than the SHE, but for which the potential *is* known with respect to the SHE.

One of the most common choices of reference electrode is the **saturated calomel electrode (SCE)**:

$$Hg_{(l)} \mid Hg_2Cl_{2(s)} \mid KCl_{(sat'd)}$$

(note that the 's' here does NOT mean 'standard' but 'solid'.) The potential of the SCE is 0.242 V relative to the SHE.

At the 'heart' of the SCE is a paste of mercury and mercurous chloride (Hg_2Cl_2, which has the old-fashioned name 'calomel'). The SCE is shown schematically in Figure 3.4.

The half-cell reaction in the SCE is as follows:

$$Hg_2Cl_2 + 2\ e^- \longrightarrow 2\ Cl^- + 2\ Hg \tag{3.7}$$

so $E_{SCE} = E_{Hg_2Cl_2,Hg}$. The mercury and calomel are both pure phases, so their activities are both unity. Accordingly, $E_{Hg_2Cl_2,Hg}$ is independent of the amounts of the two solids.

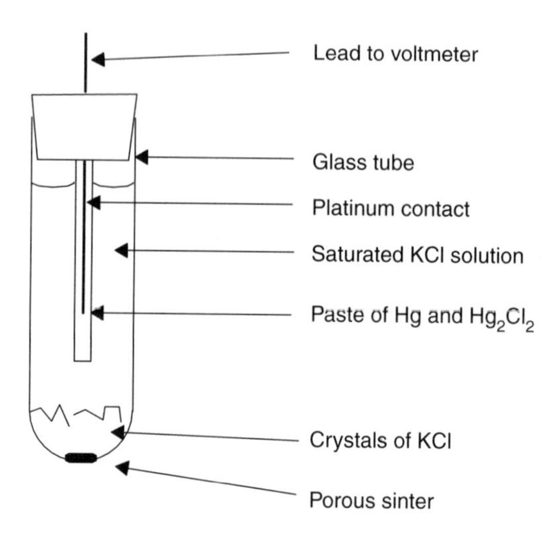

Figure 3.4 Schematic representation of the most commonly employed secondary reference, the saturated calomel electrode, (SCE). Care is needed when using this electrode to ensure that the sinter does not become blocked with recrystallized KCl – a common problem.

If the activity of the chloride ion is maintained at a constant level, then E_{SCE} will also have a constant value, which explains why this couple is the basis for a reference electrode. The activity of chloride ion is best maintained by employing a constant *surplus* of KCl crystals at the foot of the tube to ensure saturation (see Figure 3.4). An SCE not having a thick crust of KCl crystals should be avoided, since its potential might not be known.

Current through a reference electrode should always be prevented (or kept to a minimum) because current flow will probably cause $a(Cl^-)$ to alter,† itself changing $E_{Hg_2Cl_2,Hg}$. To recap, the electrode potentials $E_{O,R}$ are readily determined at constant temperature by using a voltmeter of infinite resistance, thereby minimizing the current flow.

In order to employ a reference electrode as part of an electroanalytical experiment, the analyst requires that potentials can be interconverted straightforwardly back and forth to the SHE scale. Consider the grid in Figure 3.5 which allows the interconversion of electrode potentials in aqueous systems (such a grid can be constructed quite easily).

Worked Example 3.3. What would be the potential of the Mg^{2+} | Mg couple with respect to the SCE?

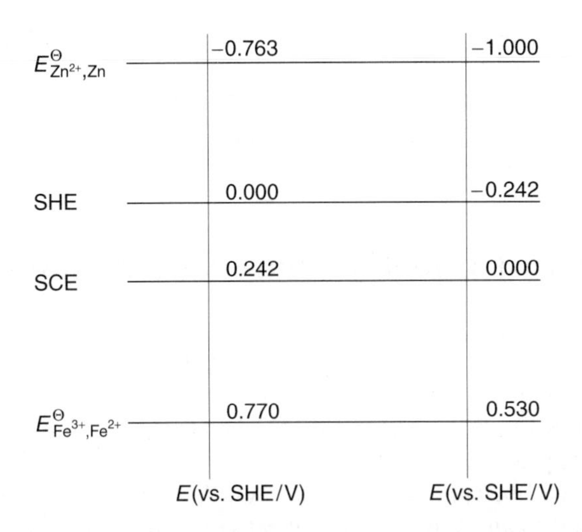

Figure 3.5 A grid which allows the interconversion of potentials between the standard calomel electrode (SCE) and the standard hydrogen electrode (SHE). In addition, the standard electrode potentials of two couples, i.e. Zn^{2+}, Zn and Fe^{3+}, Fe^{2+}, are shown for comparative purposes.

† '*a*' here means 'activity', as described in detail in Section 3.4.

Against the SHE, the value of $E^{\ominus}_{Mg^{2+},Mg}$ is −2.360 V. The value of E_{SCE} is +0.242 V with respect to the SHE, so the separation between $E^{\ominus}_{Mg^{2+},Mg}$ and E_{SCE} is (−2.360 −0.242) V, i.e. −2.602 V.

Alternatively, this same result might have been more easily obtained by using a 'number line' such as that shown in the following:

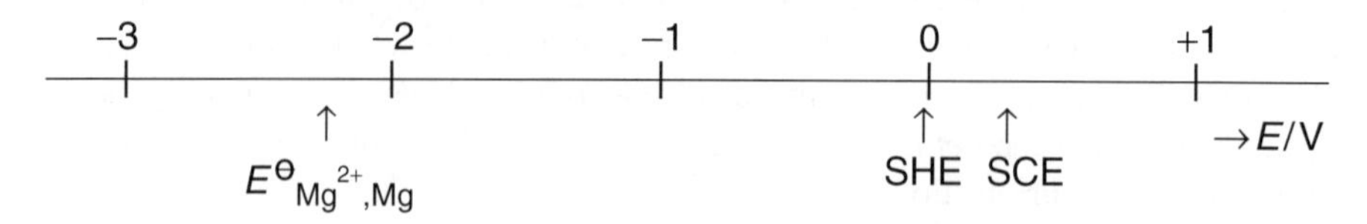

With this 'number-line' approach, it is clear that the separation between $E^{\ominus}_{Mg^{2+},Mg}$ and E_{SCE} is larger than that between $E^{\ominus}_{Mg^{2+},Mg}$ and E_{SHE}. Accordingly, using an alternative reference in this case gives a larger (i.e. more negative) value to $E^{\ominus}_{Mg^{2+},Mg}$. Note that it is always wise to prominently state when you are citing an electrode potential with respect to references other than the SHE: if no reference is cited, then the SHE is assumed.

SAQ 3.3

Draw a number line like that shown above to determine a value for $E^{\ominus}_{Mg^{2+},Mg}$ with respect to the Zn^{2+}, Zn couple.

Other reference electrodes are discussed in standard electrochemistry textbooks (see the Bibliography).

3.3 The Nernst Equation and its Permutations

Having established (i) that a meaningful cell *emf* can only be obtained at equilibrium, (ii) that the *emf* comprises two electrode potentials, and (iii) that the potential of one half cell can be defined in relation to a reference electrode, we are finally in a position to extract electroanalytical data from cells by using a potentiometric approach.

The problem will be approached as follows. We will assume that the cell is at equilibrium and the *emf* is measured by means of a high-resistance voltmeter. We will have waited for the voltmeter read-out to reach a steady value – in practice, this means that the read-out fluctuates by no more than, say, 2 mV or so. The potential of the reference electrode will be known, so we can determine accurately the electrode potential, $E_{O,R}$, of the half cell which contains the analyte of interest.

We will assume that we know the identity of the O,R couple (the analyte) and hence we know its standard electrode potential $E^{\ominus}_{O,R}$, e.g. from the tables such as those given in Appendix 3. (Note: $E_{O,R}$ represents the energy, expressed as a potential, at equilibrium, of a redox couple; $E^{\ominus}_{O,R}$ represents the *same* electrode potential but determined for the redox couple under standard conditions.)

The amounts of material in the analyte solution are related to the electrode potential $E_{O,R}$ according to the Nernst equation, as follows:

$$E_{O,R} = E^{\ominus}_{O,R} + \frac{RT}{nF} \ln\left(\frac{a_O}{a_R}\right) \tag{3.8}$$

where R is the gas constant, F is the Faraday constant, T is the thermodynamic temperature (i.e. in kelvin) and n is the number of electrons involved in the redox reaction. The terms involving a relate to activities (a type of concentration), which will be discussed in detail in Section 3.4 below. For now, we will say that the activity is the concentration which is *perceived* by the electrode.

Notice that equation (3.8) is written in terms of *natural logarithms* (ln), rather than $\log_{10}$.

Worked Example 3.4. A sample of iron-containing ore is crushed and the powder extracted to form a clear aqueous solution. A clean iron rod is immersed into the solution and the cell 'Fe | $Fe^{2+}{}_{(aq)}$ || SCE' is therefore made. The *emf* at equilibrium was measured as 0.714 V at 298 K, and the SCE was the positive electrode. What is the concentration of the iron? (Assume that all the iron exists as a simple aquo ion in the +2 oxidation state and that the solution is 'quiet'.)

Strategy. The procedure to adopt has two parts: first, we will determine the electrode potential $E_{Fe^{2+},Fe}$ from the *emf*. Secondly, the concentration will be calculated by using the Nernst equation (equation (3.8)).

Step 1. What is the electrode potential $E_{Fe^{2+},Fe}$?

From equation (3.3):

$$emf = E_{SCE} - E_{Fe^{2+},Fe}$$

and therefore:

$$\begin{aligned} E_{Fe^{2+},Fe} &= E_{SCE} - emf \\ &= 0.242 \text{ V} - 0.714 \text{ V} \\ &= -0.472 \text{ V} \end{aligned}$$

Step 2. What is the concentration, $[Fe^{2+}]$? (Note that we will actually be calculating the activity $a(Fe^{2+})$.)

We now determine $a(Fe^{2+})$ from the Nernst equation (equation (3.8)), as follows (Fe^{2+} and Fe form a two-electron couple – hence the '2'):

$$E_{Fe^{2+},Fe} = E^{\ominus}_{Fe^{2+},Fe} + \frac{RT}{2F} \ln\left[\frac{a(Fe^{2+})}{a(Fe)}\right]$$

From the appropriate tables (see Appendix 3) we can look up the value of the standard electrode potential $E^{\ominus}_{Fe^{2+},Fe}$ as -0.44 V. It is convenient to remember that the compound unit *(RT/F)* has the value of 0.0257 V.

Next, we remember that E represents a redox couple, i.e. $Fe^{2+}{}_{(aq)}$ in contact with solid Fe metal. Because the iron electrode is a solid in its normal standard state, we say that the denominator within the bracket is unity. We will see the reason for this choice of value in the next section.

Inserting values into the Nernst equation gives the following:

$$-0.472 \text{ V} = -0.44 \text{ V} + \frac{0.0257 \text{ V}}{2} \ln \left[\frac{a(\text{Fe}^{2+})}{1} \right]$$

So, by rearranging, we obtain:

$$\ln[a(\text{Fe}^{2+})] = \frac{2 \times (-0.472 \text{ V} + 0.44 \text{ V})}{0.0257 \text{ V}} = -2.490$$

and therefore:

$$a(\text{Fe}^{2+}) = \exp(-2.490) = 0.0829$$

From this single calculation, we can observe several interesting points:

(i) Notice the way that all three terms, 'E', '$E^{\ominus}$' and '$(RT/2F) \times \ln a$' have the same unit of 'volt'.

(ii) Notice also how, when rearranging to make $\ln[a(\text{Fe}^{2+})]$ the subject of the equation, these three common units of volt all cancel out to leave a logarithm term that is completely dimensionless (as is necessary from the laws of mathematics).

(iii) Finally, notice how the answer – an activity a – is therefore itself dimensionless. This leads to the general rule that **all activities are dimensionless**.

SAQ 3.4

Repeat the calculation shown above – use the same cell, 'Fe | $Fe^{2+}{}_{(aq)}$ || SCE', but this time with a different *emf* of 0.735 V. Remember that $E^{\ominus}_{Fe^{2+},Fe} = -0.44$ V. What is the new value of $a(\text{Fe}^{2+})$?

SAQ 3.5

Now perform a Nernst-based calculation in reverse: if $a(\text{Fe}^{2+}) = 10^{-5}$, what is the new value of $E_{Fe^{2+},Fe}$, and hence the new *emf*?

Types of electrodes. At this point, we need to note that, as analysts, electrodes can be grouped into classes. The division is made according to the way in which a measurement is made. In the example we have just looked at, the iron of the electrode was one part of the redox couple being studied. We see that the electrode in this system had to be iron since an electrode made of any other material would not have completed the redox couple. Electrodes of this type are called **redox electrodes**.[†]

Other commonly employed redox electrodes are metals such as copper, cobalt, silver, zinc, nickel, and other transition metals. Some p-block metals such as tin, lead and indium can also function as redox electrodes. However, s-block metals such as magnesium do not make good redox electrodes since the elemental metal is reactive and forms a layer of oxide coating, which leads to poor reproducibility, poor electronic conductivity and electrode potentials that are difficult to interpret, (see Section 3.3.1).

Redox electrodes are useful to the analyst because the metal, M, is a pure solid and invariably of a lower valence than the ionic analyte in solution, M^{n+}. This means that the denominator in the logarithmic term in equation (3.8) is always unity, i.e. the Nernst equation for a redox electrode M can be simplified to the following:

$$E_{M^{n+},M} = E^{\ominus}_{M^{n+},M} + \frac{RT}{nF} \ln a(M^{n+}) \tag{3.9}$$

Notice how this form of the Nernst equation (equation 3.9) can be thought of as a linear equation of the form '$y = mx + c$', thus allowing calibration graphs to be drawn. Figure 3.6 represents such a calibration graph for copper ions in aqueous solution. Note the important conclusion that the electrode potential decreases (i.e. becomes more negative) as the activity decreases.

The intercept of the graph shown in Figure 3.6 is the standard electrode potential, $E^{\ominus}_{M^{n+},M}$ (in this case, the intercept is $E^{\ominus}_{Cu^{2+},Cu}$ which has a value of 0.34 V). It is a good idea, when using such simple Nernst plots as an analytical method of determining an activity, to check that the intercept at $x = 0$ is indeed the standard electrode potential. (There are many compendia listed in the Bibliography at the end of this book that cite large numbers of $E^{\ominus}$ values, as does Appendix 3.)

Alternatively, if we did not have a reliable value of $E^{\ominus}$ to hand, we could have constructed a graph such as that shown in Figure 3.6 by making up solutions of known activity and then measuring the experimental values of E. The value of $E^{\ominus}$ is the intercept at $\ln a(M^{n+}) = 0$.

A graphical determination of $E^{\ominus}$ is preferred to a single calculation using equation (3.9), since random errors are more likely to be noticed and hence corrective measures can be attempted.

[†] There are a great many names given to these types of electrode. Some texts call them **non-passive** electrodes and others **metal indicator** electrodes.

Using a different convention, a simple metal in contact with its cations is also commonly termed an **electrode of the first kind**, or a **class I** or **first-order** electrode, while an electrode covered with an insoluble salt, e.g. AgCl | Ag for determining $a(Cl^-)$, is termed an **electrode of the second kind**, or a **class II** or **second-order** electrode. In this latter convention, inert electrodes for following redox reactions (cf. Chapter. 4) are somewhat confusingly termed 'redox' electrodes.

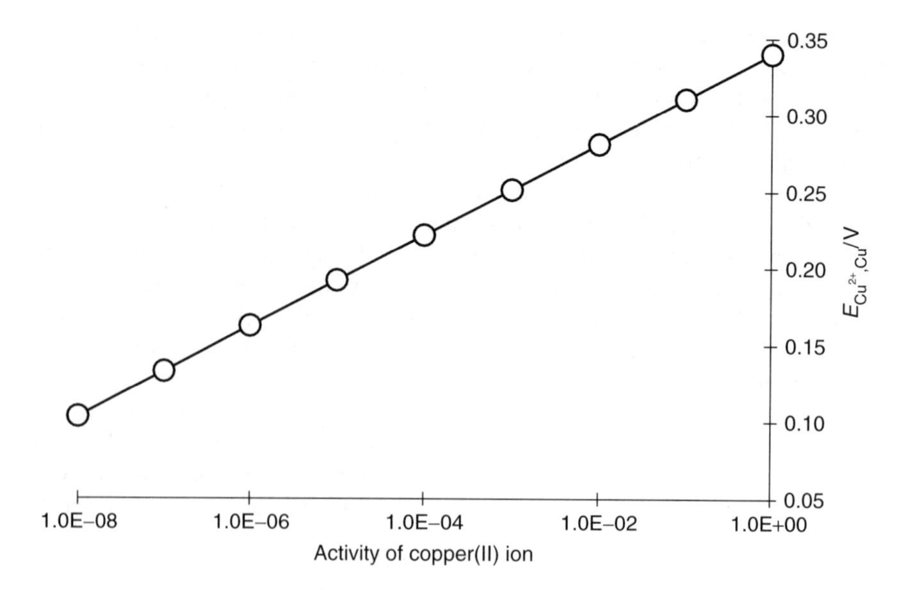

Figure 3.6 Nernst plot of $E_{Cu^{2+},Cu}$ against the logarithm of $a(Cu^{2+})$ to show the validity of the Nernst equation for a redox electrode (equation (3.9); note that the intercept is $E^{\ominus}_{Cu^{2+},Cu}$. Plots of this sort are generally a good way of obtaining values of $E^{\ominus}_{O,R}$.

If the Nernst equation (equation (3.9)) is rewritten slightly in terms of $\log_{10}$, rather than with the natural logarithm (ln), then a factor of 2.303 is introduced as follows:

$$E_{M^{n+},M} = E^{\ominus}_{M^{n+},M} + 2.303\frac{RT}{nF}\log_{10} a(M^{n+}) \tag{3.10}$$

Many analysts prefer this form because a tenfold change in the activity a causes the log term to change by '1'.

Illustrative Calculation. What is the difference between natural and base-10 logarithms when $a(Co^{2+}) = 10^{-4}$ and 10^{-5}?

(a) If $a(Co^{2+}) = 10^{-4}$, $\log[a(Co^{2+})]$ is -4, and $\ln[a(Co^{2+})]$ is -9.21.

(b) If $a(Co^{2+}) = 10^{-5}$, $\log[a(Co^{2+})]$ is -5, and $\ln[a(Co^{2+})]$ is -11.51.

We see that when using $\log_{10}$, the logarithmic terms change by '1' for every tenfold change in the activity a . In contrast, the change when using natural logarithms is an unwieldy factor of 2.303 per tenfold change.

Such a tenfold change in concentration is often described by an analyst as a **decade** (the word comes from the Latin *dec* meaning 'ten'). Many analysts say 'per concentration decade' rather than 'tenfold change in concentration'.

DQ 3.3

Since $E^{\ominus}_{M^{n+},M}$ is a constant, can we estimate the number of decades of dilution from a knowledge of $E_{M^{n+},M}$?

Answer

From the Nernst equation, (equation (3.9)), we can indeed predict how much an electrode potential will change per decade change in activity, or predict the activity from a knowledge of $E_{M^{n+},M}$. *The pre-logarithm term in equation (3.9) is a constant and (with logarithms in base-10) has the value (2.303* $\times$ RT/nF*). At 298 K, this value is equal to (0.0591/*n *V). Using slightly different units, the factor is 59.1/*n *mV.*

The term n *in this relationship is simply the number of electrons involved in the redox reaction. Accordingly, a one-electron couple will change its electrode potential* $E_{M^{+},M}$ *by 59.1 mV per tenfold change in concentration. For this reason, an analyst will say the couple shows a* **slope** *of 59.1 mV per decade (if* n = *1). The word* slope *alerts us to the fact that a* gradient *is involved. In fact, the slope in question is the gradient of a calibration curve such as that shown in Figure 3.6 when the* x*-axis is written as* $\log_{10} a(M^{n+})$.

The electrode potential $E_{M^{2+},M}$ *for a two-electron couple decreases by 29.55 mV per decade change in concentration, while* $E_{M^{3+},M}$ *for a three-electron couple decreases by 19.7 mV per decade.*

The slope is important because it gives us the opportunity to perform simple 'back of the envelope'-type calculations. All we need to know is the number of decades of concentration to which the analyte has been diluted from $a = 1$. (Remember that the Nernst equation is formulated in terms of *activity*, which is the concentration 'perceived' by the electrode.) As an illustration, we can rewrite equation (3.8) as an equation of the following type:

$$E_{M^{n+},M} = E^{\ominus}_{M^{n+},M} - \left[(\text{number of decades of dilution}) \times \frac{59.1}{n}\ \text{mV}\right] \quad (3.11)$$

Note that this equation is merely an *approximate* form of the Nernst equation. Notice also how the minus sign tells us that $E_{M^{n+},M}$ *decreases* as the solution is progressively diluted.

Worked Example 3.5. A new means of extracting nickel from its ore is being investigated. The first step is to crush the rock to powder, roast it, and then extract soluble nickel species (as Ni^{2+}) into an aqueous solution. The activity, $a(Ni^{2+})$, is monitored by a potentiometric method, where a wire of pure nickel metal functions as an electrode and is immersed in aliquot samples taken from the plant. This wire monitors the electrode potential $E_{Ni^{2+},Ni}$. If $E^{\ominus}_{Ni^{2+},Ni} = -0.230$ V, what is $E_{Ni^{2+},Ni}$ if $a(Ni^{2+}) = 10^{-6}$?

We can estimate the electrode potential by saying that the slope from the Nernst equation (equation (3.9)) is $59.1/n$ mV per decade. In this case, $n = 2$, and so the slope is 29.55 mV per decade. $E_{Ni^{2+},Ni}$ will be less than $E_{Ni^{2+},Ni}$, since the activity is less than 1 (hence the minus sign in equation (3.11)).

An activity $a(Ni^{2+})$ of 10^{-6} means that the activity is *six decades* below unit activity. Accordingly, we can then write an equation similar to equation (3.11):

$$E_{Ni^{2+},Ni} = E^{\ominus}_{Ni^{2+},Ni} - (6 \times 29.55 \text{ mV})$$

and therefore we do not need to perform a 'full-blown' calculation by using equation (3.9). In this case, we can predict that the electrode potential is 177.3 mV (or 0.177 V) *more negative* than $E^{\ominus}_{Ni^{2+},Ni}$, so $E_{Ni^{2+},Ni} = -0.407$ V.

SAQ 3.6

Silver ions are being exchanged for ammonium ions on a ion-exchange column and traces of silver ions in the eluent from the column are determined every minute when an aliquot of eluent is taken. In order to measure $a(Ag^+)$, a 'detector' of silver wire is immersed in the liquid to determine the electrode potential $E_{Ag^+,Ag}$. Assume that the silver wire only 'sees' the silver ions, i.e. that no other ions interfere.

Estimate the activity of the silver ion if the *emf* is 0.262 V. The reference electrode used was an SCE ($E_{SCE} = 0.242$ V) and $E^{\ominus}_{Ag^+,Ag} = 0.799$ V.

3.3.1 Limitations of the Nernst Equation

We shall now return to the determination of amounts of iron, e.g. as extracted from an iron-containing ore such as pyrite. Iron can exist in solution in many possible forms, with the most obvious being the ferrous and ferric aquo ions. Metallic iron is normally a good choice of redox electrode for determining the amounts of these ions in solution.

We must be careful, though: the two aquo ions are different in an electrochemical sense, and so have different standard electrode potentials with iron, i.e. $E^{\ominus}_{Fe^{3+},Fe} = -0.04$ V, while $E^{\ominus}_{Fe^{2+},Fe} = -0.44$ V. We therefore need to be certain which iron redox state is involved in the measurement.

We also need to note that if a redox electrode made of iron is not clean but is rusty, then another redox couple is possible, i.e. the couple:

$$Fe(OH)_{2(s)} + 2e^- \longrightarrow Fe_{(s)} + 2OH^-$$

which has a standard electrode potential of −0.88 V. Iron(III) oxides, hydroxides and oxyhydroxides on the surface of the electrode can cause yet more further possibilities for confusion. It should be emphasized that *measurements made with a dirty electrode give meaningless results*! For this reason, the state of the surface of a redox electrodes should be known. If possible, the electrode should be cleaned rigourously and regularly – before each and every measurement clearly being the ideal. Many redox electrodes do indeed require regular cleaning in this way before measurement's can be made. Electrode preparation is discussed in further detail in Chapter 9.

To add to the complexity yet further, we note that the solution obtained from the iron-containing ore could itself undergo reaction. Ferrous ion can be readily oxidized by oxygen from the air, so a solution that originally contained just Fe^{2+} could also contain Fe^{3+} if left standing for some time, where the oxidation process depletes the amount of Fe^{2+} in solution. In generating Fe^{3+}, such aerial oxidation has also caused yet another redox couple to form: in this case, $E^{\ominus}$ for the Fe^{3+}, Fe^{2+} couple is 0.77 V.

A redox couple that is wholly in solution can be analysed without recourse to a redox electrode – indeed, in the example given here, analysis with an iron rod would *complicate* the situation since the Fe^{3+}, Fe^{2+} system itself obeys the Nernst equation (equation (3.8)).

The electrode potential of such a solution-phase system is best followed with an **inert** electrode such as platinum or gold. An 'inert' electrode is so called because it is not involved in the redox reaction except as a 'probe' of the electrode potential E. An inert electrode is also called a **passive** electrode, **flag** electrode or **indicator** electrode.

Analyses using inert electrodes are experimentally identical to those using redox electrodes but are less useful in practice since we usually want to know how much metal is in solution, and use of the Nernst equation (equation (3.8)), e.g. for the case of Fe^{3+} and Fe^{2+}, will merely tell us the **ratio** of the respective redox states in solution.

Worked Example 3.6. An old medicine bottle is discovered that once contained tincture of iodine. The bottle still contains a trace of solid iodine. Some of the solid is dissolved in an aqueous solution of potassium iodide of concentration 0.1 mol dm^{-3}. The electrode potential of the I_2, I^- couple was determined at a platinum electrode immersed in the solution. If $E^{\ominus}_{I_2,I^-} = -0.54$ V and E_{I_2,I^-} for the resultant solution is -0.60 V, what is the concentration of the iodine? Ignore the presence of the brown $I_3{}^-$ ion.

The Nernst equation for the couple is as follows:

$$E_{I_2,I^-} = E^{\ominus}_{I_2,I^-} + \frac{RT}{2F} \ln \left[\frac{a(I_2)}{a(I^-)^2} \right]$$

(Note that here we have squared the activity of the iodide ion since the balanced redox half-cell reaction is $I_2 + 2e^- \rightarrow 2I^-$. This redox stoichiometry also explains the factor of $n = 2$ in the denominator.)

Inserting the appropriate values gives:

$$-0.60 \text{ V} = -0.54 \text{ V} + \frac{0.0257 \text{ V}}{2} \ln\left[\frac{a(I_2)}{(0.1)^2}\right]$$

(Note here how we have assumed that concentration and activity are the same thing: this assumption is why the units of concentration have been omitted – see Section 3.4.)

Rearranging now gives:

$$\frac{2 \times [-0.60 - (-0.54)] \text{ V}}{0.0257 \text{ V}} = \ln\left[\frac{a(I_2)}{10^{-2}}\right]$$

so:

$$-4.69 = \ln\left[\frac{a(I_2)}{10^{-2}}\right]$$

and therefore:

$$\exp(-4.69) = \left[\frac{a(I_2)}{10^{-2}}\right]$$

The exponential term equates to 9.38×10^{-3}. Subsequent cross-multiplying shows that $a(I_2) = 10^{-2} \times (9.38 \times 10^{-3})$, so $a(I_2) = 9.38 \times 10^{-5}$ mol dm^{-3}.

By knowing how much KI solution was employed as solvent, it is a simple matter to back calculate (from the definition of concentration) to find out how much iodine was in the bottle.

We need to appreciate from the above calculation that a huge difference in activity is represented by a very small difference between E and $E^{\ominus}$, in part because of the square term in the Nernst-equation denominator. This explains why it is essential to have a good quality voltmeter (i.e. one having a near-infinite resistance R and with the ability to display the *emf* to several significant figures) and to take readings only when true equilibrium has been reached.

Additional limitations when using the Nernst equation are treated in Section 3.6 ('The Causes and Treatment of Errors').

3.4 Differences between Concentration and Activity

3.4.1 Brief Discourse on Ion–Ion Interactions

So far, we have defined the Nernst equation in terms of activities (a) rather than concentrations (c). This representation is rather false, so we need to remedy the situation. The concept of activity was introduced because the Nernst equation cannot adequately describe the relationship between an electrode potential $E_{O,R}$

and the *concentration* of the soluble components within the redox couple that it represents. Accordingly, instead of concentration, the Nernst equation is defined in terms of **activity**.

As before, we start with the working definition that an activity is the concentration that an electrode *perceives* and, as such, is somewhat of a 'fudge factor'. More formally, activity is defined as follows:

$$a = \frac{c}{c^{\ominus}} \times \gamma \tag{3.12}$$

where c is the *real* concentration of the redox state under study and the term γ is called the **activity coefficient**. The latter term is best thought of as the ratio of a solute's 'perceived' and 'real' concentrations. Note that both the activity a and the activity coefficient γ are dimensionless, which is why the $c^{\ominus}$ ($=1$ mol dm^{-3}) term is included, i.e. to ensure that $(c/c^{\ominus})$ also becomes dimensionless.

In electrolyte solutions, we have a slightly different form of γ, which is called the **mean ionic** activity coefficient, $\gamma_{\pm}$.

The approach that we will follow is known as the **Debye–Hückel** theory. The activity laws discussed in the following are derived from a knowledge of electrostatic considerations, and apply to ions in solution that have an energy distribution that follows the well-known Maxwell–Boltzmann law. Strong electrostatic forces affect the behaviour and the mean positions of all ions in solution.

It is further postulated that the ion is surrounded by an **ionic atmosphere**, so that in the neighbourhood of any positively charged ion (e.g. a cation), there are likely to be more negative charges than positive (and vice versa). The Debye–Hückel theory states that the stronger the attractive forces, then the greater the probability of finding ions of opposite charge within the ionic atmosphere. The cation is surrounded with anions, and each anion is surrounded by cations, so the ionic atmosphere can be thought of as looking like an onion, or a 'Russian doll', with successive layers of alternate charges.[†]

Having associated with other ions, the ion is said to be **screened** from anything else having a charge (including the electrode), thus meaning that the full extent of its charge cannot be 'experienced': the cations are surrounded by anions, which are themselves surrounded by cations, etc., so the charges therefore 'cancel' each other out. In summary, the magnitude of the electrostatic interactions between *widely separated ions* will decrease. The magnitude of the screening depends on the extent of association, so we can assume that *ions are not screened at all at infinite dilution.*[‡] We give a subscript to the activity coefficient to indicate that

[†] The ionic atmosphere is not a static structure, so its composition is best treated statistically. An aggregation of ionic charges, if static, would allow for crystallization if the solution was at all concentrated. In dilute solutions, while the charges might instantaneously have a three-dimensional structure similar to that in an ionic repeat lattice, thermal vibrations soon cause such momentary interactions to break down (i.e. 'shake' free) and reform.

[‡] The term **infinite dilution** means [concentration] = 0. It is clearly nonsense to study an analyte when its concentration is 0 because there is none of it there! Parameters at infinite dilution are obtained by extrapolation from

the activity is decreased owing to ionic interactions, i.e. $\gamma_{\pm}$, and call this the **mean ionic activity coefficient**.

Since the effective charge on the ion has decreased, and since the electrode operates by detecting point charges within the interface between the electrode and the solution, the electrode perceives there to be fewer ions in the solution than are actually present. The *perceived* concentration is seen to be smaller than the *actual* concentration. From the definition of activity given above, i.e. $a = c \times \gamma_{\pm}$, we can see that $\gamma_{\pm} = a/c$, thus implying that $0 \leq \gamma_{\pm} \leq 1$ at all times. Within an analyte concentration range of $0 < c < 0.1$ mol dm^{-3}, this is generally a very good approximation. Figure 3.7 shows the relationship between $\gamma_{\pm}$ and the concentration c for a few simple electrolytes. Note how the multi-valent anions and cations cause $\gamma_{\pm}$ to vary more greatly than do mono-valent ions. As $\gamma_{\pm} = 0.035$, a solution of 0.1 mol dm^{-3} $In_2(SO_4)_3$ at an indium electrode appears to have a concentration of 0.0035 mol dm^{-3}. In this case, it should be noted that $\gamma_{\pm}$ has a large effect on the electrode potential $E_{In^{3+},In}$, i.e. the difference between values of E calculated by using concentration rather than activity would be about 30 mV.

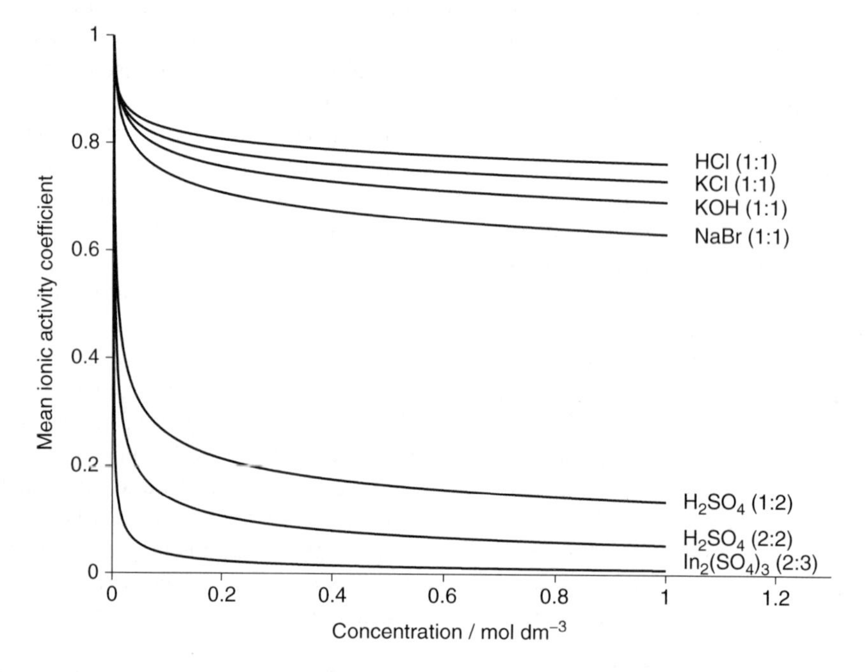

Figure 3.7 Plots of the mean ionic activity coefficient, $\gamma_{\pm}$, as a function of concentration for a variety of ionic electrolytes in water.

---

finite concentrations back to zero concentration. 'Infinite dilution' is useful to an electrochemist since all effects owing to ion–ion interactions can then be assumed to be absent.

3.4.2 Activities

3.4.2.1 The Activity of a Solid

The activity of a pure solid in its standard state is always taken as unity, so a_{Cu} or $a_{Zn} = 1$ for clean and pure Cu and Zn redox electrodes, respectively.

More complicated is the activity of an impure solid, e.g. a mixture of two metals (either as an **alloy**, or an **amalgam** with a metal 'dissolved' in liquid mercury). An alloy is indicated with a colon, so a nickel–silver alloy is written as Ni:Ag. An amalgam is indicated with the mercury in brackets, so a silver amalgam is written as Ag(Hg). Consider an alloy of cobalt (90 mol%) and silver (10 mol%): there will be two activities in such a system, i.e. one each for silver and cobalt in the Co:Ag alloy. The activity in such a situation is merely the mole fraction x, so $a_{Co} = 0.9$, and $a_{Ag} = 0.1$.

SAQ 3.7

The amalgam filling from a tooth was removed and made the positive electrode in a cell of the type SCE | Ag^+ | Ag(Hg). The cell *emf* was determined as 0.547 V. Take $E^{\ominus}_{Ag^+,Ag} = 0.799$ V and $E_{SCE} = 0.242$ V. What is the mole fraction, x_{Ag}, of silver, i.e. a(Ag), in the filling?

3.4.2.2 The Activity of a Gas

The activity of a pure gas is its pressure (in atmospheres), so $a_{H_2} = p_{H_2}/p^{\ominus}$. The activity a of pure hydrogen gas at 1 atmosphere of pressure is therefore 1.

It is not particularly common to employ pure gases, however, for safety reasons, so gases therefore tend to be mixtures. For example, hydrogen is often purchased as a mixture of 10% H_2 in a nitrogen **bath** gas (also called a 'diluent' or 'base' gas), e.g. for when constructing a safer version of the hydrogen electrode. In such a case, we can again approximate the activity to the mole fraction x: we simply say $x \equiv$ partial pressure, so $a_{H_2} = p_{H_2}/p_{TOTAL}$. Therefore, $a_{H_2} = 0.1$ if the gas is a mixture of 10% hydrogen in N_2.

SAQ 3.8

From the Nernst equation, calculate the pressure of hydrogen involved if $a(H_3O^+) = 0.011$, and $E_{H^+,H_2} = 0.008$ V. Take $p^{\ominus} = 10^5$ Pa. (Remember from the balanced half-cell reaction that $n = 2$.)

3.4.2.3 The Activity of a Solution

It is unwise to speak broadly of 'the activity of a solution' because so many different situations could be considered:

(i) The activity of a mixture of liquids: it is usually a good approximation to suggest that the activity of an impure liquid (that is, a liquid dispersed in another liquid) is equal to its mole fraction x. This situation is rare in electrochemistry, except for the case of very dilute amalgams (to a maximum mole fraction of 1% metal in Hg), so we will not consider such a situation any further in this chapter.

(ii) The activity of a solute in a solvent: when working with the very smallest concentrations (to a maximum of about 10^{-6} mol dm^{-3}), the activity a and concentration c may be considered to be wholly identical, provided that no additional solutes are present simultaneously in the same solution. Such a concentration is so tiny, however, that it gives the implication of slightly polluted distilled water.

For all other situations, we employ the **Debye–Hückel** laws, as below, to *calculate* the activity coefficient $\gamma_{\pm}$. By knowing $\gamma_{\pm}$, we then say that $a = c \times \gamma_{\pm}$ (remembering to remove the concentration units because a is dimensionless).

3.4.3 Activity Coefficients

With the above considerations in mind, Debye and Hückel devised a means of *calculating* the ionic strength I and the mean ionic activity coefficient $\gamma_{\pm}$. Before we can progress, however, we must define the extent to which a solute promotes association, and thus screening. The parameter of choice is the **ionic strength** I, which is defined formally as follows:

$$I = \frac{1}{2} \sum_{i=1}^{i=i} c_i z_i^2 \tag{3.13}$$

where z_i is the charge on the ion i, and c_i is the straightforward concentration of the ion. Equation (3.13) looks daunting, so we will consider a couple of simple examples to demonstrate the way that ionic strength is obtained.

Worked Example 3.7. What is the ionic strength I of a 1:1 electrolyte such as NaCl, at a concentration of c?

$$I = \frac{1}{2} \sum c_i z_i^2$$

so:

$$I = \frac{1}{2} \sum [[Na^+] \times (+1)^2 + [Cl^-] \times (-1)^2]$$

We next insert concentration terms, noting that one sodium and one chloride are formed per formula unit of sodium chloride (which is why we call it a 1:1 electrolyte). Accordingly, $[Na^+] = [Cl^-] = [NaCl]$, and therefore we can write:

$$I = \tfrac{1}{2}[([NaCl] \times 1) + ([NaCl] \times 1)]$$

We obtain the result that $I_{\text{NaCl}} = [\text{NaCl}]$ for this 1:1 electrolyte. Note that I has the same units as concentration.

We now consider a 1:2 electrolyte, such as $CoCl_2$. So, from equation (3.13) we obtain:

$$I = \frac{1}{2}\sum c_i z_i^2$$

and therefore:

$$I = \tfrac{1}{2}[[\text{Co}^{2+}] \times (+2)^2 + [\text{Cl}^-] \times (-1)^2]$$

Note that the calculation requires the charge *per* anion, rather than the total anionic charge. We next insert the concentrations. In this case, there are two chloride ions formed per formula unit of salt, so $[\text{Cl}^-] = 2 \times [\text{CoCl}_2]$, which gives:

$$I = \tfrac{1}{2}[([\text{CoCl}_2] \times 4) + (2[\text{CoCl}_2] \times 1)]$$

We therefore obtain the result that $I = 3 \times c$ for a 1:2 electrolyte.

Table 3.1 summarizes the relationship between concentration c and ionic strength I as a function of electrolyte type for salts $M^{x+}X^{y-}$. We will need this table often. Note that salts with higher ionic charges tend to be weak electrolytes, i.e. they do not dissociate completely. The consequences are that concentrations are too low, itself meaning that the computed ionic strength is too high. In fact, we are unlikely to know the value of I if dissociation is incomplete.

SAQ 3.9

What is the relationship between concentration c and ionic strength I for copper sulfate, a 2:2 electrolyte?

Table 3.1 The relationship between ionic strength I and concentration c for salts of the type $M^{x+}X^{y-}$. As an example, sodium sulfate (a 1:2 electrolyte) has an ionic strength that is three times larger than is c

	X^-	X^{2-}	X^{3-}	X^{4-}
M^+	1	3	6	10
M^{2+}	3	4	15	12
M^{3+}	5	15	9	42
M^{4+}	10	12	42	16

3.4.4 Revision of Debye–Hückel Theory

Knowing the ionic strength, we are now in a position to determine the mean ionic activity coefficient $\gamma_\pm$ by using the Debye–Hückel laws. There are two such laws, namely the **limiting law** and the **extended law**.

At the lowest ionic strengths ($0 < I$ (mol dm^{-3}) $\leq 10^{-3}$, which, because $I > c$, could imply a concentration as low as 10^{-5} mol dm^{-3}), we employ the **limiting** law:

$$\log_{10} \gamma_\pm = -\mathrm{A} \mid z^+ z^- \mid \sqrt{I} \tag{3.14}$$

where A is the dimensionless 'Debye–Hückel "A" constant', and has a value of 0.509 at 25°C, and z^+ and z^- are the charges *per* cation and anion, respectively. Note also that the vertical bars '|' here imply a **modulus** rather than a phase boundary. Being a modulus, the charges on the signs are treated as having magnitude alone, so we ignore the signs. From equation (3.14), we expect a plot of $\log_{10} \gamma_\pm$ against $\sqrt{I}$ to be linear (cf. Figure 3.7).

Worked Example 3.8. What is the activity coefficient of copper in a solution containing 10^{-4} mol dm^{-3} $CuSO_4$?

From Table 3.1 above, the ionic strength of such a 2:2 electrolyte is four times its concentration, so $I = 4 \times 10^{-4}$ mol dm^{-3}.

By inserting the value in equation (3.14):

$$\begin{aligned} \log_{10} \gamma_\pm &= -0.509 \mid +2 \times -2 \mid \sqrt{4 \times 10^{-4}} \\ &= -0.509 \mid 4 \mid 2 \times 10^{-2} \\ &= -4.072 \times 10^{-2} \\ \gamma_\pm &= 10^{-4.072 \times 10^{-2}} = 0.910 \end{aligned}$$

For solutions that are more concentrated (i.e. for ionic strengths in the range $10^{-3} < I$ (mol dm^{-3}) $\leq 10^{-1}$), we employ the Debye–Hückel **extended** law as follows:

$$\log_{10} \gamma_\pm = \frac{-0.509 \mid z^+ \times z^- \mid \sqrt{I}}{1 + \sqrt{I}} \tag{3.15}$$

where all terms have the same meaning as above (equation (3.14)). It is useful to note that the extended law simplifies to become the limiting law at extremely low I values since '$1 + \sqrt{I}$' tends to 1 as I tends to zero, and so the numerator will be divided by 1.

Worked Example 3.9. What is the activity coefficient of copper ion in a solution containing 10^{-2} mol dm^{-3} $CuSO_4$?

Again, we start by saying that $I = 4 \times c = 4 \times 10^{-2}$ mol dm^{-3}, for the same reasons as previously.

By inserting the values in equation (3.15):

$$\log_{10} \gamma_{\pm} = \frac{-0.509 \mid +2 \times -2 \mid \sqrt{4 \times 10^2}}{1 + \sqrt{4 \times 10^{-2}}}$$

$$= \frac{-0.509 \mid +4 \mid 0.2}{1 + 0.2}$$

$$= -\frac{0.407}{1.2}$$

$$= -\frac{0.407}{1.2} = -0.339$$

$$\gamma_{\pm} = 10^{-0.339} = 0.46$$

which is passably close to the experimental value of 0.41. We see that $\gamma_{\pm}$ can decrease substantially if the ionic strength is high.

DQ 3.4

If $\gamma_{\pm}$ depends on the ionic strength, and the ionic strength depends on the number of ions in solution, then what happens if the analyte is present in low concentration but the 'solvent' is itself an ionic solution?

Answer

This question highlights the central problem with the potentiometric approach to electroanalysis. We will illustrate this by using the following worked example.

Worked Example 3.10. *Here, we will consider a real situation. There is thought to be contamination from a zinc smelting plant, so a sample of soil from near the smelter is collected and digested in sulfuric acid (of concentration 0.01 mol dm^{-3}) in order to leach out the soluble zinc as the sulfate salt.*

The amount of zinc in the soil was determined by immersing a rod of clean, pure zinc in the solution of zinc sulfate plus sulfuric acid† *and measuring* $E_{Zn^{2+},Zn}$ *as* -0.864 *V.* What is the **concentration** of the zinc solution formed by digesting in acid and what is the **activity** of the zinc salt? *In order to answer there questions, we will need to know the standard electrode potential,* $E^{\ominus}_{Zn^{2+},Zn}$, *which is* -0.760 *V.*

---

† Here, we assume that the acid will not cause any of the zinc electrode to dissolve (which would therefore increase the amount of zinc in solution!).

Strategy. We will assume that the only solutes present are $ZnSO_4$ and H_2SO_4, and first calculate the ionic strength of the solution. (For the purposes of this calculation, we can safely assume that the amounts of zinc sulfate are negligible when compared with the amounts of sulfuric acid.) We will then calculate $\gamma_\pm$ for the zinc cation from I *(and, for simplicity, obtain $\gamma_\pm$ from the Debye–Hückel* limiting *law).*

Knowing $a(Zn^{2+})$ *from the Nernst equation and the activity coefficient $\gamma_\pm$ for the zinc cation from the Debye–Hückel law, we can therefore calculate the concentration, $[ZnSO_4]$, from equation (3.10).*

Sulfuric acid is a 2:1 electrolyte, and so (by using the data in Table 3.1) the ionic strength I *is three times the concentration, i.e.* $I = 0.03 \text{ mol dm}^{-3}$. *Next, from the Debye–Hückel extended law equation (3.15), we can obtain the mean ionic activity coefficient $\gamma_\pm$ as follows:*

$$\log_{10} \gamma_\pm = -\frac{-0.509 \mid 2 \times 1 \mid \sqrt{0.03}}{1 + \sqrt{0.03}}$$

which gives:

$$\log_{10} \gamma_\pm = -0.150 \text{ and } \gamma_\pm = 0.707$$

By using the Nernst equation, we can now calculate the activity of the zinc as follows:

$$E_{Zn^{2+},Zn} = E^{\ominus}_{Zn^{2+},Zn} + \frac{RT}{2F} \ln a(Zn^{2+})$$

By inserting the appropriate values, we obtain:

$$-0.864 \text{ V} = -0.760 \text{ V} + \frac{0.0257 \text{ V}}{2} \ln a(Zn^{2+})$$

which after rearranging gives:

$$a(Zn^{2+}) = 3.0 \times 10^{-4}$$

Finally, Knowing both $a(Zn^{2+})$ *and* $\gamma_\pm$*:*

$$[Zn^{2+}] = \frac{a(Zn^{2+}) \times c^{\ominus}}{\gamma_\pm} = 4.24 \times 10^{-4} \text{ mol dm}^{-3}$$

It is clear from this calculation that a small amount of analyte in the presence of high ionic strength from another source can significantly change an electrode potential, and vice versa.

SAQ 3.10

Repeat the calculation shown in DQ 3.4, but in this case with the zinc dissolved in 0.1 mol dm^{-3} phosphoric acid (H_3PO_4). Calculate the *concentration* of the zinc if $E_{Zn^{2+},Zn}$ is −0.880 V; take $E^{\ominus}_{Zn^{2+},Zn} = -0.760$ V.

DQ 3.5

So how much difference does it make when using activity rather than concentration in the Nernst equation?

Answer

A lot, particularly at higher concentrations and higher ionic strengths! This can be seen by considering the following worked example.

Worked Example 3.11. *We know the concentration of copper sulfate to be 0.01 mol dm^{-3} from other experiments, and so we also know (from suitable tables) that the mean ionic activity coefficient of the copper sulfate solution is 0.404. The measured electrode potential was $E_{Cu^{2+},Cu} = 0.269$ V and $E^{\ominus}_{Cu^{2+},Cu} = 0.340$ V. We will calculate the concentration of Cu^{2+} by using two different approaches, namely (i) by using the naive view that concentration and activity can be employed interchangeably, and (ii) recognizing that $a \neq c$, i.e. by also including the mean ionic activity coefficient in the calculation.*

(i) The naive view – concentration and activity are employed interchangeably.

We will start with the Nernst equation[†] *(taking '$[Cu_{(s)}]$' = a(Cu) = 1):*

$$E_{Cu^{2+},Cu} = E^{\ominus}_{Cu^{2+},Cu} + \frac{RT}{2F} \ln[Cu^{2+}]$$

Inserting the appropriate values gives:

$$0.269 \text{ V} = 0.340 \text{ V} + 0.0128 \text{ V} \ln[Cu^{2+}]$$

and therefore:

$$[Cu^{2+}] = \exp\left(\frac{0.269 \text{ V} - 0.34 \text{ V}}{0.0128 \text{ V}}\right)$$

$$= \exp(-5.547) = 0.0039 \text{ mol dm}^{-3}$$

So, the concentration calculated is far removed from the known, i.e. actual, concentration.

(ii) Including the mean ionic activity coefficient within the calculation.

[†] Note that, when using the Nernst equation in terms of concentration, we encounter the usual problem of not knowing how to deal with an electrode of solid copper metal: we will simply incorporate values for copper within the Nernst equation by saying that the denominator is unity.

Again using the Nernst equation:

$$E_{Cu^{2+},Cu} = E^{\ominus}_{Cu^{2+},Cu} + \frac{RT}{2F} \ln \left[\frac{a(Cu^{2+})}{a(Cu)} \right]$$

and inserting the appropriate values:

$$0.300\ V = 0.340\ V + 0.0128\ V \ln([Cu^{2+}] \times \gamma_{\pm})$$

Note here that we have substituted the activity a *by its constituent parts (from equation (3.12)). (The* $c^{\ominus}$ *term has been omitted for clarity, and we have also assumed that* $a(Cu) = 1$.*)*

Rearranging gives the following:

$$[Cu^{2+}] = \frac{1}{0.404} \times \exp\left(\frac{0.269\ V - 0.34\ V}{0.0128\ V}\right)$$

$$= \frac{1}{0.404} \times \exp(-5.547) = 0.01\ mol\ dm^{-3}$$

Therefore, by including the mean ionic activity coefficient into the calculation, the real *concentration is seen to be the same as the apparent concentration; without the mean ionic activity coefficient, the* apparent *(computed) concentration is about 2.4 times too small.*

From the above, it is clear that knowledge of $\gamma_{\pm}$ *is vital for an* accurate *answer. Unfortunately, calculations of this second type are often quite impossible to carry out since the mean ionic activity coefficient* $\gamma_{\pm}$ *is normally unknown.*

SAQ 3.11

The mean ionic activity coefficient, $\gamma_{\pm}$, of copper sulfate decreases to 0.158 when the concentration was raised from 0.01 to 0.1 mol dm^{-3}. By using the Nernst equation, calculate the electrode potential, $E_{Cu^{2+},Cu}$, for the following situations: (i) where $\gamma_{\pm}$ is naively assumed to be 1 (i.e. by using the Nernst equation but with concentration rather than activity), and (ii) the more realistic situation where $\gamma_{\pm} = 0.158$. Take $E^{\ominus}_{Cu^{2+},Cu} = 0.34$ V.

SAQ 3.12

When will the effect of the mean ionic activity coefficients be most likely to affect the accuracy of the concentrations calculated by the Nernst equation? (Hint – when is the value of activity most like the concentration?)

DQ 3.6

So what do we do about the problem of activity coefficients – surely we can't just 'write off' potentiometric measurements?

Answer

Variations in ionic strength are such an important concern that it is recommended for solutes to be analysed by a potentiometric procedure only *if the ionic strength is known and controlled. Furthermore, calibration steps, i.e. to determine the standard electrode potential* $E^{\ominus}$ *should also be performed in a solution of the same, known, ionic strength, e.g. in a solution of perchloric acid*† *of* $I = 1.0$ *mol dm*$^{-3}$*. Provided that* I *is always much higher than the concentration of the analyte, the latter does not contribute more than a tiny fraction of the overall ionic strength and so fluctuations in the activity coefficient* $\gamma_{\pm}$ *can be safely ignored. If calibrations are performed by using the same constant-ionic-strength solution, then activities and concentrations can be interconverted without any problems.*

To illustrate these effects, look at Figure 3.8. In this figure, (a) shows the effect that increasing the concentration has on the activity coefficient, i.e. without any additional salts being added, while (b) shows calibration graphs of $E_{Zn^{2+},Zn}$ against $[Zn^{2+}SO_4]$ as a function of the *total* ionic strength, with I being adjusted by adding inert $MgSO_4$ to the solution. It is clear that adding a second salt has significantly improved the linearity of the Nernst-based plots, and it is only at higher ionic strengths that any curvatures are apparent. Such curvatures onset at lower ionic strengths when lower concentrations of additional salts are involved, thus implying that as high a concentration of the latter as possible should be employed. In addition note that the separation between the curves in Figure 3.8(b) is decreased at higher ionic strengths, further implying that errors in the recorded *emf* will be proportionately greater.

Solutions of known ionic strength are now available commercially, as are tablets of inert electrolyte, where the latter are dissolved in a known volume of water to produce a solution of predetermined I – this is merely a volumetric procedure. Such tablets are much like buffer tablets, and are called **ionic strength adjusters**.

The following is an important point: if a constant ionic strength can be assumed, then a calibration graph can be constructed of *emf* or electrode potential against *concentration*, rather than against activity. Most commercial ion-selective electrodes (see Section 3.5) would be effectively useless without such calibration graphs.

---

† Perchloric acid is often a good choice since the perchlorate anion can be assumed to be wholly non-complexing. It should be noted, however, that anhydrous perchlorates are an explosion risk if used in conjunction with organic materials.

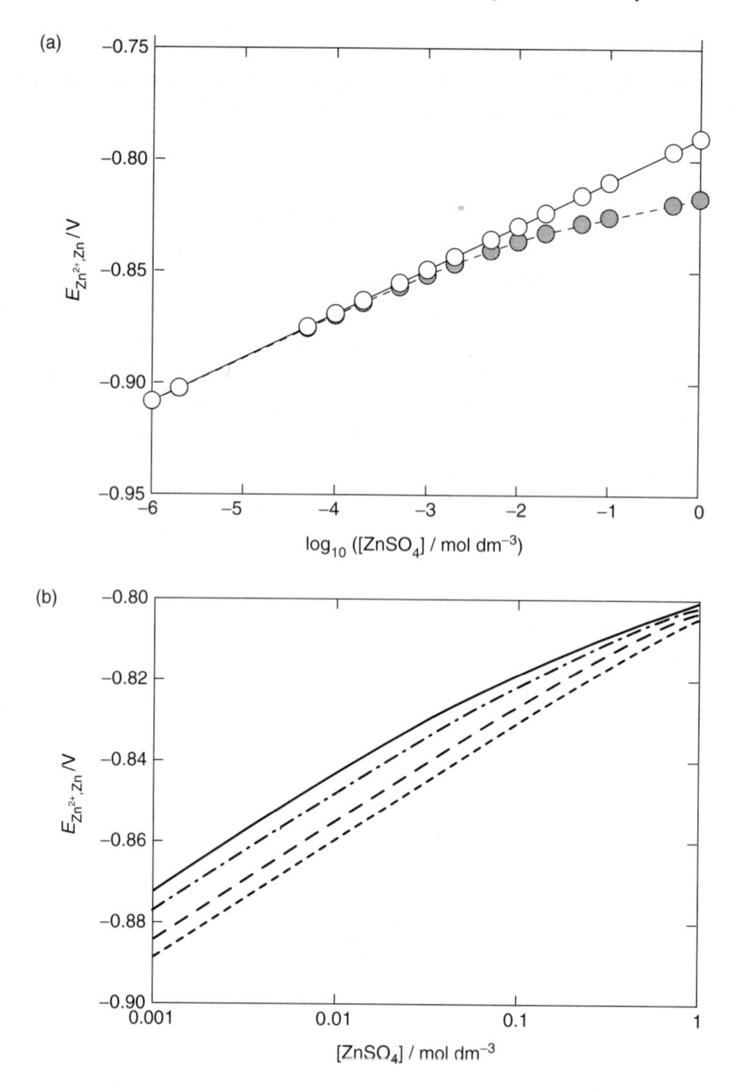

Figure 3.8 The effects of the activity coefficient $\gamma_\pm$ on $E_{Zn^{2+},Zn}$. (a) Plots of electrode potential $E_{Zn^{2+},Zn}$ against concentration: (O——) the 'perceived concentration' (activity); (●- - -) real concentration. All activity coefficients were computed using equation (3.15), while the values of $E_{Zn^{2+},Zn}$ were calculated by using equation (3.9). (b) Calibration curves of $E_{Zn^{2+},Zn}$ against [$ZnSO_4$] as a function of the total ionic strength, I, where the salt added to increase I is K_2SO_4, which is assumed to be inert: (——) 0.1; (– · – ·) 0.2; (- - - - -) 0.5; (· · · · · ·) 1.0 mol dm^{-3}. The overall ionic strength $= 4 \times$ ([$ZnSO_4$] + [$MgSO_4$]), and the values of $E_{Zn^{2+},Zn}$ were again calculated by using equation (3.9). Notice the curvatures at high concentrations and that all traces are linear at [$ZnSO_4$] $< 10^{-2}$ mol dm^{-3}.

Often, the potentiometric determination of concentration requires a preferred pH range. If pH is also important, then the ionic strength adjuster can conveniently function additionally as a pH buffer. Such tablets are called **total ionic strength adjustment buffers** (or TISABs).[†]

Incidentally, ionic strength adjusters and TISABs also decrease all junction potentials (see Section 3.6.5).

3.5 Applications Based on Calculations with the Nernst Equation

3.5.1 pH Determination and the pH Electrode

A **pH electrode**[‡] consists of a glass tube culminating in a glass bulb (see Figure 3.9(a)). The bulb is filled with a buffer containing a known concentration of chloride ion, with the pH within the bulb being typically 7 (to minimize acid or alkaline errors, as discussed below in section 3.5.2.2).[§] A silver wire is immersed in this solution, and the silver soon becomes coated with a thin film of silver chloride (AgCl), so that the solution inside the bulb then becomes saturated with AgCl. The bulb is usually made of common soda glass, i.e. glass containing a high concentration of sodium. Finally, a small reference electrode, e.g. a silver–silver chloride electrode[¶] (Ag | Ag) is positioned beside the bulb. For this reason, the pH electrode ought properly to be called a **pH combination electrode** (see Figure 3.9(b)). If the electrode does not have an additional reference, it is termed a **glass electrode** (GE) (see Figure 3.9(a)). Operation of the glass electrode requires that an additional external reference electrode be employed.

In operation, the pH electrode is immersed in an acid solution of unknown concentration. On immersion, a potential rapidly develops across the glass – hence the pH electrode's fast response. The best response is obtained if the glass is very thin, so the bulb usually has a thickness of 50 microns (μm) or so (1 micron = 10^{-3} mm = 10^{-6} m), and thus the glass electrode is particularly fragile. The glass is not so thin that it would be porous, though, so we do not need to worry about any junction potentials, E_j (see Section 3.6.5).

The non-porous nature of the glass does imply that an extremely large resistance is incurred by the cell. Accordingly, the circuitry of a pH meter must be able to operate with minute currents.

[†] The TISAB also stops complexation between fluoride and other ions in solution, particularly Al^{3+}, Fe^{3+} and silicate species.

[‡] A pH electrode is sometimes also called a 'membrane' electrode. This usage is confusing and will not be followed here.

[§] It is a commonly held misconception that the bulb contains only hydrochloric acid.

[¶] Care: some authors use the acronym SSCE to mean a sodium-chloride saturated calomel electrode, i.e. an SCE filled with NaCl rather than KCl. In order to avoid this confusion, we will avoid this acronym altogether.

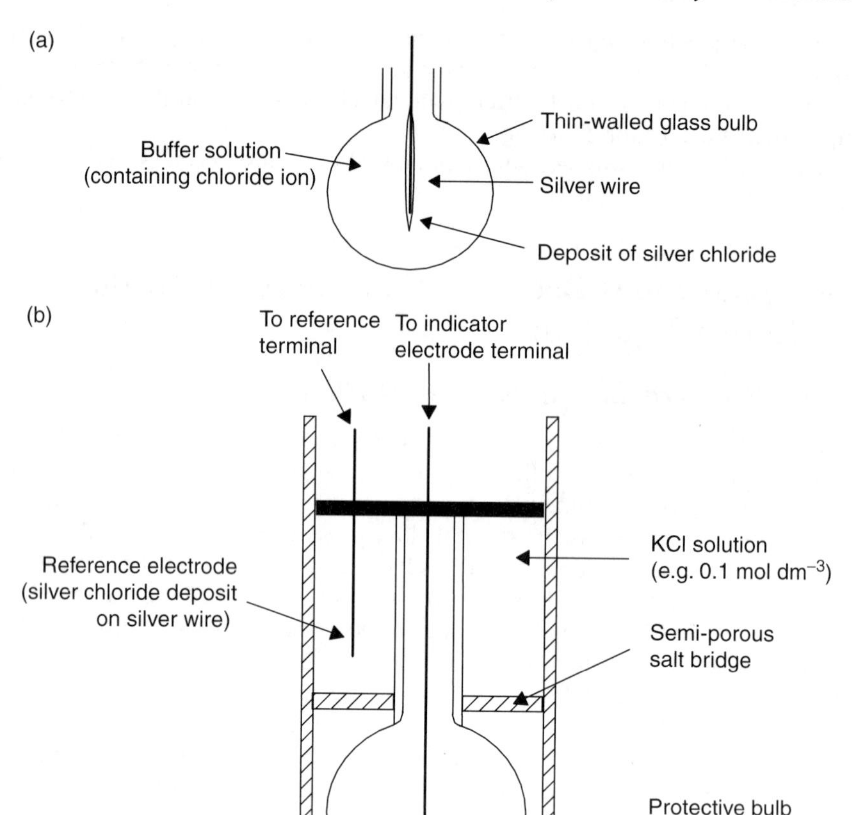

Figure 3.9 Schematic diagrams of pH electrodes, showing (a) a glass electrode, and (b) a pH combination electrode (note that the glass and the bulb are the same in both cases).

The magnitude of the potential that develops across the glass on immersion depends on the difference between the concentration of acid inside the GE (which is known) and the concentration of the acid outside the glass bulb (the analyte, which is to be measured). Provided that the internal pH does not alter, the *emf* generated is linearly related to the pH of the analyte solution – this is sometimes known as the Haber–Klemensiewicz effect (see below).

The pH dependence described above indicates that a pH meter is merely a pre-calibrated voltmeter, which converts the *emf* measured into a solution pH, according to the following relationship:

$$emf = K + \frac{2.303\ RT}{F}\text{pH} \tag{3.16}$$

where the constant K incorporates not only the $E_{\text{reference electrode}}$ but also the stresses, etc. inherent in the glass bulb. Such stresses alter all the time, thus causing K to change, and so, as a corollary, the pH electrode should be calibrated often. Since K incorporates stresses, we need to note that a pH measurement depends strongly on how 'gently' or otherwise we perform an analysis using such an electrode.

We should have noticed the remarkable resemblance between equation (3.16) and equation (3.11) presented earlier, where the latter derives from the Nernst equation. In fact, this similarity arises because of the definition of pH, i.e. pH $= -\log_{10} a(\text{H}^+)$.

Accordingly, equation (3.16) relates to an *emf* changing by a constant amount per *decade* change in activity (cf. equation (3.1)). A moment's additional reflection tells us that equation (3.16) is simply a *specific* form of the Nernst equation.

From equation (3.16), it can be readily calculated that the *emf* should change by 0.0591 V per pH unit, i.e. the electrode has a ***slope*** of 0.0591 V per *decade*. A moment's pause shows that this is actually a simple statement of the obvious–a graph of *emf* (as '*y*') against $\log_{10}[\text{H}^+]$ (as '*x*') will have a gradient of 0.0591 V, and 'per decade' because each pH unit represents a concentration change of 10 times. Therefore, a pH of 3 means $[\text{H}^+] = 10^{-3}$ mol dm^{-3}, a pH of 4 means $[\text{H}^+] = 10^{-4}$ mol dm^{-3} and a pH of 5 means $[\text{H}^+] = 10^{-5}$ mol dm^{-3}, and so on. If the glass electrode does indeed have a slope of 59 mV, its response is said to be **nernstian**, that is, it obeys the Nernst equation.

Worked Example 3.12. A pH electrode is immersed in a solution of buffer at pH 4, and the *emf* is 129 mV. What is the pH when the pH electrode is subsequently immersed in cider vinegar and the *emf* is 150 mV? Assume the behaviour to be wholly nernstian, and that I is the same for both measurements.

We will first **calibrate** the pH electrode with a buffer solution, i.e. to obtain K. Inserting the appropriate values into equation (3.16) gives the following:

$$0.129 \text{ V} = K + (0.0591 \text{ V} \times 4)$$

so:

$$0.129 \text{ V} - (4 \times 0.0591 \text{ V}) = K$$

and therefore K is -0.107 V.

Knowing K, we are now in a position to answer the question. Again, we start by inserting the relevant values into equation (3.16):

$$0.150 \text{ V} = -0.107 \text{ V} + (0.0591 \text{ V} \times \text{pH})$$

so:

$$\frac{(0.150 + 0.107) \text{ V}}{0.0591 \text{ V}} = \text{pH}$$

and we therefore calculate the pH of the cider vinegar to be 4.3.

SAQ 3.13

A pH electrode is immersed in a solution of buffer at pH 7, and the *emf* is found to be 276 mV. What is the pH when the pH electrode is subsequently immersed in a different solution and the *emf* is 502 mV? Use equation (3.16) as before.

Measurements with pH or glass electrodes are very similar experimentally, with the only real difference being the exact position of the reference electrode, i.e. either within the electrode housing (in a pH combination electrode, as shown in Figure 3.9(b)) or wherever the experimenter decides to place it (using a glass electrode).

The advantages and disadvantages of using glass and pH electrodes are summarized in Table 3.2 and below in Section 3.5.2.2.

3.5.2 Ion-Selective Electrodes

The pH electrode (and its less sophisticated parent, the glass electrode) are the most commonly encountered forms of **ion-selective electrodes** (ISEs).† Such an electrode is best defined as 'an electrode having a nernstian response to a single ion in solution' where, by 'nernstian', we again mean that the Nernst equation is obeyed. The pH electrode is an ion-selective electrode since it only responds to protons in solution (with the occasional exception of cations of the alkali and alkaline-earth metals, as discussed below).

DQ 3.7

An ion-selective electrode is said to be **selective** if it detects one particular ion only and ignores all others. Surely, the Nernst equation should only include the component parts of a redox couple – why do other ions interfere?

Answer

The Haber–Klemensiewicz effect relies on charge accumulation either side of the thin membrane of glass. Usually the proton is the only ion of suitable charge and size that can adsorb to the surface of the glass. The potential measured at the pH electrode is in fact the sum of the charges of all ions adsorbed at the glass | solution interface, so if other ions were adsorbed, the potential measured would have additional contributions, i.e. from ions other than the proton. It would be non-selective, as described in Section 3.5.2.2 below.

Ion-selective electrodes are quite commonplace now and can test for most of the ions commonly encountered in an analytical laboratory. Other than the pH

† We often see these particular ISEs described as 'solid-state ISEs'.

Table 3.2 Advantages and disadvantages of glass (GE) and the pH electrodes

Advantages

1. If recently calibrated, the GE and pH electrodes both give an accurate response
2. The response is rapid (possibly of the order of a millisecond)
3. The electrodes are relatively cheap
4. Junction potentials are absent or minimal, depending on the choice of reference electrode
5. The electrodes draw a minimal current
6. The glass is chemically robust, so the GE can be used under both oxidizing and reducing conditions, plus the internal acid solution cannot contaminate the analyte
7. The pH electrode has a very high **selectivity** – perhaps as high as 10^5:1 at room temperature – so only one foreign ion is detected per 100 000 protons (although see 'Disadvantage 6' below). The selectivity does decrease significantly above c. 35°C

Disadvantages

1. To some extent, the constant 'K' is a function of the area of glass in contact with the acid analyte. For this reason, no two glass electrodes will have the same value of K
2. For the same reason, K contains contributions from the strains and stresses experienced by the glass.
3. Following from (2) the electrode should be recalibrated often
4. The value of K, in fact, may itself be slightly pH-dependent, since the strains and stresses themselves depend on the amount of charge incorporated into the surfaces of the glass
5. The glass is very fragile and, if possible, should not be rested against the hard walls or bases of a beaker or container
6. The measured *emf* contains responses from ions other than the proton. Of these other ions, the only one that is commonly present is sodium. This error is magnified at very high pH (>11) when very few protons are in solution, and is known then as the 'alkaline error'

electrode, probably the most common are the solid-state electrodes, as typified by the fluoride electrode (described in the next section). Another common solid-state ISE is the sulfide electrode, with the 'active solid' being Ag_2S; other solid-state ISEs can test for such common ions as Cl^-, Br^-, I^-, Ag^+, Cu^{2+}, Pb^{2+} and CN^-.

In addition to solid-state electrodes, other ISEs operate with a polymer membrane, with a good example being the calcium electrode described below in Section 3.5.2.3.

Descriptions of ISEs other than the fluoride and membrane electrodes are illustrative of different types of device. Such ISEs are beyond our scope here, but are described in a voluminous literature, (see the Bibliography).

3.5.2.1 The Fluoride Electrode

We have briefly encountered the solid-state fluoride electrode, which has a fully nernstian response down to c. 10^{-5} mol dm^{-3}. The fluoride electrode is immersed in a test solution of fluoride ion (usually aqueous), and the *emf* is then determined. At its heart is a single crystal of lanthanum fluoride doped with erbium fluoride, (see Figure 3.10). Like the pH electrode, a full 'fluoride electrode' also contains a small reference electrode, meaning that a fluoride *electrode* is in reality a *cell.* The fluoride electrode does not suffer from interference from Cl^-, so an AgCl | Ag reference is the normal choice owing to its convenience and compact size.

DQ 3.8

Why does the fluoride electrode work?

Answer

Equilibrium exists between the F^- aquo ions in solution and the F^- lattice ions inside the solid LaF_3 crystal; this is true for both faces of the crystal, i.e. the face in contact with the analyte of unknown a *(F^-) and the face in contact with the internal solution of known* a *(F^-). In a similar manner to the formation of a potential across a glass membrane in the pH electrode, a potential develops across the LaF_3 crystal if the activities of fluoride*

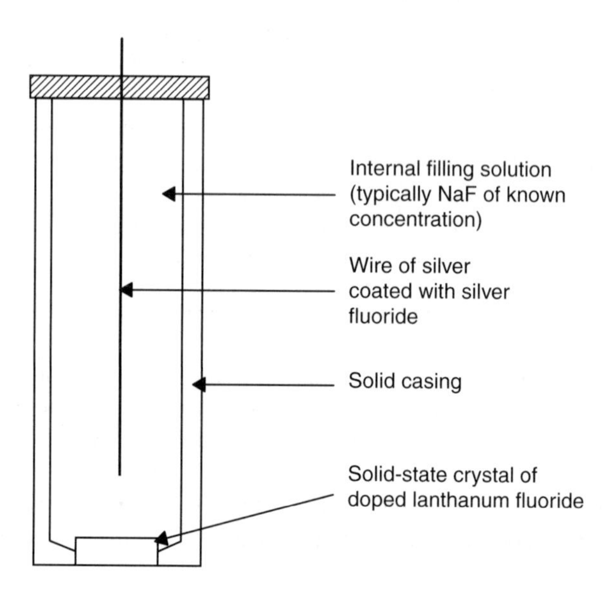

Figure 3.10 Schematic diagram of a solid-state ion-selective electrode for measuring the concentrations of aqueous fluoride ions – the so called 'fluoride electrode'. The silver wire acts as one of the electrodes, so an additional electrode is required to complete the cell.

are different on either side of it. This potential is taken as E_{ISE}, *so the measured* emf *is equal to* ($E_{ISE} - E_{reference}$).

DQ 3.9

So how do the fluoride ions enter into the LaF_3 crystal?

Answer

Ion 'hopping' is a familiar concept in the chemistry of solid-state conductors, e.g. in the semiconductor industry. In the fluoride electrode, fluoride vacancies inside the solid LaF_3 lattice allow for conduction of charge (see Figure 3.11), in turn registered by the electrode as a potential. The emf *is zero if the internal and external solutions are the same because the same numbers of fluoride ion enter the crystal from either face.*

The circuitry within a 'fluoride meter' converts the *emf* to an activity of fluoride ion, and commonly utilizes an equation that is directly analogous to equation (3.16) which allows for interconversion of *emf* and pH for the pH electrode. The pF of the solution in terms of an *emf* from a fluoride 'electrode' cell is described by the following:

$$emf = K - \frac{2.303\,RT}{F} \log_{10} a(\mathrm{F}^-) \tag{3.17}$$

where $\mathrm{pF} = -\log_{10} a(\mathrm{F}^-)$, and K incorporates the potential of the reference electrode and the standard potential of the ISE. The minus sign is needed because fluoride is an anion. Calculations using equation (3.17) are performed in an identical way to those carried out by using equation (3.16).

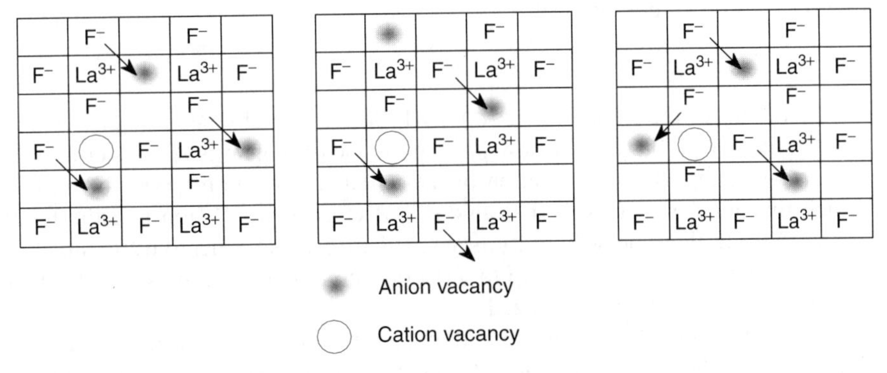

Figure 3.11 Schematic representations of the migration of fluoride ions within a crystal of doped LaF_3. In the figures shown here, the top faces are in contact with the solution of higher fluoride activity, while the bottom faces are in contact with the solution of lower activity.

SAQ 3.14

The fluoride content of a sample of toothpaste is unknown. Accordingly, a sample of the toothpaste was digested in acid solution, filtered to remove the white grit-like solid and then buffered with a total ionic strength adjustment buffer (TISAB) to pH 6. A fluoride electrode is immersed in the clear solution and the *emf* recorded when the reading was steady.

If the same fluoride ISE is immersed in a 10^{-2} mol dm^{-3} solution of sodium fluoride, the ISE gives an *emf* of 35.2 mV. What is the concentration of fluoride ion if the *emf* of the toothpaste solution is 30.2 mV? (Assume no other ions interfere and that $\gamma_{\pm} = 1$ in both cases.)

While the example given in SAQ 3.14 employed just one standard solution, it is better to construct a calibration graph with a minimum of five or six standards.

From the discussions above, we will be aware that the 'concentration' ($[F^-]$) determined will in fact be an *activity*, i.e. $a(F^-)$. It will also be apparent that adding an acid to digest the sample of toothpaste will introduce errors into the calculation since two electrolytes are involved, thereby increasing the ionic strength I (see SAQs 3.10 and 3.11). Since the preferred pH range of the fluoride electrode is 5–6, the ionic strength adjuster (TISAB) can also conveniently function as a pH buffer.

We have looked at the best way to compensate for variations in ionic strength in Section 3.4, where we saw that activity coefficients can be safely ignored if the ionic strength of all solutions are maintained throughout, thus implying that activities and concentrations can be interconverted without any problems. ISEs are sometimes supplied with a calibration graph of *emf* against concentration. Otherwise, calibration graphs need to be constructed with the ISE immersed in solutions of known composition.

3.5.2.2 Concerns about Selectivity

The selectivity of the fluoride electrode is 1000:1 in the absence of OH^-, where the **selectivity** is a measure of how precise the measurement is. Consider an ISE that is monitoring the concentration of the anion X^-: if the response of an ISE is such that nine out of every ten anions it detects is an X^- ion and the tenth ion is different, then the selectivity is 9:1. Clearly, such a selectivity is dreadful since any measurement can be up to 11% out; common sense thus dictates that the selectivity should be maximized.

If we say that an ISE responds to both potassium and sodium, with a selectivity of K^+:Na^+ = 100:1, then the Nernst equation for an ISE (cf. equations (3.16) and (3.17)) can be written in an amended form as follows:

$$emf = K + \frac{RT}{F}\left[\ln a(K^+) + \frac{1}{100}\ln a(Na^+)\right] \tag{3.18}$$

(A similar equation for anions will have a minus sign before the factor of RT/F – the difference arises from the way in which these Nernst-based equations are formulated.) The factor by which we multiply each '$\ln a$ ' term is called the **selectivity coefficient** or **selectivity ratio**. The selectivity coefficient for K^+ in the presence of Na^+ is said to be 100 because one hundred sodium ions are as 'noticeable' as one potassium ion.

Worked Example 3.13. A solution containing potassium ions is known to also contain NaCl at a concentration of 0.001 mol dm^{-3}. When 0.02 mol dm^{-3} KNO_3 is dissolved in the same chloride solution, the *emf* is 1.403 V. What is the concentration of K^+ in a new sample if the *emf* is 1.390 V?

Strategy. The first step will be to calibrate the electrode, i.e. to determine K. We will assume that activity and concentration can be interconverted (which also explains why no units are incorporated within the logarithm). Next, knowing K, we will be able to solve equation (3.18) with $a(K^+)$ as the unknown.

Calibration. Inserting the appropriate values into equation (3.18) we obtain:

$$1.403 \text{ V} = K + 0.0257 \text{ V}\left(\ln 0.02 + \frac{1}{100} \times \ln 0.001\right)$$

$$K = 1.403 \text{ V} - 0.0257 \text{ V}\left(-3.91 + \frac{1}{100} \times -6.908\right)$$

which gives:

$$K = 1.505 \text{ V}$$

Calculating for the unknown a *(K$^+$)*. Knowing K, we can then insert the relevant values into equation (3.18):

$$1.390 \text{ V} = 1.505 \text{ V} + 0.0257 \text{ V}\left[\ln a(K^+) + \frac{1}{100}\ln 0.001\right]$$

$$\frac{(1.390 - 1.505) \text{ V}}{0.0257 \text{ V}} = \left[\ln a(K^+) + \frac{1}{100}\ln 0.001\right]$$

so:

$$-4.475 = \ln a(K^+) + \left(\frac{1}{100} \times -6.908\right)$$

and therefore:

$$\ln a(K^+) = -4.406, \text{ and } a(K^+) = 0.0122 \text{ (or } 0.0122 \text{ mol dm}^{-3}\text{)}.$$

We find that the concentration of K^+ is higher when the *emf* is smaller.

SAQ 3.15

A fluoride electrode is utilized to determine the amount of fluoride ion in tap water. The water also contains chloride, with the selectivity ratio $F^-:Cl^-$ being 3000:1. An aqueous stock solution is known to contain 10^{-5} mol dm^{-3} of NaF (when acidified to pH 5.5 with hydrochloric acid), and gave an *emf* of 0.880 V. What is the concentration of NaF (in the same acid) if the *emf* changes to 0.804 V?

Assume no other ions interfere, and that all $\gamma_{\pm} = 1$.

DQ 3.10

Why is the selectivity of an ISE prone to being so poor?

Answer

The reason why the pH electrode has a poor selectivity at high pH – the so-called **alkaline error** *mentioned in Table 3.2 – arises because the soda glass from which the glass bulb is constructed contains a high mole fraction of sodium ions, so Na^+ is fairly mobile within the glass. Sodium also replaces protons from the surface of the silicate lattice of the glass, according to the following:*

$$-SiO^-H^+ + Na^+ \longrightarrow -SiO^-Na^+ + H^+$$

Such alkaline errors mean that high numbers of Na^+ ions in solution are best avoided when an electroanalytical measurement requires the use of a pH or glass electrode; a high $[Na^+]$ during pH measurement is usually due to having a high [NaOH] – hence the name 'alkaline' error. Figure 3.12 shows plots of the deviations in *emf* against pH for solutions of metal hydroxide, where the pH was determined with a standard glass electrode. An 'acid' error can also be problematic at negative pH, i.e. $a(H+) > 1$ mol dm^{-3}.

The selectivity of the fluoride electrode is usually superb – perhaps the best for any of the solid-state ISEs. The reason why hydroxide ought to be avoided in tandem with a fluoride ISE is because the OH^- and F^- ions have a very similar ionic radius, and so move through the doped LaF_3 lattice with a similar velocity and activation energy, i.e. the OH^- and F^- ions are virtually indistinguishable by this method.

We should be aware that many analytical samples are unlikely to be completely 'clean', that is, contain a single analyte. Table 3.3 lists a series of precautions by which we can hope to overcome the problem of interference when the selectivity is known to be poor, i.e. the selectivity coefficient is numerically small.

Probably the best way to overcome the problem of poor selectivity is via the **multiple standard addition** approach in which known amounts of analyte are added to the sample in increasingly high concentrations. Using this method, a

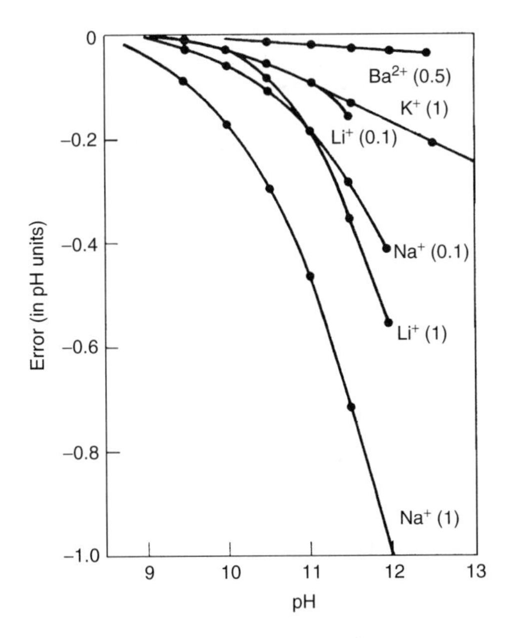

Figure 3.12 Demonstration of the extent of the 'alkaline' error, showing plots of the deviations in pH measurement under aqueous alkaline conditions as determined with a standard glass electrode (the 'Corning 015'). The figures in parentheses represent the concentrations of the metal hydroxide salt in mol dm^{-3}. From Christian, G. D., *Analytical Chemistry*, 5th Edn, © Wiley, 1994. Reprinted by permission of John Wiley & Sons, Inc.

portion of the sample has a known amount of standard added to it, and the increase in signal noted. In this way, the standard is subjected to the same chemical environment as the analyte. Perhaps a **Gran plot** will be drawn also (see Section 4.4 below).

The multiple standard addition method is time-consuming although it usually yields superior results to those of the other methods outlined in Table 3.3.

3.5.2.3 The Liquid-Membrane Electrode

The liquid-membrane electrode is another important type of ion-selective electrode. The internal filling solution contains a source of the ion under investigation, i.e. one for which the ion exchanger is specific, while also containing a halide ion to allow the reference electrode to function. The physico-chemical behaviour of the ISE is very similar to that of the fluoride electrode, except that E_{ISE} and the selectivity are dictated by the porosity of a membrane rather than by movement through a solid-state crystal.

There are two general types of liquid-membrane ISEs, namely one which involves liquid-phase ion exchange, with the response being selective to the anion or cation under scrutiny (generally polyvalent ions), while the other type involves

Table 3.3 Standard procedures used to compensate for poor selectivity when using an ion-selective electrode (ISE) as an electroanalytical tool

Method	Comment
1. If the identity and activity of the interfering ion is known, then compensate during the calculation of a by using equation (3.18)	Not very accurate as the selectivity coefficient can vary with ionic strength. Ions in the ionic strength adjuster can in fact make the problem worse
2. Prepare a calibration graph in which the analyte of interest and the interferant are present at the same activity. The level of interference will then be constant and not affect the accuracy	This method assumes that the ratio of analyte to interferent is constant and known
3. Add additional analyte to the solution to 'swamp' the interferent. The ratio of analyte to interferent will therefore increase, thus improving the precision of the measurement. This method is termed **multiple standard addition**	Time consuming but probably the best method available
4. Remove the interfering ion prior to electroanalytical measurement, e.g. by precipitation, dialysis, etc.	This method is to be used only if the two ions are distinct in terms of the parameter utilized to effect the separation. Otherwise, the analyte will be removed as well as the interferant. Avoid this method if in doubt

a selective complexation between a univalent cation and a neutral (generally macrocyclic) ligand. The most commonly used ligands are cyclic polyethers or antibiotics such as valinomycin .

In both types of liquid-membrane ISEs, the 'membrane' acts as an immiscible phase boundary between the aqueous and non-aqueous solutions inside the ISE (see the schematic diagram presented in Figure 3.13). In order to minimize mixing, the liquid membrane is held in place by an inert, porous material such as a rigid glass frit or a flexible synthetic polymer – the choice will depend on the manufacturer rather than on experimental considerations.

One of the most important examples of a liquid-membrane ISE is the calcium-selective electrode. The salt utilized as the ion exchanger is the calcium salt of dodecylphosphoric acid, e.g. dissolved in di-(n-acetylphenyl) phosphonate. The sensitivity of the electrode depends on the solubility of the ion exchanger in the test solution. The electrode response is generally nernstian down to a concentration of 10^{-5} mol dm^{-3}. In the preferred pH range of 5.5–11, the selectivity of

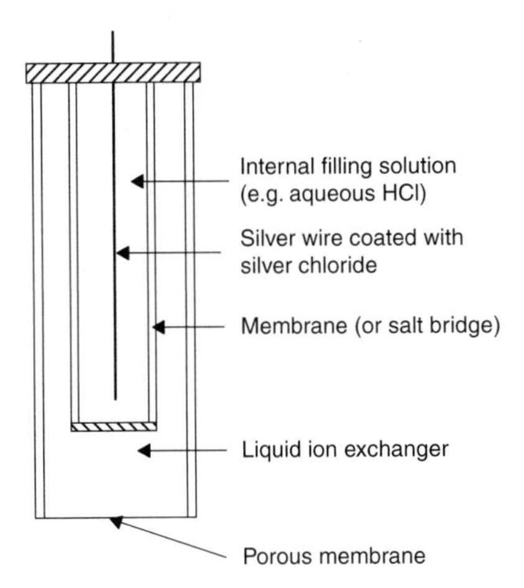

Figure 3.13 Schematic diagram of a liquid-membrane ISE. The silver wire at the top acts as one of the electrodes, and so an additional electrode will be required to complete the cell.

calcium is 2000 with respect to sodium or potassium, 200 over magnesium and about 70 over strontium.

The selectivity is severely impeded if phosphate-based buffers are employed, since calcium–phosphate complexation occurs. In principle, the selectivity of a membrane-based ISE is constant, but contamination and degradation of the membrane's polymer component means that the selectivity decreases with usage. Contamination with biological samples can also be problematic.

Other electrodes of the liquid-membrane type can be used for the determination of the ions such as Cl^-, ClO_4^-, NO_3^-, Cu^{2+}, Pb^{2+} and BF_4^-.

Calculations for the liquid-membrane electrode are performed in exactly the same manner as those used for the pH or fluoride electrode, i.e. with an equation of the following type:

$$emf = K + \frac{RT}{nF} \ln a(\mathrm{X}) \tag{3.19}$$

cf. equations (3.16) and (3.17) if contaminant ions can be assumed to be absent, or with its more complicated analogue, i.e. equation (3.18), if contaminant ions are known to participate in the potentiometric response. The manufacturer of the ISE will also be able to provide the relevant selectivity coefficients for all common contaminant ions.

3.5.3 Potentiometric Determination of Solubility Constants

Some ionic salts dissolve only imperceptibly, e.g. 10^{-13}–10^{-7} mol dm^{-3}. A simple measure of how much material actually enters solution is the **solubility constant** K_s (also called the **solubility product**, and sometimes given the symbol K_{sp}). An expression for K_s may be formulated simply by multiplying the activity of each ion in solution, and raising each to its respective stoichiometric power. As an example, the K_s for 'insoluble' copper hydroxide, $Cu(OH)_2$, is given by the following expression:

$$K_s = a(Cu^{2+})\, a(OH^-)^2 \tag{3.20}$$

Electroanalysis is a powerful means of obtaining solubility constants for solutes in which the amount of one or more of the ions can be determined electrochemically. To obtain a(ion), we will either employ an ISE or perform a calculation with the Nernst equation.

Worked Example 3.14. Excess copper (II) hydroxide solid was placed in sodium hydroxide solution (pH 9, 25°C) and left until equilibrium was reached. A copper wire was immersed in the solution and $E_{Cu^{2+},Cu}$, determined as 0.200 V; $E^{\ominus}_{Cu^{2+},Cu} =$ 0.340 V. What is the solubility constant K_s for $Cu(OH)_2$?

Strategy. $E_{Cu^{2+},Cu}$ is to be measured with a copper wire as the redox electrode, thereby allowing the activity of the copper ion to be calculated with the Nernst equation (equation (3.9)): as follows:

$$E_{Cu^{2+},Cu} = E^{\ominus}_{Cu^{2+},Cu} + \frac{RT}{2F} \ln a(Cu^{2+})$$

By inserting the appropriate values:

$$0.200 \text{ V} = 0.340 \text{ V} + \frac{0.0257 \text{ V}}{2} \ln a(Cu^{2+})$$

$$\frac{(0.200 - 0.340) \text{ V}}{0.0257/2 \text{ V}} = \ln a(Cu^{2+})$$

and hence:

$$a(Cu^{2+}) = 1.86 \times 10^{-5}$$

The concentration of hydroxide ion, $[OH^-]$, is 10^{-5} because the pH was 9, and so, from the expression for the solubility constant (equation (3.20)), we may substitute this value, yielding:

$$K_s = (1.86 \times 10^{-5}) \times (10^{-5})^2 = 1.86 \times 10^{-15}$$

An alternative means of determining $a(Cu^{2+})$ would have been from a potentiometric titration (see the next chapter for further details).

SAQ 3.16

Ferrous hydroxide is an 'insoluble salt'. A quentity of solid $Fe(OH)_2$ was placed in solution and then allowed to reach equilibrium. A wire of clean, pure iron is immersed in solution, and $E_{Fe^{2+},Fe}$ was found to be −0.576 V. Taking $E^{\ominus}_{Fe^{2+},Fe} =$ −0.44 V. What is K_s for the 'insoluble' $Fe(OH)_2$?

3.6 Causes and Treatment of Errors

There are several serious problems that we are likely to encounter when using the potentiometric approach to electroanalysis. The most serious of these derive from the mathematical components found within the Nernst equation, since an exponential function will magnify all errors. Worse, though, is the way that such magnification increases exponentially as the difference between $E_{O,R}$ and $E^{\ominus}_{O,R}$ increases; thus magnification of errors is particularly problematic when low concentrations are to be investigated or when a combination of poor experimental procedures pertain.

In this present section, we will assume that the values of $E^{\ominus}_{O,R}$ used in the calculations are wholly accurate, and that all the errors are incorporated within the electrode potential, $E_{O,R}$. The measured term in a potentiometric experiment is the cell *emf*. In the simplest systems, the *emf* is made up of electrode potentials, one per half cell (although see below). One potential is the electrode potential of interest, while the other is the potential of a reference electrode. As a crude generalization, the *emf* can be measured to within a precision of about 0.1 mV, with the exact value depending on the system under study. Fluctuations, as caused, for example, by solution stirring, may decrease the precision to as poor as 2–3 mV.

SAQ 3.17

The electrode potential $E_{Cu^{2+},Cu}$ is determined to be (0.300 ± 0.003) V at 298 K. Calculate the maximum and minimum activities of the copper ion by using the Nernst equation, taking $E^{\ominus}_{Cu^{2+},Cu} = 0.340$ V.

The first source of error will be a poor quality voltmeter, that is, a meter that fluctuates randomly owing to poor-quality circuitry. However, we can *usually* assume that all of our equipment is of a satisfactory standard.

If the voltmeter does not have a high enough resistance, then it is likely that the *emf* will not have been determined at zero current. If current passes, then the measurement is not at equilibrium, and the Nernst equation should not be employed (see Section 3.6.1).

Assuming that a true (albeit frustrated) equilibrium *is* possible, the next source of error is an incorrect reference potential, e.g. as caused by current passage through the reference electrode (see Sections 3.6.1 and 3.6.2).

A third (and usually more profound) cause of error lies in the way that the Nernst equation is formulated in terms of activities rather than concentration. Even if the *emf* and E are correct, the proportionality constant between concentration and activity (the mean ionic activity coefficient $\gamma_{\pm}$) is usually wholly unknown. Errors borne of ignoring activity coefficients (i.e. caused by ionic interactions) are discussed in Sections 3.4 and 3.6.3.

The above three sources of error all result from the fact we assume that the *emf* measured in a potentiometric experiment comprises only two terms, i.e. one per half cell. Such a situation is often simplistic. A small voltage can be induced in a cell because of the way in which an *emf* is measured – a tiny current is allowed to pass through the cell to facilitate the voltmeter to function. Such an '*IR* drop' is discussed in Chapter 2 and Section 3.6.4.

In fact, the movement of ions in and out of a half cell (for example, across a semipermeable membrane or frit) gives rise to an additional form of potential, the so-called 'junction potential'. Such potentials are discussed in Section 3.6.5.

3.6.1 Current Passage through a Cell Causing Internal Compositional Changes

All the cells that we have studied so far have been 'at equilibrium', with this term meaning in effect that a 'frustrated equilibrium' exists (see Chapter 2)

However, a true frustrated equilibrium is not attainable since a tiny charge must flow to allow measurement, so changes to the internal composition of the solution, albeit tiny, will occur during measurement. Consider the Daniell cell, Zn | Zn^{2+} || Cu^{2+} | Cu, the *emf* of which is be determined at a voltmeter. If the two electrodes were allowed to touch, however, then the cell would short out, i.e. charge would flow from the negative zinc electrode (in accompaniment with oxidative dissolution of the zinc) and charge would be accepted by the positive copper electrode (in accompaniment with reductive formation of metallic copper).

In summary, then, the amounts of material in solution (and hence their ionic activities) will alter if charge flows through the cell. The amounts of material formed and consumed can be calculated from Faraday's laws of electrolysis (see Chapter 5).

Worked Example 3.15. An electrochemical charge of 96.5 C is passed through an aqueous solution of ferric ion, thus causing the reduction reaction $Fe^{3+} + e^- \rightarrow Fe^{2+}$. The solution originally contained 0.01 mol of Fe^{3+}. How much ferric ion remains after the passage of current?

One mole of electrons has a charge of 96 500 C (a faraday) so 96.5 C represents 0.001 mol of electrons. Accordingly, 0.001 mol of Fe^{3+} will be consumed. Consideration of mass balance shows that the amount of Fe^{3+} remaining is $(Fe^{3+}_{\text{initial}} - Fe^{3+}_{\text{consumed}}) = (0.01 - 0.001)$ mol $= 0.0099$ mol.

SAQ 3.18

A large current is passed through an SHE, thus causing some of the protons to be reduced to form hydrogen gas, which decreases the activity of the proton from 1.00 to 0.95. Calculate the new value of E_{H^+,H_2}, assuming that $p_{(H_2)}$ remains as $p^{\ominus}$.

The various methods that can be used to stop such current flow have been discussed earlier in Chapter 2.

3.6.2 Current Passage through Reference Electrodes

Following on from the previous discussion, if current has been allowed to pass through a reference electrode, its internal composition will change and thus its potential will be unknown, which explains the conundrum 'when is a reference electrode not a reference electrode?' – when charge is passed through it!

DQ 3.11

Why does the potential of a saturated calomel electrode change if current is allowed to pass through it?

Answer

As mentioned above, $E_{SCE} = E_{Hg_2Cl_2,Hg}$*. The activities of mercury and calomel are each 1 since they are solids, so the Nernst equation for the SCE half cell is:*

$$E_{Hg_2Cl_2,Hg} = E^{\ominus}_{Hg_2Cl_2,Hg} + \frac{RT}{2F} \ln\left(\frac{1}{a(Cl^-)^2}\right)$$

Current passage through an SCE will cause redox change according to Faraday's laws. For convenience, if we say that an oxidative current is passed, then Hg_2Cl_2 *(calomel) is formed and Hg is lost; this formation of calomel depletes the number of chloride ions in solution. From consideration of the Nernst equation, we note that alterations to* a (Cl^-) *in an SCE itself changes* $E_{Hg_2Cl_2,Hg}$*. The activity of chloride ion is best maintained by employing a constant* surplus *of KCl crystals at the foot of the tube in order to ensure saturation. An SCE not having a thick crust of crystals should be avoided because the* E_{SCE} *might not then be known. We conclude by saying that* E_{SCE} *will go up if an oxidative current is passed through an SCE, that is, until some of the solid KCl enters solution to re-establish saturation.*

SAQ 3.19

A simple non-aqueous reference electrode is made up to contain 1×10^{-6} mol of Ag^+ in 1.0 cm^3 of MeCN, i.e. at a concentration of 1×10^{-3} mol dm^{-3}. In contact with the silver solution is a silver wire. During a coulometric experiment, 5×10^{-6} faradays of charge (i.e. moles of electrons) are passed *oxidatively* through the solution.

Calculate the number of moles of Ag that are consumed and the number of moles of Ag that are formed. Calculate the electrode potential $E_{Ag^+,Ag}$ both before and after current passage.

Assume that activities and concentrations can be used interchangeably, and take $E^{\ominus}_{Ag^+,Ag} = 0.56$ V for the silver couple in MeCN.

3.6.3 Determination of Concentration when the Mean Ionic Activity Coefficient is Unknown

The thermodynamic equations that we have been using so far, such as the Nernst equation, are written in terms of the activity a rather than the concentration c. The relationship between the two is given by the simple relation $a = (c/c^{\ominus})\gamma$, where $c^{\ominus}$ is needed to remove the units of concentration, i.e. $c^{\ominus} = 1$ mol dm^{-3}.

While this relationship is simple, it introduces more errors because the activity coefficient (or more normally, the mean ionic activity coefficient $\gamma_{\pm}$) is wholly unknown. While $\gamma_{\pm}$ can sometimes be calculated (e.g. via the Debye–Hückel relationships described in Section 3.4), such calculated values often differ quite significantly from experimental values, particularly when working at higher ionic strengths. In addition, ionic strength adjusters and TISABs are recommended in conjunction with calibration curves.

3.6.4 Cell Design and the Effects of 'IR Drop'

Ohm's law was discussed in some depth in Chapter 2, when we saw that a potential applied across a resistor will induce a current. Ohm's law is stated quantitatively as $V = IR$ (equation (2.3)). If the potential V is an *emf*, and the solution between the two electrodes has a resistance of R, then a current I will flow (it's an **ohmic current**) – this is often termed the ***IR* drop**. As should be clear by now, this current passage through a voltammetric cell is to be avoided wherever possible, since such passage causes compositional changes.

Worked Example 3.16. An *emf* of 1.305 V is formed across a cell having a resistance of 5×10^5 Ω. If the current drawn by the voltmeter is 10^{-10} A, calculate the measured *emf* by using Ohm's law.

Ohm's law states that $V = IR$, so $V = 5 \times 10^5\ \Omega \times 10^{-10}\ \text{A} = 5 \times 10^{-5}$ V (0.05 mV).

The *emf* is therefore (1.305 + 0.00005) V = 1.30505 V. Such a tiny increase in voltage can safely be ignored in almost any voltammetric experiment, unless R is very much larger than the value used here.

SAQ 3.20

How might a cell resistance be minimized, thereby lowering the potential induced? (Hint – consider all the possible causes of resistance, including all electrodes, solutions and interfaces.)

3.6.5 Additional Sources of Potential Owing to Ionic Transport and Junction Potentials

Transport in cells. Up until now, all of the cells that we have studied have been simple to describe mathematically because no charge was transferred between the two half cells. A state of frustrated equilibrium exists, thus allowing for straightforward thermodynamic analyses to be performed since all internal compositions in the cell remain static. Such cells are said to be **cells *without* transport**.

We will now look at systems in which the *emf* at equilibrium cannot be treated as exact because charge (in the form of ions) does transfer between the two half cells – these are described as **cells *with* transport**.

In reality, most real cells are never fully at equilibrium because the two half cells are located on either side of a *semipermeable* membrane, sinter or frit, such as a thin separator of rubber or terracotta (see Figure 3.14). Because ions transfer, and because the two half cells comprise different electrolytes (e.g. in terms of different concentration, etc.), the compositions of the two half cells change with time.

DQ 3.12

If ions can move across the membrane (in cells with transport), does the rate of movement have any implications?

Answer

Yes! The difference in potential between the two electrodes in a cell without transport is given by equation (3.3). In a cell with *transport, there is generally an additional potential known as a* ***liquid junction potential*** *or just a* ***junction potential****. We shall give the liquid junction potential*

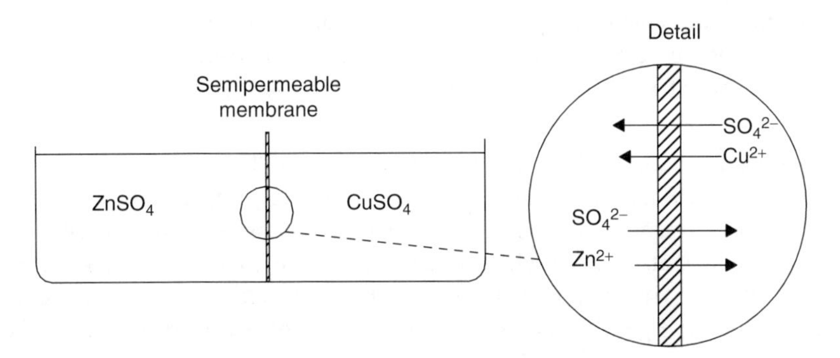

Figure 3.14 Schematic diagram showing two aqueous solutions separated by a semipermeable membrane; all ions diffuse across the membrane, but with some ions moving faster than others – hence the formation of a junction potential E_j.

the symbol E_j*, although quite a few alternative symbolisms also exist. Some texts call such a voltage a 'diffusion' potential. We define* E_j *as the potential arising from separation of charge owing to different transport rates of ions across a membrane, sinter, frit or interface.*

Up until now, all values of the *emf* have only comprised two half-cell potentials. The junction potential is an additional source of potential, so our fundamental relationship (equation (3.3)) now becomes:

$$emf = E_{\text{right-hand side}} - E_{\text{left-hand side}} + E_{\text{j}} \tag{3.21}$$

The liquid junction potential always acts in such a way that the motion of ions that move rapidly is retarded, while the motion of the slower ions is accelerated. Accordingly, equilibrium is soon attained (within a few milliseconds of the two solutions being joined via a membrane), thus allowing a constant value of E_{j} to be recorded. The absolute magnitude of E_{j} will depend on the concentrations (again, strictly the activities) of the constituent ions in the two half cells, on the charges of each ion and on the *relative* **rates** of movement across the membrane.

In common practice, liquid junction potentials are rarely large, so a value of E_{j} as large as 0.1 V should be regarded as unlikely, although not exceptional. That being so, junction potentials are a common cause of experimental error, in part because they are difficult to quantify, and in part because they are often irreproducible.

We will now look at the effects of E_{j} on thermodynamic calculations, and then decide on the various methods that can be used to minimize them. One of the most common reasons for performing a calculation with an electrochemical cell is to determine the concentration or activity of an ion. In order to carry out such a calculation, we would first construct a cell, and then, knowing the potential of the reference electrode, we would determine the half-cell potential, i.e. the electrode potential E of interest, and then apply the Nernst equation.

Worked Example 3.17. Consider the simple cell 'SCE | | $Cu^{2+}_{(aq)}$ | $Cu_{(s)}$'. From a measurement of the *emf* and a knowledge of E_{SCE}, the electrode potential, $E_{\text{Cu}^{2+},\text{Cu}}$, was determined to be 0.300 V. We will consider three possible situations, and calculate the activity of the copper ion in each case: (i) $E_{\text{j}} = 0$, (ii) $E_{\text{j}} = 30$ mV, with the junction *adding* to the electrode potential for the copper half cell, and (iii) $E_{\text{j}} = 30$ mV, with the junction *subtracting* from the electrode potential for the copper half cell.† As usual, we will assume a value of $a(\text{Cu}) = 1$ throughout and $E^{\ominus}_{\text{Cu}^{2+},\text{Cu}} = 0.340$ V.

† By 'adding to' and 'subtracting from', we are tacitly saying that the majorily of E_{j} comes from the right-hand side and left-hand sides of the cell, respectively.

(i) From the Nernst equation for this couple:

$$E_{Cu^{2+},Cu} = E^{\ominus}_{Cu^{2+},Cu} + \frac{RT}{nF} \ln \left[\frac{a(Cu^{2+})}{a(Cu)} \right]$$

Inserting the appropriate values, we obtain:

$$0.300 \text{ V} = 0.340 \text{ V} + 0.0128 \text{ V} \ln a(Cu^{2+})$$

and therefore:

$$a(Cu^{2+}) = \exp\left(\frac{0.300 \text{ V} - 0.340 \text{ V}}{0.0128 \text{ V}} \right) = \exp(-3.125) = 0.044$$

(ii) We will again start with the Nernst equation. If $E_j = 30$ mV, then the measured value of $E_{Cu^{2+},Cu}$ will be 30 mV too high, i.e. the real value of $E_{Cu^{2+},Cu}$ is 0.270 V.

Inserting the relevant values into the Nernst equation, we obtain:

$$0.270 \text{ V} = 0.340 \text{ V} + 0.0128 \text{ V} \ln a(Cu^{2+})$$

and therefore:

$$a(Cu^{2+}) = \exp\left(\frac{0.270 \text{ V} - 0.340 \text{ V}}{0.0128 \text{ V}} \right) = \exp(-5.47) = 0.0042$$

So by poor experimental design, we have allowed a liquid junction potential to form in our cell. While we have the same *emf* as in the first cell, in fact the activity of copper ion in this new cell is only 9.5% of the activity calculated for the case where no liquid junction potential was induced.

(iii) We use a similar reasoning to that above, with the E_j potential (30 mV) being subtractable. The measured value of $E_{Cu^{2+},Cu}$ will therefore be 30 mV too *low*, i.e. the real value of $E_{Cu^{2+},Cu}$ is 0.330 V.

Putting the appropriate values into the Nernst equation once more, we obtain:

$$0.330 \text{ V} = 0.340 \text{ V} + 0.0128 \text{ V} \ln a(Cu^{2+})$$

and therefore:

$$a(Cu^{2+}) = \exp\left(\frac{0.330 \text{ V} - 0.340 \text{ V}}{0.0128 \text{ V}} \right) = \exp(-0.78) = 0.458$$

So the activity of the copper ion is 9.5 times greater than in the cell where no liquid junction potential was measured.

To summarize, in this two-electron cell, a liquid junction potential of about 30 mV causes the activity of the ion to *appear* different by a factor of about 10. We have considered both of the cases where the junction was formed at the positive electrode and at the negative electrode. In fact, we will often be unsure which is the case, i.e. whether to consider:

$$E_{\mathrm{O,R}} = emf - E_{\mathrm{reference}} + E_{\mathrm{j}}$$

or:

$$E_{\mathrm{O,R}} = emf - E_{\mathrm{reference}} - E_{\mathrm{j}}$$

which means that there is a 100-fold range of possible activities in this cell.

We therefore need to be careful in terms of cell design, and must have a keen appreciation that junction potentials are always likely to exist.

SAQ 3.21

An electrode potential $E_{\mathrm{Ag^+,Ag}}$ is measured as 0.670 V. In fact, this value is too high because the value of $E_{\mathrm{Ag^+,Ag}}$ also incorporates a liquid junction potential of 22 mV. Calculate two values of $a(\mathrm{Ag^+})$, first by assuming that $E_{\mathrm{Ag^+,Ag}}$ is accurate, and secondly by taking account of E_{j}. What is the error in $a(\mathrm{Ag^+})$ caused by the liquid junction potential? Take $E^{\ominus}_{\mathrm{Ag^+,Ag}} = 0.799$ V at 298 K.

DQ 3.13

If junction potentials are so important, how do we measure their magnitude?

Answer

Part of the reason why junction potentials are so 'disastrous' experimentally is because they are usually so difficult to quantify. Accordingly, a method of determining an approximate value of E_j *is essential. Generally, we prepare a series of cells that have elements in common. It is necessary that the junction potentials are similar in magnitude, as will be shown in the following worked example.*

Worked Example 3.18. The following three cells are constructed:

(A) SCE | Ag^+(0.1 mol dm^{-3}) | Ag;
(B) SCE | Ag^+(0.01 mol dm^{-3}) | Ag;
(C) Ag | Ag^+(0.01 mol dm^{-3}) || Ag^+(0.1 mol dm^{-3}) | Ag.

What is the magnitude of E_{j}?

Cells A and B are almost identical, except that the concentration of the silver ion differs, while cell C is different. In the latter cell, the double phase boundary

'||' represents a salt bridge, which is included to decrease the magnitude of the E_j. Therefore, we can write the following:

Cell A $emf_A = E_{Ag^+,Ag(0.1)} - E_{SCE} + E_j'$

Cell B $emf_B = E_{Ag^+,Ag(0.01)} - E_{SCE} + E_j''$

Cell C $emf_C = E_{Ag^+,Ag(0.1)} - E_{Ag^+,Ag(0.01)} + E_j$

The primes are used merely to indicate that the three junction potentials are different.

The junction potentials for cells A and B can be assumed to be very similar in magnitude because their liquid junction potentials will be dominated by chloride ions diffusing *out from* the sinter at the bottom of the SCE (see Figure 3.4), rather than by silver diffusing *into* the SCE. Cell C has no sinter but a salt bridge. If we therefore consider cells A and B:

$$emf_A - emf_B = (E_{Ag^+,Ag(0.1)} - E_{SCE}) - (E_{Ag^+,Ag(0.01)} - E_{SCE})$$

Notice that the two liquid junction potentials cancel out because we have said that $E_j' = E_j''$.

By substituting the expression for emf_C into the above equation, we then obtain:

$$emf_A - emf_B = E_{Ag^+,Ag(0.1)} - E_{Ag^+,Ag(0.01)} = emf_C - E_j$$

which gives:

$$E_j = emf_C - (emf_A - emf_B)$$

SAQ 3.22

Determine the magnitude of the liquid junction potential E_j in cell C of the following, taking $E_{SCE} = 0.242$ V:

(A) SCE | Ag^+(0.1 mol dm^{-3}) | Ag ($emf = 0.498$ V);
(B) SCE | Ag^+(0.01 mol dm^{-3}) | Ag ($emf = 0.439$ V);
(C) Ag | Ag^+(0.01 mol dm^{-3}) || Ag^+(0.1 mol dm^{-3}) | Ag ($emf = 0.070$ V).

As junction potentials can have such a devastating effect on electroanalytical data, we need next to consider some means of minimizing them. There are two general methods that can be used – placing a salt bridge in the circuit (as alluded to above) or adding a swamping electrolyte to the solution.

Using a salt bridge. Following directly from the calculation above, the first method of minimizing the junction potential is to choose an electrolyte characterized by similar transport numbers and activites for its anions and cations. However, such experimental conditions are usually impracticable.

In normal electrochemical usage, the most effective way of minimizing a junction potential is with a **salt bridge** – 'bridge' because it connects the two half cells and 'salt' because it must be saturated with a strong ionic electrolyte. In practice, the salt bridge is typically a thin strip of filter paper soaked in electrolyte or a U-tube containing the electrolyte. The electrolyte is usually KCl or KNO_3 in relatively high concentrations, while the U-tube contains the salt within a gelling agent such as agar or gelatine.

In operation, each end of the salt bridge is dipped in the half-cell solutions, thereby connecting the two half cells. A typical cell might be as follows:

$$Zn_{(s)} \mid Zn^{2+}_{(aq,c_1)} \mid S \mid Cu^{2+}_{(aq,c_2)} \mid Cu_{(s)}$$

where we have represented the salt bridge by '| S |', with S being the electrolyte within the salt bridge. Note that the two half cells comprise electrolyte at different concentrations.

DQ 3.14

How does the salt bridge work?

Answer

First, we must recognize that all ionic diffusional changes involve both ends of the salt bridge. Secondly, because the electrolyte in the bridge is gel-like (usually), ionic motion into, through and from the bridge is quite slow because the viscous nature of the gel will minimize ionic diffusion. Retardation of the ionic motion will itself enable the system to settle quickly to a reproducible state. As all ionic motion is slowed, the differences *in diffusion rate are themselves minimized.*

Thirdly, and more importantly, the concentration of the salt in the bridge should greatly exceed the concentrations of analyte within each of the half cells (exceed, if possible, by a factor of 10–100). Figure 3.15

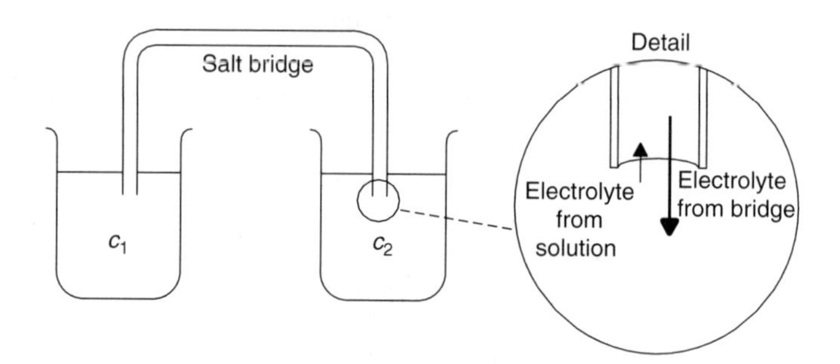

Figure 3.15 Schematic diagram showing two half cells separated by a salt bridge, where the inset is intended to illustrate the way in which more ions *leave* the bridge end than *enter* it.

illustrates how the extent of diffusion from *the bridge (represented by the large arrow) is far greater than the extent of diffusion* into *the bridge (represented by the small arrow). This aspect is crucial because junction potentials are caused by diffusional effects,* so a junction will still form at either end of the bridge.

If the electrolyte in the bridge is concentrated, however, then these two junction potentials will be dominated by the ions moving from *the bridge and so their values will be almost equal, but opposite in magnitude, thus causing them to more or less cancel each other out: with care, a junction potential* E_j *of as little as 1–2 mV can be achieved with a salt bridge if the electrolyte is concentrated.*

The concentration of the salt in the bridge has a large effect on E_j (see Table 3.4). It can be seen from this table that lower values of E_j are realized when the salt bridge is impregnated with larger concentrations of salt.

KCl or KNO_3 are the salts usually employed in the bridge because their respective transport numbers for anions and cations are almost equal:

$$\text{KCl} \quad t^+ = 0.49 \quad t^- = 0.51$$

$$\text{KNO}_3 \quad t^+ = 0.51 \quad t^- = 0.49$$

The effect of swamping electrolytes. The second method of minimizing the junction potential is to employ a swamping electrolyte, S. Addition of a high concentration of ionic electrolyte will greatly increase the ionic strength of the solution and thus all of the activity coefficients ($\gamma_\pm$) will be decreased to quite small values.

If all values of $\gamma_\pm$ are decreased, then differences between the activities also decrease – it is these differences in activity that cause the diffusion in the first place. Accordingly, after addition of a swamping electrolyte, fewer ions diffuse and so the chemical potentials equalize, with a smaller junction potential being formed.

Table 3.4 Values of junction potential in aqueous cells as a function of the concentration of inert KCl within a salt bridge

[KCl] (mol dm^{-3})	E_j(mV)
0.1	27
1.0	8.4
2.5	3.4
4.2[a]	<1

[a]Saturated system.

Summary

In this chapter, the potentiometric method of obtaining electroanalytical data has been discussed at some length.

Before electroanalytical data can be obtained from a cell, it is essential to ensure that the cell is in a state of *frustrated* equilibrium The measured *emf* represents the separation in potential between the electrode potentials of two half cells. One half cell is conveniently a reference electrode, so $E_{O,R}$ for the analyte is the only unknown. From $E_{O,R}$, the central relationship behind the electroanalytical determination of concentrations by using the potentiometric approach is the Nernst equation (equation (3.8)), which is written in terms of *activities*.

The activity a and concentration c are related by $a = (c/c^{\ominus}) \times \gamma_{\pm}$ (equation (3.12)), where $\gamma_{\pm}$ is the mean ionic activity coefficient, itself a function of the ionic strength I. Approximate values of $\gamma_{\pm}$ can be calculated for solution-phase analytes by using the Debye–Hückel relationships (equations (3.14) and (3.15)). The change of $\gamma_{\pm}$ with ionic strength can be a major cause of error in electroanalytical measurements, so it is advisable to 'buffer' the ionic strength (preferably at a high value), e.g. with a total ionic strength adjustment buffer (TISAB).

In order to obtain a *concentration* from the Nernst equation, it is necessary to obtain accurate potentiometric measurements at equilibrium, i.e. at zero current.

Having revised a few basic electrochemical ideas, such as the nature of reference electrodes, the standard hydrogen electrode and the $E^{\ominus}$ scale based on it, we next looked briefly at thermodynamic parameters such as the electrode potential E, the standard electrode potential $E^{\ominus}$ and *emf*, and then discussed how $\Delta G'$, $\Delta H'$ and $\Delta S'$ (where the prime indicates a frustrated cell 'equilibrium') may be determined.

A major part of this chapter considered the ways used to calculate concentrations or activities by using variants of the Nernst equation. For example, we saw how the solubility constant K_s can be measured potentiometrically.

We then described how ion-selective electrodes (ISEs) function by means of generating an *emf* that is related to the activity of a single ion. The simplest ISEs are glass electrodes for pH determination, and the related pH combination electrodes. Other solid-state ISEs, such as the fluoride electrode, function in the same way.

Unfortunately, most ISEs must be selective if the measurements obtained are to be useful, so the Nernst equation is therefore adapted to take account of the ISE selectivity by using the concept of a selectivity coefficient (or 'ratio'). The problems of selectivity can be overcome by using a standard-addition method, such as drawing a Gran plot.

When considering potentiometric errors, it is necessary to appreciate how a liquid junction potential, E_j, arises, and appreciate how such potentials can lead to significant errors in a calculation. In addition, we saw how the IR drop can affect a potentiometric measurement (and described how to overcome this). Finally, we discussed the ways that potentiometric measurements are prone to errors caused by both current passage through the cell, and by the nature of the mathematical functions with which the Nernst equation is formulated.

Chapter 4

Potentiometry: True Equilibrium and Monitoring Systems with Electron Transfer

Learning Objectives

- To appreciate that potentiometry can be applied to systems in which electrons are transferred, provided that the redox reagents are allowed to mix, so solutions ought to be stirred.
- To appreciate that the concept of 'frustrated' equilibrium does not apply when solutions of redox species are allowed to mix, so any equilibrium in this case is a *true* equilibrium.
- To note that potentiometric titrations are always zero-current measurements.
- To learn that the end point in a redox titration is obtained as the point of inflection in the S-shaped graph produced by plotting '*emf*' against 'volume of redox reagent'.
- To be aware that the redox reagents must be chosen with care: for complete oxidation or reduction of the analyte, the equilibrium constant of the redox reaction, $OX_1 + RED_2 \rightarrow RED_1 + OX_2$, must exceed about 10^6, so the separation between $E^{\ominus}$ for the two couples must exceed about 0.35 V for a one-electron couple.
- To learn that the solubility constants (products), K_s, of sparingly soluble salts can be obtained from a potentiometric titration: the activity of one constituent ion is determined directly from the *emf* at the end point, and the salt stoichiometry then allows K_s to be calculated.

- To learn that the end point can also be indicated by the colour of a redox indicator, that is, by adding a small amount of a redox couple that has different colours for its two principal redox states. (In addition, to appreciate that the redox indicator is a *probe* only, and does not participate in the redox reaction.)
- To learn why the ideal choice of a redox indicator is one which has a standard redox potential as close as possible to the end point of the titration system.

4.1 Introduction to Potentiometry

We have seen already that the root of the word 'potentio-' means that we are looking at a potential, so the crucial piece of apparatus in a potentiometric experiment is a potentiometer (which, in practice, will almost certainly be marketed and labelled as a 'voltmeter'). In the previous chapter, we looked at the simplest forms of potentiometric experiment in which the two half cells were physically separated, one from the other, in order to prevent electron transfer (et) occurring.[†] To recap, such a separation is a key requirement for ensuring that a **frustrated equilibrium** holds.

In this present chapter, we will be looking at a slightly more complicated situation, i.e. one in which the contents of two redox half cells are not separated but are allowed to mix. Because mixing occurs, redox chemistry can occur, i.e. electron-transfer reactions are not forbidden. Any electrochemical equilibrium attained is thus a *genuine* **thermodynamic** equilibrium and is *not* 'frustrated'.

Although electron-transfer reactions occur, the electrode in no way supplies or conducts away the electrons – it is merely a **probe** of the potentials of the constituent redox couples in solution. Accordingly, potentiometric titrations are always *zero-current* measurements.

4.1.1 Redox Titrations: End Points and Shapes of Curves

DQ 4.1

What happens during such a potentiometric experiment, i.e. what are the implications of the electron-transfer reaction?

Answer

One of the simplest potentiometric experiments in this context is the ***redox titration****. We shall consider a real example, i.e. the determination*

[†] While we say that no electron transfer takes place in a cell under frustrated equilibrium, we should really say that no *net* electron transfer occurs. This point will be explored further in Chapter 7.

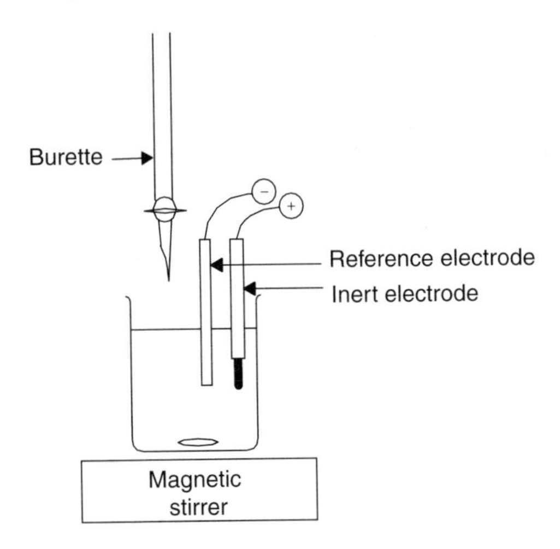

Figure 4.1 Schematic diagram of the apparatus required when monitoring a redox reaction via a potentiometric titration; while a burette is depicted here, the apparatus may be contained within an autotitrator.

of the concentration of ferrous ion (Fe^{2+}) with ceric ion (Ce^{4+}) of known molarity. In this system, we shall consider that a known volume of the Fe^{2+} analyte solution is placed in a beaker, and is titrated in a conventional manner with a Ce^{4+} solution of known concentration.

The redox reaction occurring on addition of ceric ion to the solution of ferrous ion is as follows:

$$Fe^{2+}{}_{(beaker)} + Ce^{4+}{}_{(burette)} \longrightarrow Fe^{3+} + Ce^{3+} \qquad (4.1)$$

Figure 4.1 shows a simple schematic diagram of how such a titration would be performed experimentally. In essence, we carry out a normal titration (just as though we were titrating an acid with an alkali), with the determination of the end point being the only important difference. Instead of using indicators such as litmus or phenolphthalein, which both rely on colour changes, in this case we immerse two electrodes – one of which is a reference electrode – into the solution and follow the emf *during the course of the titration as a function of the volume of redox reagent that is added. It is because the course of the redox titration is followed by a voltmeter that allows us to call this a* potentiometric *titration. (In practice, we might employ an autotitrator as a means of speeding up the procedure.)*

In order to ensure that both species do not precipitate or form oxo ions during their redox reactions, all solutions are prepared with fairly concentrated acid, thus causing all of the ionic strengths I *to have high values.*

SAQ 4.1

Consider the redox reaction shown in equation (4.1). From a potentiometric titration, it is found that 12.5 cm^3 of Ce^{4+} (0.01 mol dm^{-3}) will completely oxidize 25.0 cm^3 of Fe^{2+} solution. What is the concentration of the ferrous iron? (Hint – remember the equation, $c_1V_1 = c_2V_2$, from acid–base titrations.)

DQ 4.2

If we follow the potentiometric titration with a voltmeter, what is the source of the *emf*?

Answer

As with all electrochemical cells, the emf *represents the separation between two half-cell potentials (see equation (3.3)). In accepting this statement, we are implying that the cell has neither a junction potential nor any* IR *drop. The latter assumption is valid since the solution in our example is acidified and hence its ionic strength* I *is high. Solutions with high ionic strengths can be assumed to have a minimal solution resistance* R.

The assumption that junction potentials are absent is also valid, since our usual practice during a potentiometric titration is to stir the solution vigorously.[†] *Such stirring causes rapid mixing, thus obviating any worries about poor mass transport. In addition, note that junction potentials are caused by differing rates of ionic movement across a membrane. No such membrane is incorporated anywhere in our cell here.*

The cell *emf* comprises two half-cell potentials. One of these half cells will be a standard reference, such as a saturated calomel electrode (SCE), while the other will be an inert electrode such as platinum or gold.[‡]

There is no redox couple in solution at the start of the ferrous–ceric titration because the solution contains only Fe^{2+}. The oxidation of ferrous to ferric occurs as soon as an aliquot of ceric ions enter the solution to effect the redox reaction shown in equation (4.1). The bulk of the initially present ferrous ions remain, with the ferric products of the redox reaction residing in the same solution, i.e. a Fe^{3+}, Fe^{2+} redox couple is formed.[§] This couple has the electrode potential $E_{Fe^{3+},Fe^{2+}}$.

Note that all of the ceric ions in the aliquot are converted (in this case, reduced) to form cerous ions, i.e. there is only one redox form of the cerium in solution.

[†] Solution stirring is necessary in order to effect efficient mixing and hence rapid electron transfer. Care is needed, however, because if the mixing is too vigorous (e.g. causing a vortex to form) then the reading on the voltmeter will fluctuate, thus leading to imprecisions in the measurements.

[‡] Care is needed with notation at this point: some texts call an inert electrode operating in this context a 'redox' electrode. However, we will stick with the terminology developed over the previous chapters.

[§] Remember that we should not write the couple as 'Fe^{3+} | Fe^{2+}' because both redox states exist in the same solution, and so there is no phase boundary (cf. the rules for writing a cell schematic given in Appendix 2).

Therefore, there is no Ce^{4+}, Ce^{3+} *couple* present during the first part of the titration. Accordingly, the *emf* can be represented by the following:

$$emf = E_{Fe^{3+},Fe^{2+}} - E_{\text{reference electrode}} \tag{4.2}$$

As the volume of ceric ion solution increases, so the ratio of ferrous ion to ferric ion will decrease, and hence the electrode potential $E_{Fe^{3+},Fe^{2+}}$ changes during the course of the titration. This is why a potentiometric titration can be followed by using the *emf* as the reaction variable.

SAQ 4.2

The cell SCE || A^{n+}, A^{m+} | Pt is constructed to use in a potentiometric titration. At the equivalence point,† the end point *emf* is 0.308 V. Taking $E_{SCE} = 0.242$ V, calculate the electrode potential $E_{A^{n+},A^{m+}}$.

DQ 4.3

So how is the end point determined?

Answer

As we have just seen, there are no redox couples – either Ce^{4+}, Ce^{3+} *or* Fe^{3+}, Fe^{2+} *– in solution before the titration is started. As soon as ceric ions are added to the solution, the* Fe^{3+}, Fe^{2+} *couple forms. At no point* before *the end point will any ceric ions remain in solution, i.e. within a reasonable time scale – because they react – and so there is no* Ce^{4+}, Ce^{3+} *couple in solution.*

After *the end point, all of the ferrous ions will have been consumed and converted to ferric ions, i.e. there is no longer a* Fe^{3+}, Fe^{2+} *couple in solution. However, after the end point, the ceric ions added to the solution will not react because there is nothing left to oxidize, so* Ce^{4+} *and* Ce^{3+} *reside together as a new, different redox couple with its own electrode potential,* $E_{Ce^{4+},Ce^{3+}}$.

The end point of the titration is seen as the transition from one electrode potential to the other, as shown below in Figure 4.2.

An 'S'-shaped curve is produced (this shape arises from the logarithmic dependence of *emf* on activity, as discussed below). The curve is symmetrical if the reaction is 1:1, 2:2, etc., i.e. both redox couples involve the same number of electrons. The **equivalence** point is obtained as the **point of inflection** of such symmetrical curves. If the reaction has a different stoichiometry, then the trace is not symmetrical, and ascertaining the equivalence point can be considerably more difficult. For this reason, it is advantageous to know the stoichiometry beforehand and thus choose the couples to be employed accordingly.

---

† The equivalence point, e.g. in a titration, occurs when there are equal amounts of the electroactive analyte in the beaker and the oxidant or reductant region in the burette.

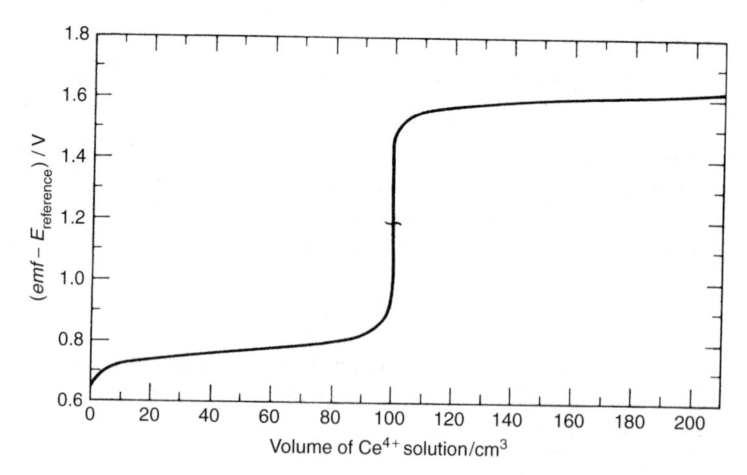

Figure 4.2 Plot of the variable ($emf - E_{SHE}$) against the volume of ceric ion solution during a potentiometric determination of [Fe^{2+}]. The end point is clearly shown by a sharp transition from the standard electrode potential of the analyte couple, $E^{\ominus}_{Fe^{3+},Fe^{2+}}$, to that of the titrant couple, $E^{\ominus}_{Ce^{4+},Ce^{3+}}$ (cf. equation (4.1)).

SAQ 4.3

200 mg of dishwashing liquid were dissolved in water, and titrated against hexadecylpyridinium chloride, with the *emf* being recorded as a function of volume (see Figure 4.3 below). Use this figure to determine a rough value for the *emf* at the end point.

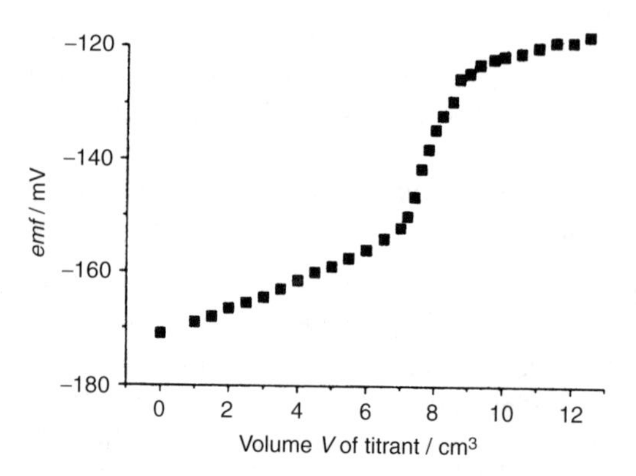

Figure 4.3 Potentiometric titration curve obtained by dissolving 200 mg of dishwashing fluid in water and titrating against hexadecylpyridinium chloride solution, employing a zeolite–polydimethylsiloxane (NaY–PDMS) modified electrode (cf. SAQ 4.3). From Matysik, S., Matysik, F.-M., Mattusch, J. and Einicke, W.-D., *Electroanalysis*, **10**, 98–102, (1998), © Wiley-VCH, 1998. Reproduced by permission of Wiley-VCH.

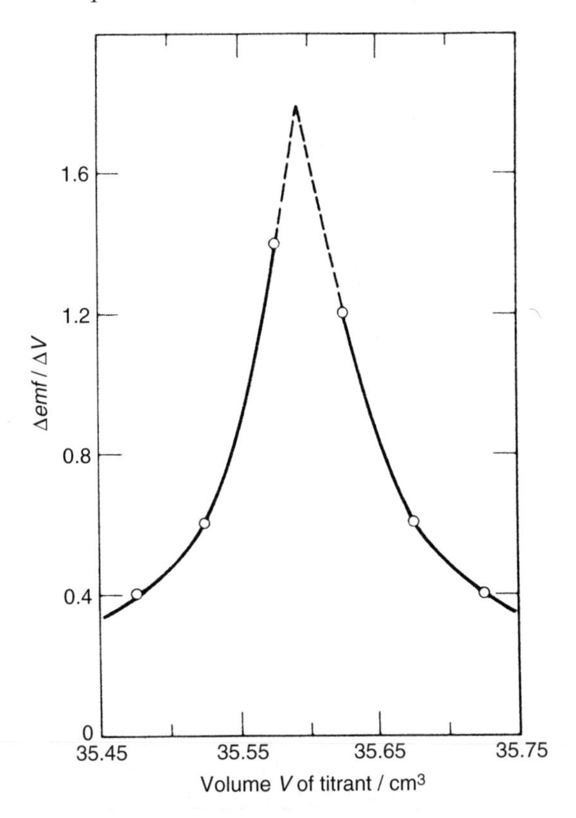

Figure 4.4 Plot of $\Delta emf/\Delta V$ against volume, i.e. the first-derivative curve of the plot shown in Figure 4.2, for the redox reaction represented by equation (4.1).

The error in determining the end point potentiometrically is probably about 2%, with this error arising principally from the reading of the volume. Such an error will be about the same as that found in a conventional acid–base titration. Errors of this type can be decreased by first titrating against a known standard or by ensuring that larger volumes of redox reagents are involved.

Conversely, a better approach involves drawing a plot of Δemf against ΔV, i.e. obtaining the first-derivative curve (as shown in Figure 4.4), where the end point is now given by a peak.[†]

Derivative plots such as that shown in Figure 4.4 can greatly increase the accuracy of the end point determination, provided that a sufficient amount of data is obtained. We need to note that the rate of change of *emf* with volume V is often very large near the equivalence point, and so it is easy to miss some of the data.

[†] A superior end point is obtained if the effects of dilution are compensated for by using an abscissa axis of $V + \Delta V/2$.

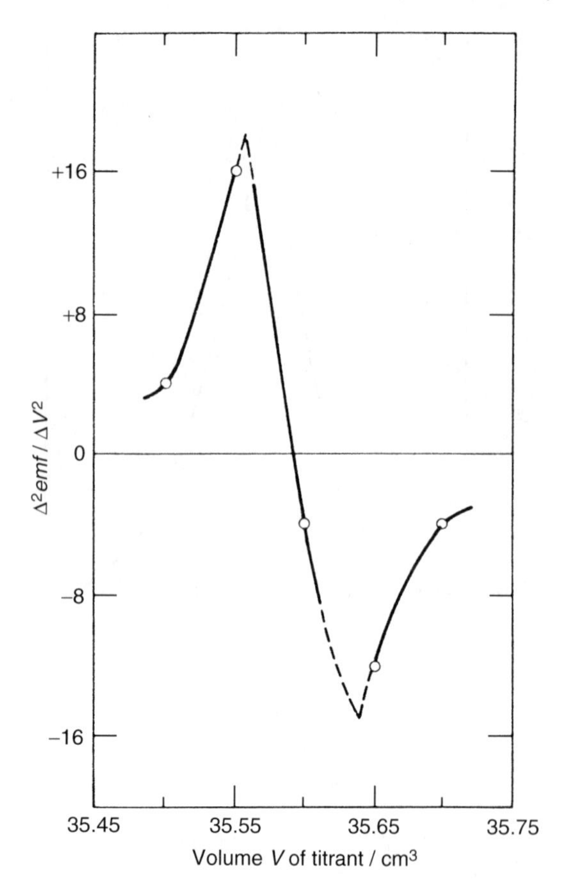

Figure 4.5 Plot of $\Delta^2 emf/\Delta V^2$ against volume for the redox reaction represented by equation (4.1), i.e. the second-derivative curve of the plot shown in Figure 4.2 (as well as being the first-derivative curve of that shown in Figure 4.4).

This is why a dashed line has been drawn near the peak in Figure 4.4. For this reason, a second-derivative curve is often produced, in which case the end point is given by the point of **inflection** (see Figure 4.5). For a 1:1 reaction (such as that represented by equation (4.1)), the inflection occurs at $\Delta^2 emf/\Delta V^2 = 0$, and the end point can then be taken as the volume at which the line crosses the axis.

In summary, we see that redox systems can indeed qualify as volumetric reagents for quantitative analysis.

DQ 4.4

Are there any restrictions on the reagents that can be used in a potentiometric titration?

Answer

We need to think in terms of both kinetics and thermodynamics when considering restrictions on the choice of redox systems. For simplicity, we will continue to employ the case where ferrous ion is titrated with ceric ion.

The main **kinetic** *consideration is the* **time-scale** *of the redox reaction: if the relevant electron-transfer reaction is slow, then we run the risk that measurements are taken before a true equilibrium has been attained after the addition of each aliquot. In practice, however, most analytes are oxidized or reduced within a very short time-scale – probably within microseconds if mixing is efficient, and provided that the oxidant (e.g.* H_2O_2, MnO_4^-, Ce^{4+} *or* $Cr_2O_7^{2-}$*) or reductant (e.g. chromous ion,* Cr^{2+}*, dithionite,* $S_2O_4^{2-}$*, or thiosulfate,* $S_2O_3^{2-}$*) is sufficiently powerful. Note that while oxidation of* Cr^{2+} *to form* Cr^{3+} *is fast, the respective reductive electron-transfer reactions* of Cr^{3+} *(to form* Cr^{2+}*) are too slow to employ in a redox titration of this sort.*

There are two **thermodynamic** *factors that must be considered. The first thing to remember is that the product of the redox reaction must be soluble and electroactive. This is an obvious point, since both the oxidized and reduced forms of the titrant need to comprise a redox couple. Ceric ion,* Ce^{4+}*, is therefore a good choice of oxidant, since* Ce^{4+} *and its* Ce^{3+} *product are both completely electroactive. However,* H_2O_2 *is not a good oxidant in the context of potentiometric titrations, since the product of the reaction is water. Even if the* H_2O_2–H_2O *couple was completely electroactive, fast and nernstian (see Chapter 3) in behaviour – and it is not – most of these systems are prepared in aqueous solution, and so the* amounts of water formed as the product of oxidation *would be very difficult to determine.*

Secondly, if the volume of ceric ion is to be an acceptable measure of the quantity of ferrous material in solution, then the reaction needs to go to **completion**, *that is, the equilibrium constant* K *of the redox reaction (e.g. equation (4.1)) must be high.*

SAQ 4.4

The value of $E^\ominus$ for the quinone–hydroquinone couple is 0.699 V. Look at the list of electrode potentials given in Appendix 3 and decide which redox couples would be powerful enough to oxidize hydroquinone to quinone.

Worked Example 4.1. What does 'completion' mean in the context of a potentiometric titration?

The equilibrium constant of the redox reaction represented by equation (4.1) is given by the following:

$$K = \frac{a(Fe^{3+})a(Ce^{3+})}{a(Fe^{2+})a(Ce^{4+})} \tag{4.3}$$

Note that we can safely assume that all of the mean ionic activity coefficients are essentially the same because I is so high. In fact, we do not even need to know this 'common' value of the $\gamma_{\pm}$, since all of the coefficients cancel out in equation (4.3). For this reason, we can safely write K in terms of concentrations rather than activities (a).

For all practical purposes, we require a minimum extent of reaction, ξ, of 99.9% (or 0.999 mol of reaction). If ξ is 99.9%, then we have the following:

$$\frac{[Fe^{3+}]}{[Fe^{2+}]} = \frac{99.9}{0.1} \approx 10^3 \text{ and } \frac{[Ce^{3+}]}{[Ce^{4+}]} = \frac{99.9}{0.1} \approx 10^3$$

and so by substituting into equation (4.3), we find that K is 10^6.

DQ 4.5

If a 'complete' reaction implies a value of $K \approx 10^6$, which systems are suitable for redox titrations of this type, i.e. what is the relationship between $E_{O,R}$ of the two half cells and K?

Answer

The actual values of $E^{\ominus}_{O,R}$ *(and hence* $E_{O,R}$*) are irrelevant – what matters is the* difference *between the* $E_{O,R}$ *values for the left-hand and right-hand sides of the cell, i.e. the magnitude of the* emf *that is possible.*

From the Nernst equation (equation (3.8)), we can determine the equilibrium constant K *from the difference between the two values of* $E^{\ominus}$ *as follows:*

$$E^{\ominus}_{RHS} - E^{\ominus}_{LHS} = \frac{RT}{nF} \ln K \tag{4.4}$$

It is very common to see the left-hand side of this equation written as $E^{\ominus}_{cell}$ *or even* $emf^{\ominus}$*. The latter usage is not recommended, however, since confusion is likely to occur – equation (4.4) does not rely on the* emf *in any way.*

Worked Example 4.2. So, what is the value of K for the cell represented by equation (4.1)?

From the appropriate tables (e.g. Appendix 3), we find that:

$$E^{\ominus}_{RHS} - E^{\ominus}_{LHS} = E^{\ominus}_{Ce^{4+},Ce^{3+}} - E^{\ominus}_{Fe^{3+},Fe^{2+}} = (1.610 - 0.770)\ V = 0.840\ V$$

and by inserting the relevant values into equation (4.4), we obtain:

$$0.840\ V = 0.0257\ V \times \ln K$$

and so:

$$\frac{0.840 \text{ V}}{0.0257 \text{ V}} = \ln K$$

After taking the exponential:

$$K = \exp(32.68) = 1.56 \times 10^{14}$$

Note how both units of 'volt' cancel out, thus allowing us to take this exponential.

As we have seen above in Worked Example 4.1, the *minimum* acceptable value of K is 10^6 to a good first approximation.

SAQ 4.5

Taking $n = 1$, what is the minimum separation between the standard redox potentials that will yield a value of K of 10^6?

From SAQ 4.5, we see that the first criterion for suitability when titrating two redox systems is that their electrode potentials must be separated by a relatively small potential when $n = 1$.

SAQ 4.6

By using a similar approach to that used for SAQ 4.5, show that the separation is 0.26 V when $n = 2$.

Note that a separation of 0.2 V or less will generally yield a graph that is not worth using, because the separation will not then be clear, i.e. the error in measuring ΔE will be too great.

Aside *The Shape of Potentiometric Titration Curves*

Suppose that the amount of Fe^{2+} in a sample is f. As a quantity, x, of Ce^{4+} is added, so the amount of Fe^{2+} decreases to $f \times (1 - x)$. Simultaneously, the amount of Fe^{3+} increases to $x \times f$, where, again, x relates to the amount of Ce^{4+} added.

If the equilibrium constant of the reaction, K, is infinite, then the product (xf) would be equal to the amount of oxidant added. It follows that at an intermediate stage in the titration, the electrode potential is the ratio of fx and $f(1 - x)$.

The fraction, '$x/(1 - x)$', is the cause of the titration curve's 'S' shape (cf. Figure 4.3 above).

4.1.2 Determination of Equilibrium Constants and Solubility Constants

We looked earlier (Section 3.5.3) at sparingly soluble salts. In Worked Example 3.13, we examined the case where excess solid AgCl was placed in pure water at 25°C, and then left until equilibrium was reached. The activity $a(\mathrm{Ag}^+)$ was obtained via the electrode potential as $E_{\mathrm{Ag}^+,\mathrm{Ag}} = emf + E_{\mathrm{SCE}}$, and the Nernst equation (equation (3.9)) was then used. Such a method is prone to systematic errors, e.g. 'What if E_{SCE} is inaccurate?', or 'What if the salt and solution have not reached equilibrium?'. Accordingly, we can now see that the experimental determination of a solubility constant, K_s, via a potentiometric titration is superior to that involving a simple cell under frustrated-equilibrium conditions (see Worked Example 3.14).

DQ 4.6

How is the solubility product of silver chloride determined via a potentiometric titration?

Answer

We can employ the following technique. A clean silver electrode is immersed in a solution of chloride ion of known concentration, while a calomel electrode (SCE) is immersed in a second solution of saturated KNO_3. The two containers are connected via a salt bridge (to minimize junction potentials, E_j), thus constructing the circuit shown in Figure 4.6.

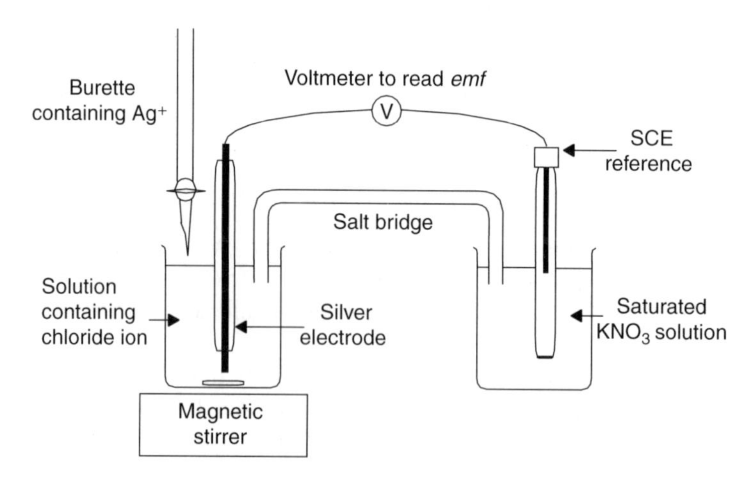

Figure 4.6 Schematic representation of the apparatus required when monitoring a precipitation process via a potentiometric titration. The salt bridge is impregnated with a saturated solution of KNO_3.

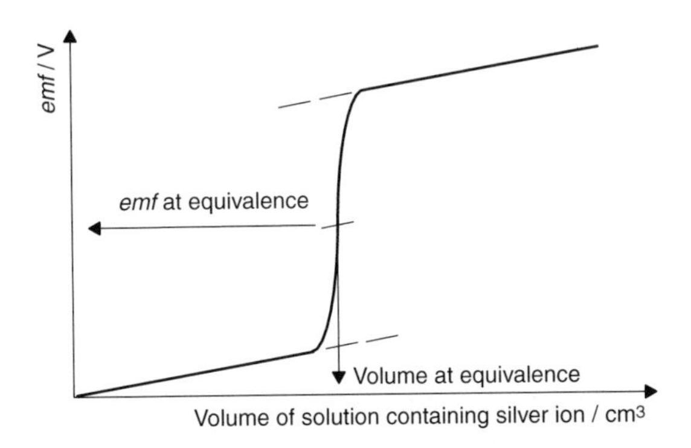

Figure 4.7 Plot of *emf* ($= E_{\mathrm{Ag^+,Ag}} - E_{\mathrm{SCE}}$) against the volume of silver ion solution added during a titration against chloride ion, as followed by a potentiometric procedure. The point of inflection is the end point, which can be conveniently determined as follows. Draw a line that is parallel with, and exactly midway between, the two extrapolants (the outermost dashed lines). The end point is indicated by the volume at which this line crosses the *emf*–volume curve (as indicated by the downward arrow). The equivalence *emf* is also indicated.

The cell schematic in this case may be written as follows:

$$SCE \mid KNO_{3\,(saturated\ solution)} \mid Ag^{+}{}_{(aq)} \mid Ag$$

where the central portion represents the salt bridge.

Silver nitrate solution (again of known concentration) is added to the chloride solution in small, precisely controlled aliquots, with the emf *being determined after each addition, thus enabling a graph of* emf *against volume to be drawn (see Figure 4.7). The point of inflection on this graph is the point of equivalence.*

Knowing the emf *at the end point, we can then determine the electrode potential* $E_{Ag^+,Ag}$ *(via the relationship,* emf $= E_{Ag^+,Ag} - E_{SCE}$*), and hence use the Nernst equation in a manner analogous to that adopted earlier in Worked Example 3.13.*

Worked Example 4.3. The cell, SCE | $KNO_{3(\mathrm{saturated\ solution})}$ | $Ag^{+}{}_{(aq)}$ | Ag, was constructed as described above, titrating Ag^+ with the Cl^- initially present. A graph of *emf* against the volume of added $AgNO_3$ was then plotted. The *emf* at the end point was 0.270 V. If we take $E^{\ominus}_{\mathrm{Ag^+,Ag}} = 0.799$ V and $E_{\mathrm{SCE}} = 0.242$ V, what is K_s?

Strategy. (i) We measure the electrode potential of the silver ion from the cell *emf* at the end point, (ii) from the value of $E_{\mathrm{Ag^+,Ag}}$, the activity of the silver ion is calculated by using a form of the Nernst equation (equation (3.9)), and (iii) now knowing $a_{\mathrm{Ag^+}}$, K_s may then be determined, since $K_s = a_{\mathrm{Ag^+}} \times a_{\mathrm{Cl^-}}$, and at equivalence, $a_{\mathrm{Ag^+}} = a_{\mathrm{Cl^-}}$, and so $K_s = (a_{\mathrm{Ag^+}})^2$.

(i) In order to measure $E_{Ag^+,Ag}$, we first employ the equation defining the *emf* (equation (3.3)), by saying:

$$emf = E_{\text{positive electrode}} - E_{\text{negative electrode}}$$

so:

$$emf = E_{Ag^+,Ag} - E_{SCE}$$

and therefore:

$$E_{Ag^+,Ag} = 0.270 \text{ V} + 0.242 \text{ V} = 0.512 \text{ V}$$

(ii) Next, from the Nernst equation:

$$E_{Ag^+,Ag} = E^{\ominus}_{Ag^+,Ag} + \left(\frac{RT}{F}\right) \ln a(Ag^+)$$

Rearrangement then gives:

$$a(Ag^+) = \exp\left(\frac{E_{Ag^+,Ag} - E^{\ominus}_{Ag^+,Ag}}{RT/F}\right) = \exp\left(\frac{0.512 \text{ V} - 0.799 \text{ V}}{0.0257 \text{ V}}\right)$$

and so:

$$a(Ag^+) = 1.41 \times 10^{-5}$$

(iii) Finally, by using the relationship, $K_s = a(Ag^+) \times a(Cl^-)$, and knowing that a_{Ag^+} and a_{Cl^-} are equal at the equivalence point, we obtain:

$$K_s = [a(Ag^+)]^2 = (1.41 \times 10^{-5})^2 = 2 \times 10^{-10}$$

Note how we have assumed throughout this calculation that concentration and activity can be employed interchangeably.

SAQ 4.7

A known volume of KBr solution is pipetted into a clean beaker and a freshly polished silver electrode is then immersed in this solution. After a potentiometric titration with $AgNO_3$ solution at 298 K, the electrode potential at the equivalence point ($E_{Ag^+,Ag}$) is determined to be 0.441 V. What is K_s for the pale yellow AgBr that is formed?

SAQ 4.8

A cobalt electrode is immersed in a solution of Co^{2+}, and is then titrated with hydroxide solution at 298 K to form $Co(OH)_2$. What is the solubility product of $Co(OH)_2$ if $E_{Co^{2+},Co} = -0.41$ V at the equivalence point? Take $E^{\ominus}_{Co^{2+},Co} = -0.28$ V.

DQ 4.7

What is the physico-chemical basis for the point of inflection being taken as the equivalence point?

Answer

For the purposes of this discussion, we will stay with the example of an Ag^+ solution (as $AgNO_3$) added to a solution of chloride. This chloride solution contains no silver at the beginning of the titration, so the silver redox electrode is not a part of a couple. Therefore, the electrode does not measure anything meaningful at all before the start.

As soon as the first aliquot of Ag^+ is added to the chloride solution, a white precipitate of 'insoluble' AgCl will form. Some of this AgCl, however, will remain in solution. Accordingly, the silver couple will form, and its existence will be registered at the silver electrode as $E_{Ag^+,Ag}$. The amount of silver remaining in solution will be minute, and so the emf *will increase by only a tiny amount. Note that $[Cl^-]$ will not change much after the addition of only one aliquot, so $[Cl^-] \ggg [Ag^+]$.*

We will now look at the relative amounts of ions in solution at the very end of the titration, i.e. after the end point has been reached and surpassed, and note that 'all' of the chloride ion will have been consumed by the precipitation of solid AgCl, and that the addition of further $AgNO_3$ solution to the reaction solution merely dilutes it. At the end of the titration, we see that $[Cl^-] \lll [Ag^+]$.

Clearly, between these two extremes, the conditions are such that $[Cl^-] = [Ag^+]$. The shapes of the emf–V *curve on either side of the point of inflection are the same – a mirror image in fact – implying that the two regimes, above and below $V_{inflection}$ represent deviations from a common position. The only situation existing between the two extremes that could have any physico-chemical basis is that where $[Cl^-] = [Ag^+]$.*

An additional point worth mentioning is that the potentiometric method can monitor several partially soluble salts at once. For example, if a solution contains chloride, bromide and iodide ions, then a plot of *emf* against the volume of cation (e.g. Ag^+) will contain three inflection points (see Figure 4.8), one for each of the three silver halides. K_s for AgI is smaller than that for AgCl, while K_s (AgBr) has an intermediate value, so the first inflection point represents the precipitation of AgI, the second represents formation of AgBr and the third represents the formation of 'insoluble' AgCl.[†]

These three silver halides all have different colours, so the successive formation of the three salts can easily be followed by the naked eye.

[†] Care should be taken when determining K_s for salts such as these since their photostability is poor, cf. the standard use of such salts in black and white photography. Therefore, the experiment is best performed in the dark, or by employing light of low photon energy (e.g. using red filters).

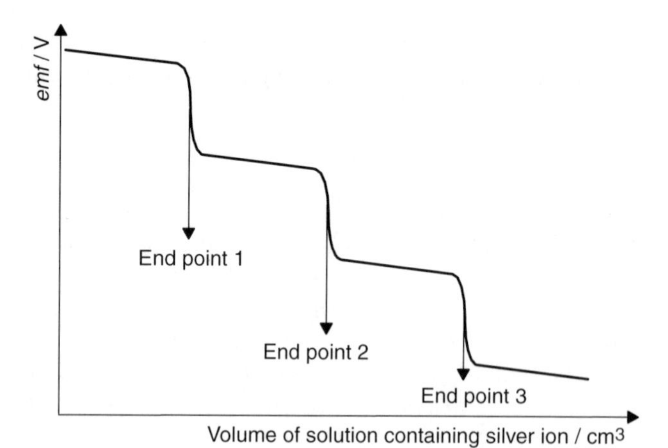

Figure 4.8 Plot of *emf* ($= E_{Ag^+,Ag} - E_{SCE}$) against the volume of silver ion added during a mixed halide (Cl^-, Br^- and I^-) titration, as followed by a potentiometric procedure. The three points of inflection, on the curve determined by using the experimental set-up shown in Figure 4.6, represent the three end points, thus allowing for the simultaneous determination of the three values of K_s. Points (1)–(3) relate to the precipitation of AgI, AgBr and AgCl, respectively.

4.2 Redox Indicators

The end point of a redox titration can be detected by use of a **redox indicator**, that is, a redox couple (O,R) in which the oxidized and reduced forms of the couple have distinctly different UV–visible spectra and hence, present different colours to the naked eye:

$$\begin{array}{ccc} O + ne^- & = & R \\ \text{colour 1} & & \text{colour 2} \end{array}$$

In some contexts, an indicator (dye) such as this is said to be **electrochromic**, that is, the potential determines the colour (the name comes from the Greek word *chromos*, meaning 'colour'). A few representative redox indicator systems are shown in Table 4.1.

DQ 4.8

How is a redox indicator used?

Answer

Suppose we add a small quantity of redox indicator to a relatively large volume of our redox titration mixture, e.g. aqueous ferrous and ferric ions. In addition, suppose that the standard electrode potential of the indicator couple (we will call it $E^{\ominus}_{in}$*) is greater than that of the ferrous–ferric*

Table 4.1 Some commonly used redox indicator systems

Name	Structural formula[a] and colour change (written as a reduction reaction)	$E^{\ominus}$ (V (vs. SHE))
Diphenyl benzidine[b] (DPB)	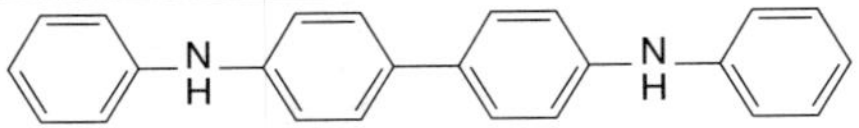 violet + $2H^+$ + $2e^-$ → colourless	+0.76
Iron tris(*o*-phenanthroline)[c] ('ferroin')	pale blue + e^- → red	+1.11
Indigo Carmine[c]	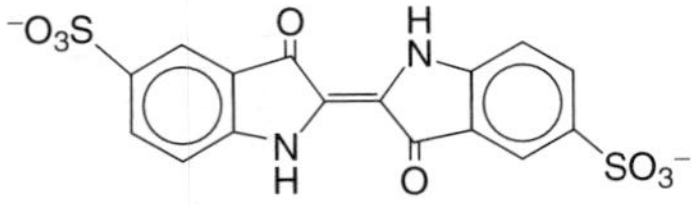 ^-O_3S … SO_3^- blue + $2e^-$ + $2H^+$ → red	+0.70
Methylene Blue (MB)[c]	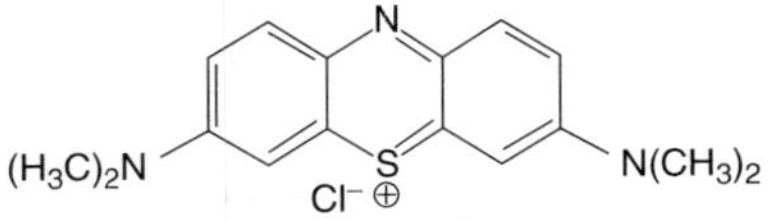 $(H_3C)_2N$ … $N(CH_3)_2$, Cl^- blue + e^- → colourless	+0.53
Methyl viologen	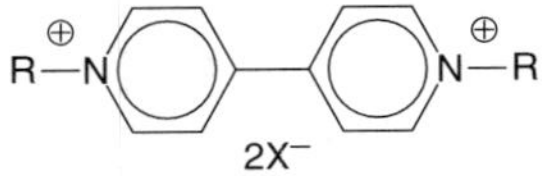 R—N⊕ … ⊕N—R, $2X^-$ colourless + e^- → blue	−0.446

[a]Most of these colour changes arise from changes in the conjugation. Accordingly, only the structure of the stablest of the redox states is shown.
[b]Diphenylamine is added to the solution as a precursor.
[c]In acid (1 mol dm^{-3}).

system. Ferrous ion is in excess, and so a tiny amount of the indicator will be oxidized by the titrant, as follows:

$$Fe^{2+} + O_{in} \longrightarrow Fe^{3+} + R_{in}$$

(We will treat this system as one which undergoes a one-electron transfer.)

If the quantity of indicator is minute, then its presence will not affect the $a(Fe^{2+})/a(Fe3^{+})$ *ratio – the indicator takes no part in the reaction, and acts merely as a* ***probe****. Conversely, addition of iron will completely shift the* $a(O_{in})/a(R_{in})$ *ratio in one direction, so only one of the two colours possible within the indicator system will be visible. The potential of the system can be assumed to have remained constant.*

The average human eye can detect absorbance changes for the indicator colour between about 1:10 and 10:1. This means that if the indicator is completely reduced, we won't know until about 10% of it has been re-oxidized by the addition of ceric ion.

The potential of the system is given by the following:

$$E_{in} = E^{\ominus}_{in} + \frac{RT}{nF} \ln\left(\frac{1}{10}\right)$$

After 90% of the indicator has been re-oxidized, the eye will then again detect no further changes, and so the upper potential limit is therefore given by:

$$E_{in} = E^{\ominus}_{in} + \frac{RT}{nF} \ln\left(\frac{10}{1}\right)$$

The (electrode) potential range over which the eye can observe colour changes in a redox indicator is therefore given by the following expression:

$$E = E^{\ominus}_{in} + \frac{RT}{nF} \ln(100)$$

By assuming that $n = 1$*, the limits for this particular indicator are* $E^{\ominus}_{in} \pm 0.059 \ln\ (100)$ *V, i.e. a working range of c. 120 mV. The ideal choice of indicator is seen to be one which is characterized by a standard redox potential as close as possible to the end point of the titration system.*

SAQ 4.9

An analyst has poor eyesight, and consequently can only detect (oxidation/reduction) absorbance changes between 1:5 and 5:1. What is the minimum separation between E and $E^{\ominus}_{in}$?

4.3 Treatment of Errors

4.3.1 Routine Errors

Potentiometric titrations are usually more accurate than simple acid–base titrations since we are not looking here for a *visual* change in colour or intensity. Furthermore, the potentiometric approach means that we can follow a titration without having to add an additional chemical to the analyte solution – always a good idea if complexation is a possible side reaction.

An additional benefit of the potentiometric titration is that we are looking for *changes* in a variable (where the variable here is *emf*). Accordingly, the potential of the reference does not need to be known. This latter consideration is particularly pertinent if derivative curves, such as those shown in Figures 4.3 and 4.4, are to be constructed.

When considering errors, we will assume that the concentration of the titrant is known accurately, e.g. as determined by prior titration against various standards. We also assume that possible 'problems' with volumetric glassware, the correct reading of burettes, handling of pipettes, etc., can be ignored or are wholly absent. In such a situation, the most common error encountered in a potentiometric titration is an incorrect determination of the end point, owing to the graph of *emf* against volume displaying a distortion from the ideal shape shown in Figure 4.2.

The simplest case in which we can overcome such distortion is when we initially have insufficient data to enable us to correctly ascertain the position of the end point. Figure 4.4, for example has some data missing, which explains why the peak is 'pencilled in' with dashed lines. In such a situation, we simply repeat the titration, taking care this time to obtain more data at volumes close to the suspected equivalence point.

Another common reason for distortion is that the two values of $E^{\ominus}_{\mathrm{O,R}}$ – for the analyte and for the titrant – are too close together, implying that $K < 10^6$. A more careful choice of redox reagent should alleviate this problem.

Alternatively, if the titration is performed too quickly, then it is possible that a true equilibrium has not been achieved. A simple test of whether distortion is caused by slow electron transfer is to repeat the titration, but in this case, waiting between the addition of each aliquot.

4.3.2 Errors Caused by Low Precision of Volume Determination: Use of Gran Plots

We sometimes find that the end point is still elusive, even with due care and patience. In such cases, a **Gran plot** may help – a plot of $\log_{10}$ (*emf*) against volume will be linear (or at least, passably linear), and the volume at equivalence is then given by the intercept on the x-axis.

Consider the case in which the solubility of silver chloride is being measured by using a silver ion-selective electrode (ISE). The ISE is immersed in a solution

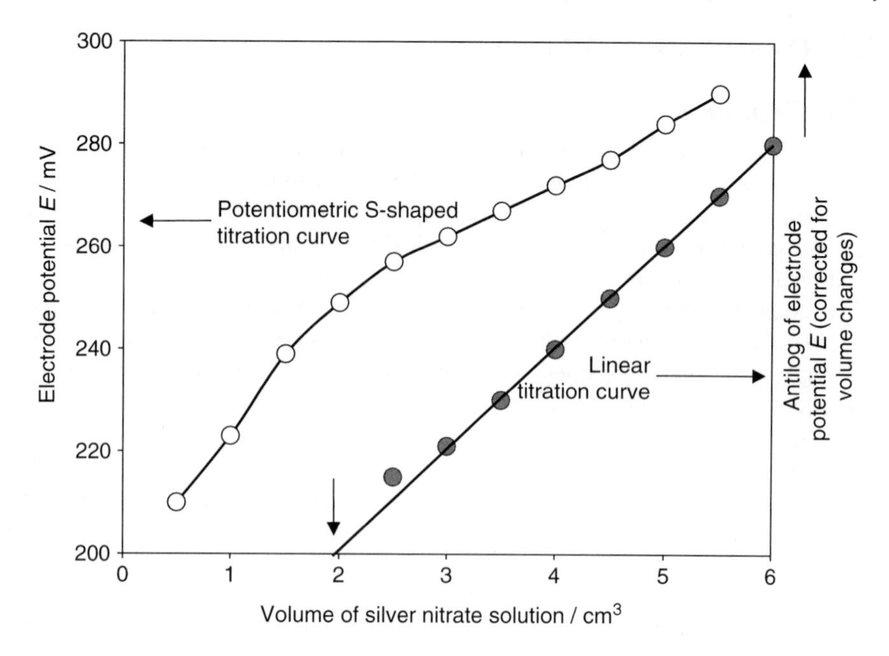

Figure 4.9 Gran plot of electrode potential E against volume of silver ion (O), used to determine the solubility constant K_s of AgCl. E is obtained from the *emf* of the cell, 'reference electrode | Ag^+ | ISE', where the ion-selective electrode is a solid-state ISE based on Ag_2S. The ISE is immersed in a stirred solution of KCl (of initial concentration 5×10^{-5} mol dm^{-3}); the volume values on the x-axis relate to the aliquots of added $AgNO_3$ (of concentration 2×10^{-3} mol dm^{-3}). The antilog trace $E(\bullet)$ is clearly proportional to the volume of $AgNO_3$ added. From Christian, G.D., *Analytical Chemistry*, 5th Edn, © Wiley, 1994. Reprinted by permission of John Wiley & Sons, Inc.

of KCl, and aliquots of $AgNO_3$ are then added. The activity, $a(Ag^+)$, increases as the amount of silver is added, so the *emf* of the cell, 'reference electrode | Ag^+ | ISE', increases accordingly (see Figure 4.9). After many such additions, the Gran plot becomes linear and so extrapolation to lower volumes is valid. The intercept† is indicated with an arrow, and represents the true end point volume.

Note how the Gran plot in Figure 4.9 is curved at intermediate and small addition levels, thus making adjustment quite difficult. Such curvature is best taken to indicate that a 'complication' is present, e.g. that the precipitate has an appreciable solubility (when determining K_s) or that dissociation of a complex has occurred (when trying to measure $K_{complexation}$).

† This intercept relates to a position of the x-axis which is determined by running a 'blank' experiment, that is, by adding the same $AgNO_3$ solution to water. The value of the *emf* when $[AgNO_3] = 0$ is then taken to represent the correct position of *emf* = 0.

4.3.3 Errors Caused by Poor ISE Selectivity

We saw earlier in Section 3.5.2.2 how the **selectivity** of an ion-selective electrode (ISE) can be occasionally problematic, e.g. the activity of fluoride ion when determined by a fluoride ISE can be significantly in error if the solution also contains sufficient hydroxide ion to raise the pH above about 8.

Employing the method known as **multiple standard addition** helps when trying to discern the effects of the sample *matrix* in which an analyte is dissolved. In this technique, the *emf* is determined as a function of the amount of standard solution added to the sample.

Figure 4.10 shows a Gran plot which can be used for such a situation. In this figure, the addition values on the right-hand side of the abscissa represent the volumes (i.e. the *known additions*) of contaminant brought about by adding, e.g. small amounts of hydroxide to a solution in which a fluoride ISE is immersed. The intercept of the plot on the left-hand side represents the equivalent amount of hydroxide within the sample prior to the multiple standard addition process. This intercept† tells us the concentration of hydroxide. Therefore, knowing $[OH^-]$, we

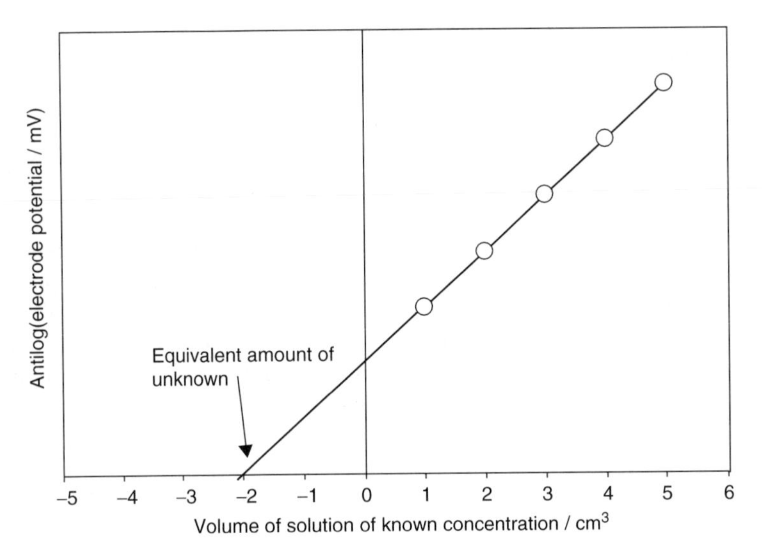

Figure 4.10 Gran plot of $\log_{10}$ *emf* against volume of 'contaminant'. The equivalent amount of contaminant in the analyte is obtained as the intercept on the x-axis. Such plots are particularly useful when determining concentrations with ion selective electrodes in the presence of ions or substances known to have poor selectivity. From Christian, G.D., *Analytical Chemistry*, 5th Edn, © Wiley, 1994. Reprinted by permission of John Wiley & Sons, Inc.

† The correct value of y at which to draw the x-axis in the plot shown in Figure 4.10 must first be obtained by performing an experiment with the same set of electrodes immersed in a solution known to have a zero amount of contaminant, but otherwise identical to that under study.

can then use expressions such as equation (3.18) to determine an accurate value of $a(F^-)$.

Summary

Many redox systems are suitable for use as volumetric reagents for quantitative analysis provided that (i) both states within the oxidized and reduced forms of the redox-active titrant comprise a fast nernstian couple, (ii) all redox states are soluble in the solutions employed, and (iii) the separation between the standard electrode potential for each of the constituent half cells is $0.35/n$ V (where n is the number of electrons in the titrant couple).

During a redox reaction, a potentiometric titration can be employed to determine a *concentration* of analyte rather than an *activity*, since we are only using the *emf* as a reaction variable in the accurate determination of an end point *volume*. For this reason, an absolute value of $E_{\text{reference electrode}}$ need not be known, as we are only concerned with *changes* in *emf*. It is, however, advisable to titrate at high ionic strength levels in order to minimize fluctuations in the mean ionic activity coefficients.

During a precipitation reaction, a potentiometric titration can also be employed, but here we generally determine an activity since the *emf* is related to the Nernst equation. For this reason, an absolute value of $E_{\text{reference electrode}}$ should be known in this case.

The end point on a potentiometric plot can often be determined merely by *inspection* of the *emf* at the point of inflection. Precision can be improved (when determination of the end point is difficult) by constructing derivative plots.

When the electrode is an ISE, poor selectivity can impair the final value of the concentration that is determined, in which case a Gran plot is constructed in order to correct for the way that the *emf* comprises various contributory components.

Chapter 5
Coulometry

Learning Objectives

- To learn that coulometry is the study of charge flow, usually for a charge Q, indicating that the potential has been shifted away from its equilibrium value.
- To appreciate that the charge Q is often determined in terms of the current I.
- To appreciate that charge flow indicates that one material is formed simultaneously at the expense of another. Since such materials are generally in solution, coulometry is a means of following concentration changes.
- To learn that the amounts of material formed and the electrochemical charges Q that form them are related by Faraday's laws.
- To learn that the proportionality between the amount of material formed and Q is the Faraday constant F, where F is the charge carried by one mole of electrons.
- To appreciate that not all of the charge that flows is useful – the component that has formed material is termed 'faradaic' while the remainder is called 'non-faradaic'. The efficiency of the charge flow relates to that fraction of the overall charge which is faradaic. All electroanalytical methods require the non-faradaic charge to be minimized, with the ideal faradaic efficiency being 100%.
- To appreciate that a majority of non-faradaic currents are caused by the effects of adsorption, capacitance and the electrode double-layer, or by competing side reactions such as solvent splitting.
- To appreciate that the area A of an electrode is often slightly larger than its geometric area: this different A is termed the 'electrochemical' area.

- To learn that one of the most important areas of coulometry for the electroanalyst is 'stripping', in which analyte is allowed to accumulate on the surface of, e.g. a hanging mercury-drop electrode ('pre-concentration'), and then electro-oxidized ('stripped').
- To appreciate that analyte pre-concentration followed by stripping allows for the quantitative determination of analytes otherwise too dilute for accurate detection, i.e. stripping improves the sensitivity.
- To learn that the use of microelectrodes is a valuable means of decreasing the capacitive effects of the electric double-layer.
- To appreciate why using microelectrodes can improve the accuracy of the measurement, and how an assembly of microelectrodes can further improve the sensitivity of analysis.
- To learn that mediators are redox species used to effect redox chemistry on biological species that are electroactive, yet inert at most conventional electrodes: the charge employed in electro-converting the mediator is the same as would have been used in converting the analyte if it was electroactive.
- To appreciate that mediators can be electro-reduced or electro-oxidized, yet will effect a *chemical* redox reaction with the analyte.

5.1 Introduction to Coulometry and Faraday's Laws

In Chapter 1, we saw that electrochemistry is the branch of chemistry employed by an analyst when performing electroanalytical measurements, while in Chapter 2, we saw that electrochemical measurements fall within two broad categories, namely determination of a potential at zero current, and determination of a current, usually by careful variation of an applied potential. These two branches of electroanalysis are bridged in this present chapter by showing – on an elementary level – why charge flows, and also explaining how an analyst can interpret and thus process quantitative data during charge flow.

The SI unit of charge is the **coulomb** (which has the symbol 'C'), which is named after the eighteenth-century French scientist Charles Augustin de Coulomb. The modern analyst understands the word root 'coulo-' to indicate that aspects of charge flow are studied, so that '**coulometry** is the study of charge flow'. Note that sometimes we speak of the *charge* Q, while at other times we discuss the *charge flow* in terms of its first derivative, i.e. the current I. The following equation from Chapter 2 gives the relationship between Q and I:

$$I = \frac{\mathrm{d}Q}{\mathrm{d}t}$$

In the electroanalytical techniques discussed in previous chapters, care was taken to emphasize that no charge flowed. Indeed, Section 3.6.1 showed briefly how charge flow could so hamper the interpretation of a potentiometric measurement that the *emf* data obtained were effectively worthless.

DQ 5.1

What happens when charge flows through a cell?

Answer

Let us revisit the electrochemical cell shown earlier in Figure 3.1. In this figure, two redox electrodes are immersed in solutions of their respective ions, with the half cells being connected by a salt bridge. If we were to connect an infinite-resistance voltmeter between the cells, then it would be possible to perform potentiometric experiments such as those described in the previous chapter. One electrode would be positive with respect to the other, with the separation in potential between the two electrodes being the emf *– but only if the measurement was performed at equilibrium. (As before, we take the word 'equilibrium' to imply that no charge flows.)*

Now consider a new situation, in which we have removed the voltmeter and directly joined the two electrodes. As seen before, ***R****eduction occurs at the* ***R****ight-hand electrode (notice the alliteration), while oxidation occurs at the left, i.e. the right-hand electrode is positive and the left-hand electrode is negative. These potentials are not absolute – the terms 'negative' and 'positive' are relative, and relate to the potential of one electrode with respect to the other.*

Charge flows through the cell as soon as the wires of the electrodes touch each other, with electrons† *travelling from the more negative to the more positive electrode. It is these electrons which allow the reduction reaction to proceed at the right-hand side of the cell, where they are consumed as part of the electrode reaction. Since the electrons from the left 'fuel' the reaction at the right, we see that the extent of electrochemical reaction must be the same at both sides of the cell. We say that these reactions are* ***complementary****.*

SAQ 5.1

Consider a cell made up of two half cells, where one contains the Fe^{3+}, Fe^{2+} couple and the other the Cu^{2+}, Cu couple. By looking up the respective values of the standard electrode potentials $E^{\ominus}$ given in Appendix 3, deduce the spontaneous cell reaction that would occur if the leads connecting the two half cells were allowed to 'short' by touching.

Such complementarity goes further. The charge flowing at the left-hand electrode is the same in magnitude as the charge flowing through the right-hand

† In common with most electrochemical work, the electroanalyst assumes that electrons are *particles* rather than *waves*. While recognizing that the energy of the electron is quantized, the electroanalyst says that the differences in energy between quantum levels are so small that a **continuum** of energies can safely be assumed.

electrode:

$$Q_{\mathrm{LHS}} = -Q_{\mathrm{RHS}} \tag{5.1}$$

The minus sign in equation (5.1) reminds us that reduction occurs at the right while the opposite redox reaction, i.e. oxidation, occurs at the left of the cell. From equation (2.1) (see earlier), we can write an analogous form of equation (5.1) in terms of the current, as follows:

$$I_{\mathrm{LHS}} = -I_{\mathrm{RHS}} \tag{5.2}$$

Both of these equations are important relationships, so we will refer to them often.

If current is conducted (we often use the words **drawn** or **passed**) through the wires, then the same current must be drawn through the solutions of the cell. We must recognize that current is carried through the wires by electrons and through the solution by ions, both **anions** (negative) and **cations** (positive).† It is these ions which move from the **bulk** of the solution to the electrode where they undergo electrochemical modification. We say that they 'undergo electrode reaction' or are 'electromodified'.

Important	Current is carried through the **wires** by **electrons**
	Current is carried through the **solution** by **ions**

The amount of ionic charge carried through the solution is the same as the electronic charge carried through the electrode. Similarly, we could say that the amount of ionic current carried through the solution is the same as the electronic current carried through the electrode, as follows:

$$I_{\text{ions through solution}} = I_{\text{electrons through the external circuit}} \tag{5.3}$$

If we were to place a zero-resistance meter between the two electrodes, we could monitor the amount of charge that flows. Such a meter would be called an **ammeter** if it measured the current, or a **coulometer** if it measured the charge. (In practice, most modern meters are multi-function devices and can measure both, changing from one function to another at the flick of a switch.)

A meter placed in the circuit (see Figure 5.1) tells us the charge (or current) that flows through the electrodes. Following from equation (5.3), though, we see that in fact an ammeter (or coulometer) also tells us how much charge (or current) has flowed *through the solution of the cell*. In other words, the meter tells us how many electrons have been consumed by electrode reactions in solution. We see

---

† Sometimes the anions and cations move to differing extents, as defined by the transport number t. In fact, there is no need for both anions and cations to move appreciably at all, i.e. $t^+ \gg t^-$ or $t^- \gg t^+$. We shall ignore this possibility by simply saying that ions move through the solution.

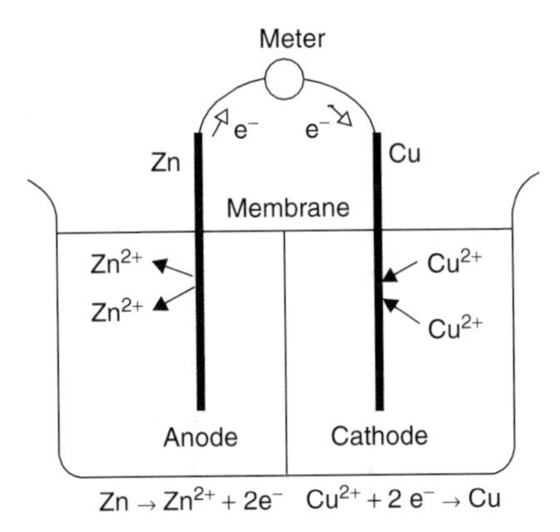

Figure 5.1 Schematic representation of a cell during discharge, showing a meter reading the charge or current passed. A coulometer placed in series will measure the charge passed, while an ammeter in series will measure the current passed.

that the reading on the meter is, in effect, quantifying the amount of reaction occurring at the electrodes. We are now in a position to discuss **coulometry**.

DQ 5.2

What if the reaction we want does not occur naturally?

Answer

Up until now, we have been considering the somewhat simplistic situation where the driving power for the reaction is provided by the energy released by the components within the cell as they react, i.e. when undergoing redox reactions. A battery, e.g. for powering a torch or watch, is an application for such a cell during its discharge.

As an electroanalyst, however, we need to note that the overwhelming majority of cells are not of such a simple nature. The electrode reactions in such cells usually have to be forced, that is, an external power pack is incorporated into the circuit in order to make sure that the intended electrode reactions will occur.

An analogy from everyday life might be the motion of a car. For example, a car will roll *down* a hill just as a battery will spontaneously discharge – the impetus in both cases is to attain a position of lower energy. If we wish the car to go *up* the hill, we then need to supply it with energy (in practice, we turn on the engine to release, by combustion, the energy stored within its petroleum-based fuel), and, in a similar sense, we need an external source of energy to effect many cell reactions. For now, we will call such a supply a **power pack**.

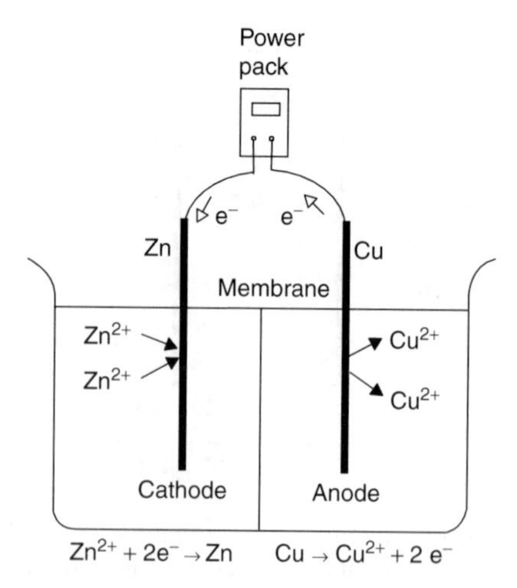

Figure 5.2 Schematic representation of a cell with a power pack (in series) to force electron-transfer reactions to occur; also indicated on the circuit are the anode (the positive electrode at which oxidation occurs) and the cathode (the negative electrode at which reduction occurs). Note that this figure would be equivalent to Figure 5.1 if the power pack was to drive the electrons in the opposite direction.

When a reaction is forced, we call the electrode that is made positive the **anode**, since anions are attracted to it. Oxidation reactions occur at the anode. We then term the negative electrode the **cathode**, since cations are attracted to it. Reduction reactions occur at the cathode. Figure 5.2 shows a schematic representation of such a situation.

Note that the terms 'anode' and 'cathode' are not at all useful when discussing cells in a state of frustrated equilibrium, since no charge flows.

DQ 5.3

How are the readings on the ammeter related to the 'amounts' of electrode reaction?

Answer

We will start by saying that an ammeter measures current and not charge, but (as discussed below) the amount of charge passed can be obtained as the integral of current with time.

We will define the topic of coulometry from within the context of ***Faraday's laws*** *(which were formulated in 1834 by the great Victorian scientist Sir Michael Faraday), as follows:*

1. *The number of moles of a species formed at an electrode during electrolysis is proportional to the electrochemical charge passed, where* $Q = It$.

2. *A given charge liberates different species in the ratio of their relative formula masses, divided by the number of electrons in the electrode reaction.*

One mole of electrons carries a charge of L × e, *where* e *is the charge on a single electron and* L *is the Avogadro constant.*[†] *This quantity of charge* (L × e) *has a value of 96 487 C, and is known as the* **Faraday** *constant* F, *and so we talk in terms of 'faradays of charge'. The basic reaction at an electrode is electron transfer to effect reduction (the analyte gains electrons) or oxidation (the analyte loses electrons), as follows:*

$$O + \mathrm{n}e^-{}_{(Pt)} \longrightarrow R \qquad (5.4)$$

The subscript 'Pt' is inserted to remind us that the electrons come from a conductive electrode. Platinum (Pt) is usually an excellent choice of electrode, while other common electrode materials are gold, mercury and glassy carbon (GC).

From Faraday's laws, we can see that:

*1*F *liberates 1 mol of Ag (via the reaction* $Ag^+{}_{(aq)} + e^-{}_{(Pt)} \longrightarrow Ag_{(s)}$*);*

*1*F *liberates 1/2 mol of Cu (via the reaction* $1/2\ Cu^{2+}{}_{(aq)} + e^-{}_{(Pt)} \longrightarrow 1/2\ Cu_{(s)}$*), i.e.*

2F *liberates 1 mol of Cu (via the reaction* $Cu^{2+}{}_{(aq)} + 2e^-{}_{(Pt)} \longrightarrow Cu_{(s)}$*);*

*1*F *liberates 1/3 mol of Al (via the reaction* $1/3\ Al^{3+}{}_{(aq)} + e^-{}_{(Pt)} \longrightarrow 1/3\ Al_{(s)}$*), i.e.*

*3*F *liberates 1 mol of Al (via the reaction* $Al^{3+}{}_{(aq)} + 3e^-{}_{(Pt)} \longrightarrow Al_{(s)}$*).*

Note that all *cations are initially in solution and will certainly be solvated to some extent. In addition, notice that the symbol for the electron is again subscripted to show that charge comes from an electrode rather than from a homogeneous electron-transfer reaction in solution (cf. the potentiometric titrations we discussed in the previous chapter).*

SAQ 5.2

How many moles of electrons are required to reduce 0.05 mol of Sn^{2+} to form Sn^0?

[†] Note that the Avogadro constant is also commonly given the symbol N_A. We will use L in this present text.

For basic reactions, **coulometry** of this sort is simple and straightforward, although we need to be careful in practice since some reactions can proceed with different stoichiometries:

$1F$ generates 1 mol of Fe^{2+} (from $Fe^{3+}{}_{(aq)} + e^{-}{}_{(Pt)} \longrightarrow Fe^{2+}{}_{(aq)}$);
$2F$ generates 1 mol of Fe (from $Fe^{2+}{}_{(aq)} + 2e^{-}{}_{(Pt)} \longrightarrow Fe_{(s)}$);
$3F$ generates 1 mol of Fe (from $Fe^{3+}{}_{(aq)} + 3e^{-}{}_{(Pt)} \longrightarrow Fe_{(s)}$).

Worked Example 5.1. 100 coulombs of charge are passed reductively through a solution of silver ions. What mass of silver is formed?

Strategy. Before we start, we need to know (i) what is the electrode reaction. Next, we need to determine (ii) the number of moles (or fractions thereof) of charge which flows through the cell. This is Faraday's first law in action. Knowing the number of moles, we then invoke Faraday's second law and decide from the reaction stoichiometry (iii) how many moles of metal are formed. Finally (iv), now knowing the number of moles of metal, we can calculate the mass of metal from the known atomic mass. The following procedure is therefore adopted:

(i) We will consider that the reaction is the simple one-electron reduction, $Ag^{+}{}_{(aq)} + e^{-}{}_{(Pt)} \longrightarrow Ag_{(s)}$.

(ii) One mole of electrons has a charge of 96 485 C, so 100 C = 100 C/96 485 C. Therefore, we perform 1.036×10^{-3} mol of electrode reaction.

(iii) We have caused 1.036×10^{-3} mol (worth) of electrode reaction to take place, and from the reaction stoichiometry, one electron generates one atom of silver metal, so 1.036×10^{-3} mol of silver are formed by 1.036×10^{-3} mol (worth) of the electrode reaction.

(iv) Finally, the atomic mass of silver is 107.9 g mol^{-1}, and as we have formed 1.036×10^{-3} mol of silver, the mass of silver is therefore $(107.9 \text{ g mol}^{-1}) \times (1.036 \times 10^{-3} \text{ mol}) = 0.112$ g.

SAQ 5.3

2734 C of charge effects the reduction reaction, $Zn^{2+}{}_{(aq)} + 2e^{-}{}_{(Pt)} \longrightarrow Zn_{(s)}$. What mass of zinc metal is formed?

SAQ 5.4

What charge was passed if 0.075 g of silver are formed by electrochemical deposition?

5.1.1 Faradaic and Non-Faradaic Charge

We have talked about charge being passed. In reality, any current will comprise two components, i.e. **faradaic** and **non-faradaic**. Faradaic charge is that component of the overall charge which can be said to follow Faraday's laws, i.e. is linked directly with the sum of the electron-transfer reactions effected. The remainder of the current does not follow Faraday's laws, and hence it is said to be '*non*-faradaic'. To summarize, we could say that:

$$I_{\text{overall}} = I_{\text{faradaic}} + I_{\text{non-faradaic}} \quad (5.5)$$

The existence of a non-faradaic component to the overall current explains why the amount of material formed by electrochemical formation will generally be less than the theoretical amount, since the 'theoretical amount' relates only to I_{faradaic}. Clearly, the coulometric **efficiency** should be maximized, i.e. $I_{\text{non-faradaic}}$ should be minimized by careful choice of experimental design, reagents and apparatus. Note that coulometric efficiency is also called **faradaic efficiency**.

SAQ 5.5

Elemental iodine can be formed from aqueous iodide ion (by means of the oxidative electrode reaction, $2I^- \rightarrow I_2 + 2e^-$) by passing a charge of 8.04×10^4 C. The amount of iodine was subsequently determined (by titration with aqueous thiosulfate) to be 0.239 mol. What is the electrolytic efficiency?

SAQ 5.6

Charge is forced through a cell to perform the electrode reaction, $Zn^{2+} + 2e^- \rightarrow Zn^0$. Previous experiments revealed the reaction has a faradaic efficiency of 77.1%. Given this efficiency, calculate the overall charge passed, by considering that 10^{-4} mol of zinc metal forms.

DQ 5.4

What is the cause of $I_{\text{non-faradaic}}$, i.e. why is the electrochemical 'efficiency' lower than the 100% theoretical maximum?

Answer

Electrode reactions occur when they are energetically favourable and do not occur if the thermodynamics of an electrode reaction imply that it is non-spontaneous. The simplest cause of $I_{\text{non-faradaic}}$ *arises from competing electrode reactions, i.e. two electrode reactions can occur in tandem if their energies are similar. Probably the most common of these reactions is electrolytic side reactions such as solvent splitting.*

Worked Example 5.2. During the reductive formation of copper metal (from an aqueous solution of Cu^{2+}), it is noticed that hydrogen gas is formed at the negative cathode, that is, in addition to the formation of a layer of fresh, pink copper metal. The volume of the gas at STP is 2.24 dm^3, and the overall electrochemical charge passed was 1.40×10^5 C. What is the electrolytic efficiency?

Strategy. We will need to decide first (i) the identity and stoichiometry of the second electrode reaction. Then, we will work out (ii) how much of the charge was required to form the hydrogen by this route. Therefore, knowing the overall charge and the amount consumed in the side reaction, (iii) we can work out the faradaic fraction utilized to form copper metal.

(i) The reaction at the electrode that forms the hydrogen gas is splitting of the solvent (here water), i.e. $2H_2O_{(aq)} + 2e^-{}_{(Pt)} \longrightarrow H_{2(g)} + 2OH^-$, which is a two-electron process.

(ii) From the ideal gas equation, 0.1 mol of hydrogen gas occupies a volume of 2.24 dm^3. Since the formation of H_2 is a two-electron process, Faraday's second law tells us that 0.1 mol of the gas was formed by consuming 0.2 mol of electrons (i.e. $0.2F$). A charge of $0.2F$ equates to $0.2 \times 96\ 485$ C = 19 297 C.

(iii) If 1.93×10^4 C of charge formed the H_2 gas, and 1.4×10^5 C was the overall charge passed, then the fraction of charge passed in forming the hydrogen was 0.138 or 13.8%.

The faradaic efficiency of the cell under these conditions is therefore (100 – 13.8)% = 86.2%.

In addition to simultaneous electrode reactions, there are many other reasons why the efficiency can be poor, as will be discussed in subsequent sections.

DQ 5.5

How accurate is the measurement of Q, and how much of the difference between moles of charge passed and moles of material formed is caused by the measurement?

Answer

Coulometry carried out with a proper *coulometer is usually highly accurate. We need to note, however, that it is an extremely common experimental practice to obtain* Q *as the integral of current* I *with time* t.

An even more approximate method is to draw a graph of current (as 'y') against time (as 'x') and then ascertain the area under the curve. If the current is constant, then charge is obtained as $Q = I \times t$. *This latter method is likely to introduce errors of as much as 5% to a measured charge.*

5.1.2 The Effects of Absorption, Capacitance and the Electrode Double-Layer

Solvent splitting and electrolytic side reactions are an extremely common contribution to $I_{\text{non-faradaic}}$. We will look in this present section at additional components of the observed non-faradaic charge.

In solution, all electrodes are surrounded by a **layer** of water molecules, ions, and other atomic or molecular species. We will not look in depth at this topic, except to refer to the two principle layers, which are named after one of the original pioneers of electrochemistry, namely the nineteen-century great, Hermann Helmholtz. The two Helmholtz layers are often said to comprise the **electrode double-layer** (or 'electric double-layer').

The extent to which ions, etc. **adsorb** or experience an electrostatic ('**coulombic**') attraction with the surface of an electrode is determined by the material from which the electrode is made (the **substrate**), the chemical nature of the materials adsorbed (the **adsorbate**) and the potential of the electrode to which they adhere.[†] Adsorption is not a static process, but is dynamic, and so ions etc. stick to the electrode (adsorb) and leave its surface (**desorb**) all the time. At equilibrium, the rate of adsorption is the same as the rate of **desorption**, thus ensuring that the fraction of the electrode surface covered with adsorbed material is constant. The double-layer is important because faradaic charge – the useful component of the overall charge – represents the passage of electrons *through* the double-layer to effect redox changes to the material in solution.

In elementary physics, a **capacitor** is usually depicted as comprising two layers (or 'plates') which are separated by a distance and bear different amounts of charge (Figure 5.3), as seen in practice by the formation of a difference in potential between the two plates. The electric double-layer around the electrode will similarly behave as a capacitor since the inner layer is different from the outer layer in terms of the numbers of ions adsorbed, and hence the total amount of charge it comprises (Figure 5.4).

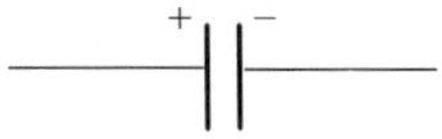

+ −

Figure 5.3 Schematic representation of a typical capacitor.

[†] While we will not discuss them further, we need to recognize that other factors will also effect the extent of adsorption, such as the choice of solvent, temperature and ionic strength, as well as other possible adsorbate species in solution, where the latter can then compete for adsorption sites on the electrode surface.

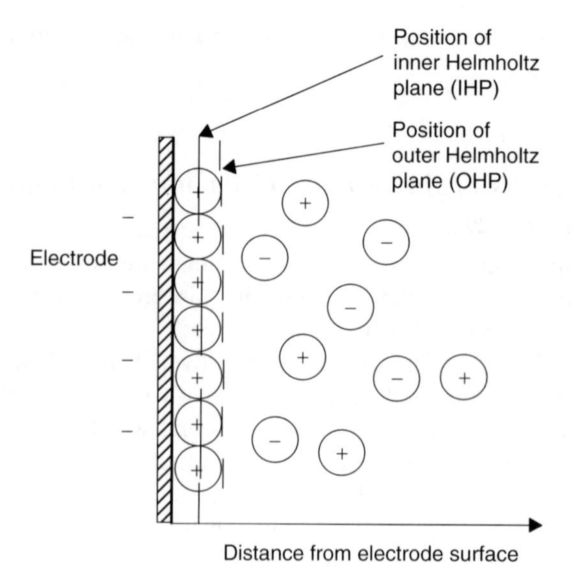

Figure 5.4 Schematic representation of the double-layer around an electrode, showing the positions of the inner and outer Helmholtz planes, and the way that ionic charges are separated. The circles represent solvated ions.

DQ 5.6

Clearly, 'charging' a capacitor requires charge, so how does charging the double-layer around the electrode affect a coulometric measurement?

Answer

The area of an electrode is finite and essentially constant.† *Similarly, the thickness of the electric double-layer does not vary by a large amount. As an empirical rule, we find that the* ***double-layer capacitance*** *has a value in the range 10–40 μF cm^{-2}, where F is the SI unit of capacitance, the* ***farad****. Note that a capacitance without an area is not particularly useful – we need to know the* complete *capacitance.*

To reiterate, the amount of charge consumed during charging the double-layer is a function of potential, so as soon as coulometry commences, the charge held within the double-layer will change as ions adsorb and/or desorb in response to the change in potential. For convenience, we will discuss these processes in terms of a reduction reaction, with electrons constantly leaving the electrode.

† The area of an electrode grows with time if a conducting material is electro-deposited on to it, e.g. during the electro-deposition of a metal. We will ignore this complication here.

SAQ 5.7

A copper electrode is made from flat copper sheet. The electrode is 3 cm long and 1 cm wide (ignore its thickness), and has a capacitance of 25 μF cm^{-2}. What is its capacitance?

During reduction, some of this charge will be 'captured' within the double-layer in order to alter the amounts of material adsorbed, and to readjust the positions of ions undergoing electrostatic interactions. The remainder will conduct *through* the double-layer to be taken up by analyte residing in, or just beyond, the layer, i.e. during a reduction reaction. In summary, we can say:

$$Q_{\text{overall}} = Q_{\text{double-layer charging}} + Q_{\text{for electroanalytical redox reaction}} \quad (5.6)$$

This equation can be viewed as a rewrite of equation (5.5).

DQ 5.7

If $Q_{\text{non-faradaic}}$ (i.e. $Q_{\text{double-layer charging}}$) is such a problem, then how is coulometry made practicable experimentally?

Answer

While $Q_{\textit{double-layer}}$ *is a function of potential, etc., we have seen that its value is known to be about 10–40* $\mu F\ cm^{-2}$. *This, in tandem with the simple relationship given in equation (5.6), indicates that the first and easiest method of increasing the coulombic efficiency is to pass a large charge. By this means, the second term on the right-hand side of equation (5.6) is made so much larger than the first that* $Q_{\textit{double-layer charging}}$ *can be more or less ignored.*

Worked Example 5.3. A coulometric experiment is performed to determine the faradaic efficiency of an electrode. The faradaic electrode reaction, '$Sn^{2+}{}_{(aq)} + 2e^{-}{}_{(Pt)} \rightarrow Sn_{(s)}$', occurs with no electrolytic side reactions. The electrode has an area of 9.8 cm^2 and the double-layer is charged with 20 mC cm^{-2}. What is the faradaic efficiency when (i) a total charge of 1.0 C is passed, and (ii) a total charge of 320 C is passed?

The proportion of the overall charge needed to charge the double layer is readily computed as 'charge per area × area', so $Q_{\text{double-layer charging}} = 0.196$ C.

(i) The efficiency if an overall charge of 20 C is passed is (1 – 0.196 C/1.0 C) × 100 = 80.4%.

(ii) The efficiency if an overall charge of 3.2 × 10^3 C is passed is (1 – 0.196 C/320 C) × 100 = 99.94%.

We see that, to a good approximation, the non-faradaic component is not a significant contribution if the overall charge is large, but $Q_{\text{double-layer charging}}$ *is* significant to Q_{overall} if the overall charge is relatively small.

SAQ 5.8

Clay is just one example† of a material used to modify the electrochemical properties of electrodes to form a **chemically modified electrode** (CME) (as described below). A porous-clay CME has an area of 5 cm^2, and charging the double-layer requires a charge of 1.43 C per square centimetre. Repeat the calculations shown above in Worked Example 5.3 to determine the respective faradaic efficiencies.

We will see from the answer to part (i) of SAQ 5.8 that if the charge required by the double-layer is enormous, then *all* of the charge passed is non-faradaic and thus electroanalysis using such an electrode is impossible.

DQ 5.8

Following on from SAQ 5.8, why would anyone employ a clay-modified electrode for an electrochemical analysis?

Answer

At first sight, a clay-modified electrode is an unlikely choice since its capacitance is so large. However, the area of such electrodes is usually huge (see below), and hence the large capacitance.

We need, however, to recognize that a professional electroanalyst must perform many analyses every day. One of the factors which limits an analyst's time will be the speed at which each coulometric measurement can be performed. Electrodes of this sort are explicitly intended to help speed up such measurements, as follows.

Faraday's first law says that (for faradaic charges) Q $\propto$ *amount of material electro-modified, so the overall amount of charge to be passed during an analysis is a constant. In order to see how large-area electrodes provide this additional speed, we first remember that the amount of charge passed,* Q*, is often gauged as the product of time* t *and current* I *(*$Q = I \times t$*). If the time available is to be shortened, then the current should be increased.*

So we should look at the current. To increase the current through an electrode of area A *requires pushing more charge through it per unit time. We shall see in the next chapter that a higher current usually requires a more extreme potential,* V*, and that a more extreme potential is likely to cause electrode side reactions such as solvent splitting. In short, forcing*

† Other popular choices of CMEs are electrodes coated with redox polymers, conducting polymers, or inorganic films (see later).

a higher current, while saving time, will usually decrease the efficiency of the electrode reaction and decrease the accuracy of the coulometric measurement – which is clearly counter-productive.

A superior means of passing more charge per unit time is to maintain the same current density i *(current per unit area), but with a larger area* A. *In this way, the charge passed per unit area stays constant (thereby keeping the potential more manageable, and decreasing the likelihood of side reactions). However, the overall charge passed has increased in direct proportion to the increase in electrode area (see equation (1.1) earlier). We see that if the area of an electrode is made larger, then more analyte can be electromodified (i.e. undergo a redox reaction* in a controlled way*) than at a smaller electrode.*

In summary, large-area electrodes are a good way of increasing the speed at which a coulometric analysis can be performed, and are especially useful if large charges need to be passed, e.g. if concentrations of analyte are high.

DQ 5.9

So why have a *clay-modified electrode* at all? Why not just have an ordinary electrode, but one which is made of a fine wire mesh?

Answer

Many analyses are *performed with mesh electrodes, thereby increasing the surface area. However, there is an additional reason for using a clay-modified electrode. We saw above (in Worked Example 5.2) that electrode side reactions are a common enemy of accurate electroanalytical work. One of the best ways to stop such side reactions, even when working with high currents, is to coat the electrode with a layer of a specially formulated thin film. The electrode plus its layer is termed a* **chemically modified electrode (CME)***. In a similar manner to a catalyst, the layer is designed to alter the energetics of the electrode reactions, although in this case the layer is put there to* stop *reactions at the electrode. Electrodes that are completely passive to water splitting, for example, are a major goal of research at the moment, and will save a lot of money when perfected since the aim is for* $I_{side\ reaction}$ *to be wholly absent.*

While chemically modified electrodes are excellent for stopping side reactions, they tend not to possess smooth continuous layers but, rather, they are often porous or so rough as to be virtually three-dimensional (Figure 5.5). The electrode surface is often said to be **fractal** for this reason.

DQ 5.10

If the surface of an electrode is fractal, how can we know its area?

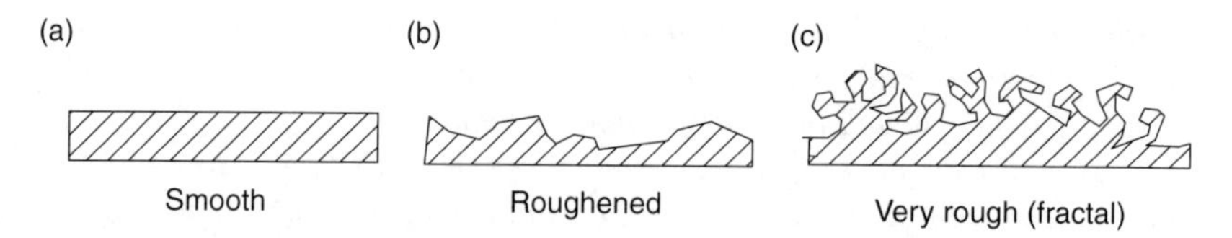

Figure 5.5 Schematic representations of electrode surfaces: (a) a smooth surface; (b) a roughened surface; (c) a surface so uneven that it may be considered as fractal.

Answer

In fact, as a good generalization, we usually do not need to know the area of an electrode when performing coulometry – it is sufficient to determine only the charge passed.

When we do need to know the area of an electrode, we talk about its ***electrochemical area****, i.e. the area that the electrode is perceived to have. The electrochemical area is generally greater than the actual, i.e.* ***geometric****, area.*

SAQ 5.9

An electrode has a capacitance of 21 μF. Calculate the electrochemical area if the material has an intrinsic capacitance of 17.2 μF cm^{-2}.

Determination of the electrochemical area is best achieved by measuring the current at the electrode under study during a voltammetric experiment with an analyte of known concentration. (This is briefly described in the following section, with a more detailed treatment being given in Chapter 6).

5.2 Stripping

In the previous section, we introduced the way that coulometry can be employed as an analytical tool, looking specifically at some simple forms of the technique. We saw that the charge passed was a simple function of the amount of material that had been electromodified, and then looked at ways in which the coulometric experiment was prone to errors, such as non-faradaic currents borne of electrolytic side reactions or from charging of the double-layer.

In this present section, we will look at more specific forms of coulometry. Other than electrolysis of a bulk solution, the most common technique is **stripping**, which is the method of choice when only a tiny amount of analyte is in solution.

DQ 5.11

During normal coulometry, how does the precision of the ammeter affect the ranges of concentrations that can be analysed?

Answer

If the concentration of the analyte is minute, then the magnitude of the current may itself be smaller than the precision of the ammeter employed. In fact, a ***hanging mercury-drop electrode (HMDE)*** *is typically employed, although it cannot accurately determine a concentration smaller than about* 10^{-6}–10^{-7} *mol dm*$^{-3}$ *because the non-faradaic component of the overall charge (e.g. charging of the double-layer) is comparable with the faradaic component of the charge (charge consumed by redox reactions involving analyte).*

Imagine a beaker containing 100 cm^3 *of Cu(II) ions at a concentration of* 10^{-6} *mol dm*$^{-3}$*. We calculate that the beaker contains* 10^{-7} *mol of* Cu^{2+}*. A typical coulometric determination would be inaccurate (if not worthless) since the concentration,* $[Cu^{2+}]$*, is too small.*

This simple demonstration shows that coulometric measurements are often not valid when determining trace or ultra-trace amounts of analyte, and some form of pre-concentration procedure needs to be employed in order to increase $c_{analyte}$ *to a more reasonable level.* ***Stripping*** *is one means of achieving such pre-concentration.*

Imagine a working electrode of liquid mercury. A small drop of mercury is suspended from the end of a capillary, and then made sufficiently negative so that any Cu^{2+} ions impinging on it will be reduced straightaway to form Cu^0 (i.e. to form an amalgam) (see Figure 5.6(a)). Atoms of copper metal will quickly 'dissolve' in the mercury to form a surface layer of Cu(Hg) amalgam.

After a period of time, a steep concentration gradient forms around the electrode since the solution immediately adjacent to the HMDE is entirely depleted of Cu^{2+}. In response, (solvated) Cu^{2+} analyte ions from the **bulk** of the solution will diffuse† toward the HMDE and themselves be reduced. After a further period of time, all of the copper ions will have been removed from solution and accumulate on the surface of the drop (Figure 5.6(b)). Here, we say that we have ***exhausted*** the solution.‡

DQ 5.12

How do we know that all the copper has been removed?

† An alternative way of performing the experiment is to *gently* stir the solution in a controlled, reproducible manner, i.e. replenishing the solution around the HMDE by *convection*. Extreme care is needed when stirring, however, in order to avoid the viscous drag of the solution causing the mercury drop to fall (drag is inevitable during stirring). Once fallen, any copper within the amalgam is effectively lost from the analysis, so we would then have to start the analysis again.

‡ Confusion can arise in some texts when the depletion of analyte from solution (i.e. pre-concentration) is termed 'stripping'. In such cases, removal of the material pre-concentrated on the electrode (i.e. *after* the deposition process) is termed 'redissolving'.

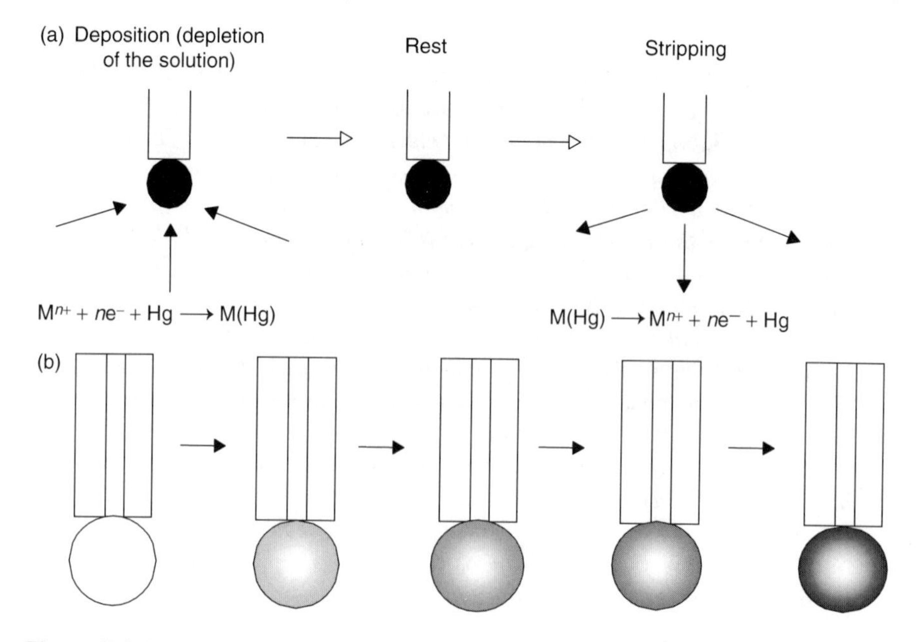

Figure 5.6 Schematic representation of stripping when using a cathodically biased hanging mercury-drop electrode (HMDE) suspended in a solution of reducible ions: (a) the sequence of processes occurring during the stripping process; (b) the accumulation of copper at the surface of the HMDE (the increasing surface concentration of Cu^0 is indicated by the intensity of colour).

Answer

Copper was removed from solution by making the hanging mercury-drop electrode (HMDE) sufficiently cathodic, thereby reducing Cu^{2+} to form Cu^0. The electrons required for reduction are registered by a coulometer or ammeter in the circuit as charge or current, respectively. When the ammeter read-out says zero (or at least when the coulometer read-out shows that the overall charge passed is constant at a very small level and has stopped increasing), then it is assumed that ***exhaustive electrolysis*** *(or 'deposition') is complete, i.e. we say the solution is 'exhausted'.*

DQ 5.13

Can we just take the coulometer reading when determining the overall amount of copper(II) in solution?

Answer

Occasionally, it is *possible to obtain the number of moles of copper,* n(Cu^{2+})*, in solution by directly measuring* $Q_{exhaustive\ electrolysis}$ *and then* *applying Faraday's laws. The calculation of concentration is simple once* n(Cu^{2+}) *is known.*

SAQ 5.10

All the copper in a solution was removed by exhaustive electrolysis. The volume of the solution was 450 cm^3, and the stripping current was 3×10^{-4} C. What was the concentration of the copper? Assume 100% electrolytic efficiency.

The usual situation is that $Q_{\text{exhaustive electrolysis}}$ is considered to be inaccurate since the charge needed to achieve double-layer charging comprises this parameter as a significant, but almost certainly unknown, component.

The way that the stripping analysis is performed in practice is much more sophisticated. After the solution has been exhausted, all of the copper (as $Cu^0(Hg)$ amalgam) resides on the surface of the mercury drop. The potential of the drop is changed from cathodic to anodic (we say that we have **stepped** the potential), and the copper is all oxidized back to Cu^{2+} and the charge determined – as $Q_{\text{oxidation}}$. The potential chosen should be more positive than $E^{\ominus}$ for the analyte couple by at least 0.2 V.

$Q_{\text{oxidation}}$ is the analytical signal employed in calculations of [Cu^{2+}]. We know when all of the copper has been oxidized as the current decreases to zero (this follows because there is nothing left *to* oxidize – all of the copper on the electrode has been consumed).

Because the copper resides *within* the surface amalgam of the electrode, its concentration is high. Accordingly, all of the copper is oxidized at once, and the charge which is consumed by charging the double-layer can safely be ignored. Taking this into account, the number of moles of copper in the HMDE is obtained directly from $Q_{\text{oxidation}}$ from Faraday's laws.

5.3 Microelectrodes

An electrode of surface area 100 μm^2 or less is called a **microelectrode** and provides a means of decreasing the double-layer capacitance which can affect our coulometry experiments so badly. Microelectrodes are also useful when the cell considered is also tiny, as, for example, is the case when performing *in vivo* voltammetry (see next chapter) with biological samples. For example, a nerve ending is typically 10–100 pm in diameter, so electroanalytical experiments using a conventional electrode would be impossible.

The advantages of such electrodes goes beyond the simple restrictions borne of size. The major advantages of microelectrodes follow from their geometry – the electrode is in fact a fine wire encased within an unreactive sheath of plastic, thus ensuring that only one of its end faces is in contact with the solution. This end is typically flat and circular with a diameter of 0.5–50 μm.

DQ 5.14

It was said earlier that the current was proportional to electrode area. Surely a microelectrode will then only pass a tiny current?

Answer

It is true that only a minute amount of electrolysis can occur at the electrode because of its size. However, because the electrode is small, the layer around the electrode that is depleted of analyte is quite thin – certainly far thinner than that around a conventional electrode.

SAQ 5.11

The maximum electrochemical current that a platinum electrode can pass is about 0.2 A cm^{-2}. What is the maximum current that can pass through a circular microelectrode of radius 5 μm?

A thinner depletion layer itself implies a steeper concentration gradient between the face of the electrode and the solution bulk, so in consequence the rate of mass transport to the microelectrode is considerably higher than to normal-sized electrode.

A fast rate of mass transport is useful to the electroanalyst because the faradaic component of the charge is made greater, while the non-faradaic current is not affected. In addition, note that $I_{\text{non-faradaic}}$ will be small anyway since it is in proportion to electrode area.

We saw earlier that the limit of the electrode sensitivity was the ratio of faradiac to non-faradaic currents. The net result of using a microelectrode is to increase this ratio, and thus allow the analyst to analyse solutions of lower concentration – perhaps as low as 10^{-8} mol dm^{-3}.

It is also possible sometimes to dispense with the need to add unreactive electrolyte to solution if the electrode being used is a microelectrode.

A new approach is to assemble a large number of microelectrodes together. Studies and applications of such **micro-arrays** are a growth area at present. In these assemblies, if each electrode is polarized to a different potential, then (in principle, at least) each one could then monitor the amounts of different analytes.

Alternatively, an assembly of microelectrodes can alleviate some of the problems associated with the individual microelectrodes. Such a **random array of microelectrodes (RAM)** comprises about 1000 carbon fibres (each of diameter 5–7 μm) which are embedded randomly within an inert adhesive such as an epoxy resin. (The ends of the fibres need to be widely spaced.) The net result is to generate an electrode system with a superior response time and a current which is 1000 times that of a single microelectrode. By increasing the current in this way, the sensitivity of measurement is further increased.

5.4 Introduction to Electron Mediation

Occasionally, the analyst wishes to coulometrically quantify the amount of a redox-active material that is in solution, but finds that the material itself is electro-*inactive*. A good example of this is provided by enzyme electrochemistry.

Enzymes are the naturally occurring macromolecular species within a cell or organism that catalytically facilitate reaction. A great many enzymes will catalyse electron-transfer reactions, yet are wholly unreactive at straightforward electrodes. In such cases, we perform the redox reaction 'one step removed' and *chemically* effect the redox change at the molecule of interest. If redox change is wanted, then a **mediator** must be included in the electrochemical system.

A mediator is a redox-active species in the same phase as the biological analyte. Electrochemical redox changes to the mediator generate either a reducing or an oxidizing agent which may then *chemically* effect an electron-transfer reaction to the biomolecule. The process of mediation is represented schematically in Figure 5.7, although note that in many situations, it is the reverse which occurs, i.e. oxidation rather than reduction.

In consequence, it is the mediator which effects the redox changes at a redox site of the enzyme.

Worked Example 5.4. Consider the example of methyl viologen (its structure is given earlier in Table 4.1). At about −0.45 V (vs. SHE), the MV^{2+} dication is reduced by the electrode to form a radical cation ($MV^{+\bullet}$), as follows:

$$MV^{2+} + e^{-}_{(Pt)} \longrightarrow MV^{+\bullet} \tag{5.7}$$

Once formed, the $MV^{+\bullet}$ radical acts as a reducing agent and *chemically* reduces the redox centre of a biological molecule (illustrated here with the example of the electro-inactive enzyme, cytochrome-*c*), as follows:

$$MV^{+\bullet} + \text{cyt-c}_{(ox)} \longrightarrow MV^{2+} + \text{cyt-c}_{(red)} \tag{5.8}$$

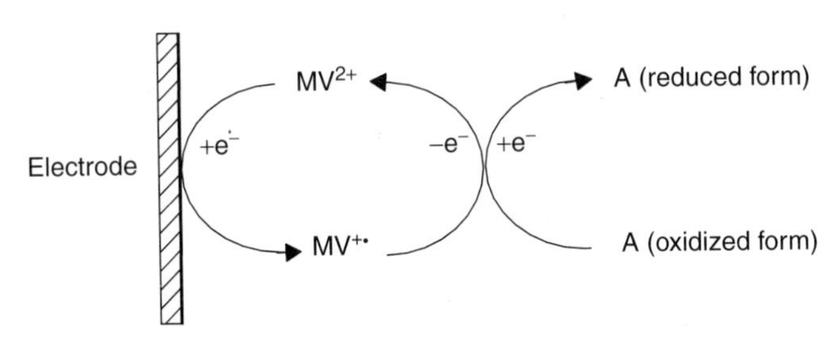

Figure 5.7 Schematic representation of the way that methyl viologen, as a mediator, allows for the electroanalytical determination of an electro-inactive material A.

Table 5.1 The five simple tests that need to be satisfied in order for a species to be considered as suitable for use as a mediator (after Steckhan, E. and Kuwana, T., *Ber. Bunsenges. Phys. Chem.*, **78**, 253–259 (1974)

1. The electron-transfer reaction should be simple, fully reversible and preferably involve the transfer of only *one* electron
2. Both redox states of the redox couple should be chemically stable and undergo little or no structural or bonding changes during the conversion between the two redox states
3. Mediators will need to be in solution, and usually buffered at pH 7 (to stop denaturing of the natural macromolecules)
4. They must be able to undergo a reasonably fast electron-transfer reaction with the biological redox site
5. The electron-transfer reaction between the mediator and the electrode ought to be fast

We see that a good choice of mediator requires the electrode potential of the mediator to 'couple' with the redox state of the analyte.

Mediators are occasionally termed **mediator titrants** or **auxiliary redox couples**, while mediation is sometimes called **indirect coulometry**. The efficiency of the reaction shown in equation (5.8) is assumed to be 100%, and so, in practice, the coulometric experiment is performed in the same way as if direct electron transfer occurred – the charge required to reduce (or oxidize) the mediator is determined, and Faraday's laws are then applied.

SAQ 5.12

Methyl viologen (MV) is used as a mediator during the one-electron reduction of an enzyme. The concentration of the enzyme is thought to be about 10^{-6} mol dm^{-3}. How long will it take the MV to reduce all the enzyme if added to 1 dm^3 of enzyme solution, and the maximum current passed is 1 μA?

A species must fulfil five simple tests for it to be considered as a suitable choice of mediator (see Table 5.1). The most common mediators for biological systems are methyl viologen, benzyl viologen and hydroquinone. Each of these mediators obey all five of the criteria presented in this table.

5.5 Treatment of Errors

The most common type of errors found during coulometry is the incorporation of non-faradaic charge within the overall charge measured, e.g. as caused by double-layer charging or electrolytic side reactions. These aspects of coulometry have been discussed above.

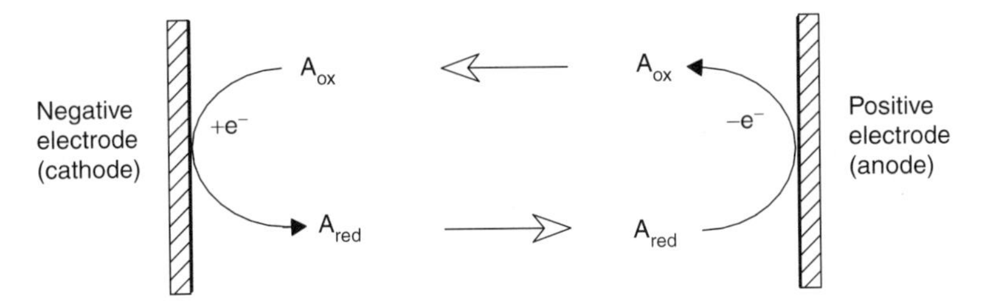

Figure 5.8 Schematic representation of a coulometry cell, showing how a single molecule of analyte A can be repeatedly oxidized and reduced, thereby giving an overly high charge Q. The subscripts 'ox' and 'red' relate to the 'oxidized and 'reduced' forms, respectively, while the open-headed arrows represent movement of species A through the solution.

An alternative error arises from the geometry employed during coulometry: consider the coulometric cell shown in Figure 5.8. In Section 5.1 above, we saw that oxidation and reduction reactions always occur in accompaniment in a cell, with $I_{\text{oxidation}} = -I_{\text{reduction}}$ (see equation (5.2)).

Reduction of analyte occurs at the cathode (on the right-hand side of the cell). Once formed, however, the reduced form of the analyte couple diffuses across the cell – it may also be swept along by the stirred solution – and/or be re-oxidized again at the anode. Clearly, a single molecule of analyte could be oxidized and reduced many times, thus leading to an artificially high charge at the coulometer. For this reason, the two halves of the coulometry cell should be separated if possible, e.g. with a semipermeable membrane or frit, or we should ensure that the product of electron transfer should be a solid, i.e. it is immobilized as soon as it is formed.

Summary

Coulometry is the study of electrochemical charge passage. The overall charge passed has two components, namely faradaic (which follows Faraday's two laws of electrolysis) and non-faradaic (which does not).

Non-faradaic charge can arise in many different ways. The first is the occurrence of electrode side reactions such as solvent splitting (which also follow Faraday's law, but decrease the faradaic efficiency of the analyte of choice). The second is the existence of the electric double-layer, which is charged during electroanalytical processes at an electrode. The best way to avoid the former is to pass a smaller current, while the best way to avoid the latter is by the use of a microelectrode. Microelectrodes allow charge to flow through the electrode in such a way that $Q_{\text{double-layer}}$ contributes only a small proportion of the overall charge that is passed.

While forcing a higher current through an electrode will save time, we usually expect a large I to also decrease the efficiency of the electrode reaction by promoting side reactions, which decreases the overall accuracy of the coulometric analysis. For this reason, the use of large-area electrodes is a good way to increase the speed at which a coulometric analysis can be performed, and is particularly advantageous if larger charges need to be passed, which is the case when the concentrations of analyte are high.

Mediators may allow for coulometry to be carried out when an analyte is not itself electroactive. Such mediators are often aromatic organic molecules which can interconvert reversibly between several oxidation states. The mediator is itself electromodified at an electrode, and then effects a *chemical* redox charge to the analyte in the same solution. Coulometry is then possible, provided that the transfer of charge from the mediator to the analyte is 100% efficient.

Chapter 6

Analysis by Dynamic Measurement, A: Systems under Diffusion Control

Learning Objectives

- To appreciate that 'dynamic' electrochemistry implies that concentration changes occur in response to redox reactions at the electrode of interest.
- To appreciate that dynamic electrochemical and electroanalytical experiments usually require three electrodes.
- To learn that the changes in concentration caused by current flow will follow Faraday's laws, so the analytical variable during measurement is current, where $I \propto c_{\text{analyte}}$.
- To appreciate that to minimize non-faradaic currents, the polarography solution should be purged of oxygen and contain a surfactant depolarizer in low concentration.
- To learn that addition of a swamping electrolyte to a still solution of analyte ensures that the principal form of mass transport is diffusion.
- To learn that the working electrode during polarography is a constantly replenished mercury drop, while the working electrode during voltammetry is usually a solid electrode such as platinum, gold or glassy carbon.
- To learn that in polarography, the magnitude of the diffusion current I_d is proportional to analyte concentration according to the Ilkovic equation.
- To learn that numerical values of the polarographic half-wave potential $E_{1/2}$ are a characteristic of the analyte.

- To learn that accurate values of $E_{1/2}$ may be obtained from the intercept of a Heyrovsky–Ilkovic graph, while the slope of such a graph indicates the extent of electrode irreversibility.
- To learn how equilibrium constants of association, K, may be obtained polarographically by analysing shifts in the half-wave potential $E_{1/2}$ as a function of complex concentration.
- To learn how to test for electro-reversibility from simple diagnostic tests concerning the cyclic voltammogram (CV) peak currents and potentials.
- To learn that the magnitude of the current peak in cyclic voltammetry (after suitable correction for baseline drift, where applicable) is proportional to analyte concentration according to the Randles–Sevčik equation.
- To appreciate how the time-scale of the cyclic voltammogram, τ, is a function of the scan rate ν; variation of ν therefore allows insights into the kinetics and mechanisms of electrode reactions .
- To appreciate how the analytical sensitivity of polarography and voltammetry can be enhanced by sampling the current, or by pulsing the potential in normal pulse, differential pulse and square-wave pulse methods to attain a lower concentration limit of about 10^{-8} mol dm^{-3}.
- To learn how the analytical sensitivity can be further enhanced to about 10^{-11} mol dm^{-3} by pre-concentration of the analyte in the technique of stripping voltammetry.

6.1 Experimental Introduction to Dynamic Electrochemistry

We will assume for all of the techniques discussed in this chapter that the analyte solution is quiet (that is, still and unstirred) in order to ensure that mass transport by convection is absent. Furthermore, we will also assume that an excess of ionic electrolyte has been added to the solution to ensure that mass transport by migration is also absent. We see that the only form of mass transport remaining is *diffusion*, and hence the subtitle to this chapter.

We now introduce **voltammetry** and the subset technique of **polarography**. The word root 'voltam-' tells us straightaway that we are looking at both potential ('volt-') and current ('am-'). We will see that during any voltammetry experiment, the potential of an electrode is varied while we simultaneously monitor the current that is induced. This occurs as a result of the electrode being **polarized**, that is, its potential is forced away from its equilibrium value.

The concept of polarization is so important that many aspects of this chapter will be discussed in terms of it. Indeed, the root 'polar-' of polarography implies polarization. As a working definition, we say that polarization represents 'the deviation of the potential of an electrode from its equilibrium value'. The magnitude of the deviation is termed the **overpotential**, η, and is defined according to the following equation:

$$\eta = E_{\text{electrode}} - E_{\text{equilibrium}} \tag{6.1}$$

Note that, from the way equation (6.1) is written, η can be either positive or negative.

We have seen already that an absolute potential at an electrode cannot be known, so, in accord with all other electrochemistry, it is the potential *difference* between two electrodes which we measure. However, if the potential of the electrode of interest is cited with respect to that of a second electrode having a known, fixed potential, then we can know its voltage via the concept of the standard hydrogen electrode (SHE) scale (see Section 3.1). We see that a reliable value of overpotential requires a circuit containing a reference electrode.

We call the electrode of interest – that at which the electrochemical changes of interest occur – the **working electrode** (WE). When we cite an overpotential, we cite the potential of the working electrode with respect to the potential of the reference. The overpotential η is seen to be positive during anodic electrochemistry and negative during cathodic electrode processes.

A common way of indicating that 'equilibrium' is involved is to say that the electrodes are 'open-circuit', that is, the power source (or whatever is driving the charges that move) is absent or remote from the circuit.

In most voltammetry, the power source is a **polarograph** or **potentiostat**. The name potentiostat comes from the roots 'poten-' (voltage) and 'stat-', implying 'holds steady'. We will see later why such a steady potential is essential. A 'polarograph' is an instrument (hence '-graph', in this context) that polarizes.

DQ 6.1

It was stated in Chapter 5 that equal and opposite amounts of charge will flow through the two electrodes (equations (5.1) and (5.2)). However, in Chapter 3, it was also shown why it is inadvisable to pass charge through a reference electrode. Accordingly, if the potential of the working electrode is to be determined with respect to a reference electrode, can we trust its potential?

Answer

It is certainly true that a non-zero overpotential implies that current is flowing. In order to alleviate the apparent paradox of wanting current passage at the working electrode (WE) while the reference electrode (RE) is kept at equilibrium, we need to employ a third electrode, i.e. the ***counter electrode*** *(CE). The latter is only included in the circuit to facilitate current passage at the WE, so it is rare to monitor the potential*

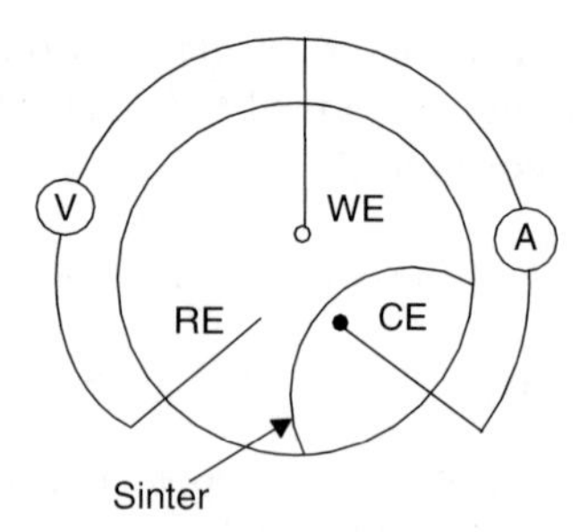

Figure 6.1 Schematic representation of a cell used for voltammetry, in which the potential of the WE is measured with respect to the RE and the current through the WE is determined with respect to the CE. Wherever possible, the spatial separation between the WE and RE is minimized.

of the CE. Figure 6.1 shows a simple diagram of a cell used for voltammetry, in which the potential of the WE is measured with respect to the RE and the current through the WE is determined with respect to the CE. In order to maintain this potential, the cell should ideally be thermostatted, since it should be noted that the potential of a redox potential is temperature-dependent (see the Nernst equation, equation (3.8)).

DQ 6.2

However, if charge passes through the counter electrode, what about the products of the electrode reactions that occur there?

Answer

If current flows through the working electrode, then it also flows through the counter electrode. From Faraday's laws, such a current flow implies that material must have been formed at the counter electrode (CE).

It is unwise to assume that the products of electrode reaction at the CE are benign, so the CE should be excluded from the solution bulk. In practice, a satisfactory extent of exclusion is achieved by placing it in a separate compartment within the electrochemical cell, but with electrolytic contact between the CE compartment and the main body of the cell being achieved via a sinter or frit.

DQ 6.3

Why does current flow when the potential is altered, i.e. what is the cause of the voltammetric current?

Answer

We now have a new definition of when an electrochemical measurement is non-equilibrium (or 'dynamic'), i.e. those for which $\eta \neq 0$.

In the simplest case, we can say that no current passes through the cell during an equilibrium *electrochemical experiment, and the activities of the electroactive species in solution will obey the Nernst equation.*

Equilibrium is disrupted when we force charge to flow through an electrode. Such charge flow is accompanied by electron uptake (reduction) or electron loss (oxidation) by the electroactive analyte at the electrode, thus causing the concentrations to change – and hence the activities will change also. (These changes explain why voltammetry is a ***dynamic*** *experiment.)*

Dynamic electrochemistry is seen to alter the ratio of a*(O) to* a*(R) for the redox couple at the surface of the working electrode (i.e. at the* ***electrode | solution interface****). Note that this alteration occurs* during *electrolysis, such that the electrode potential* $E_{O,R}$ *can shift according to the Nernst equation.*

We can state this argument in reverse – alteration of the potential at the electrode | solution interface will itself cause the ratio of a*(O) to* a*(R) to alter to that dictated by the Nernst equation, and the conversion of material from its reduced to its oxidized forms (or back) requires the production (or consumption) of charge. In fact, we can write a variant*[†] *of the Nernst equation (equation (3.8)), as follows:*

$$E_{working\ electrode} = E^{\ominus}_{O,R} + \frac{RT}{nF} ln \frac{a(O)_{surface}}{a(R)_{surface}} \qquad (6.2)$$

where 'surface' here relates to the surface of the working *electrode alone.*

This idea of current flowing as a function of polarizing the electrode (shifting its potential away from equilibrium) lies at the very heart of voltammetry. Note that the magnitude of the current – and current is a derivative quantity (equation (2.1)) – tells us the *rate* at which the electrochemical conversion occurs.

DQ 6.4

If the a(O)/a(R) ratio depends on the potential of the working electrode, and we measure the current through this electrode as a function of that potential, how is it possible to simultaneously monitor both the current and potential through the *same* electrode?

Answer

In fact, we don't monitor two variables through the same electrode. Instrumentally, we employ a feedback loop within the voltage source

---

[†] Note that the true Nernst equation (equation (3.8)) has an electrode potential on the left-hand side, i.e. a potential that is *measured* (at equilibrium). The variant here does not have an electrode potential at all: the term on the left-hand side is an *imposed* potential, i.e. it is applied through a power source such as a potentiostat.

(that is, the potentiostat or polarograph). The potential at the working electrode is maintained, by the circuitry of the voltage source, at its desired value by causing charge to flow. The potential is a function of the ratio of oxidized and reduced forms of the electroactive material, according to the above Nernst equation (equation (6.2)).

Experimentally, then, the potential of the working electrode is monitored with respect to the reference electrode, while the current measurement is actually performed via the counter electrode.

6.2 Chronoamperometry: Current Determined with Time

In the chronoamperometric experiment, we measure current (hence 'amp-') as a function of time ('chrono-'). It is usual to commence with the solution around the electrode containing only one redox form of the analyte.

We will use the example of thallium ion. The potential of the working electrode will be **stepped** from a potential at which only Tl^+ is the stable form to a potential at which only Tl^{3+} is the stable form. Figure 6.2(a) shows a plot of potential against time – note that the rise in potential here is essentially vertical. It would be completely vertical but for the requirement to charge the double-layer around the electrode. The potential before the step is, e.g. 0 V, i.e. it is well cathodic (negative) of $E^{\ominus}_{Tl^{3+},Tl^+}$ (= 1.252 V), so Tl^+ is the only stable redox form, and no Tl^{3+} will form. The potential after the step is, e.g. 1.6 V, i.e. it is well anodic of $E^{\ominus}_{Tl^{3+},Tl^+}$, so Tl^{3+} is the only stable redox form here, thus causing Tl^+ to oxidize to Tl^{3+}.

Figure 6.2(b) shows a trace of current against time in response to the potential step. The trace shows a rapid rise in current, with this rise requiring perhaps as long as a few thousands of a second (i.e. milliseconds). The time between the potential step and the maximum is known as the **rise time**. The current trails off smoothly after the rise time until, eventually, it reaches zero. Such plots are often termed **transients** to emphasize their pronounced time dependence.

DQ 6.5

Why do 'double-layer effects' cause a longer time delay in the current trace than in the potential trace?

Answer

The delay in reaching the maximum voltage (Figure 6.2(a)) is caused by charging *of the double-layer, i.e. merely adjusting the positions of the ions. In contrast, the time delay in reaching the peak in the current trace (Figure 6.2(b)) is caused by electrochemical changes to the analyte within the double-layer, according to Faraday's laws.*

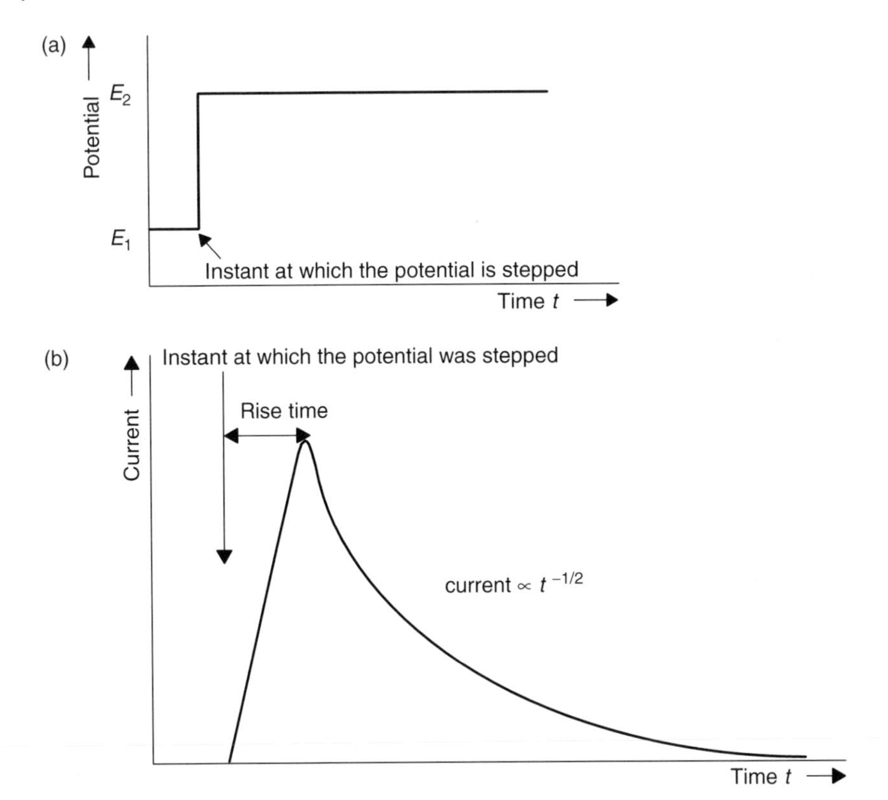

Figure 6.2 During the chronoamperometry experiment, the potential is stepped from an initial potential E_1 (at which no electromodification occurs) to a second potential E_2 at which the electrode reaction is complete, i.e. the current is limiting: (a) trace of potential against time; (b) trace of current against time. In both cases, the only form of mass transport is diffusion.

6.2.1 The Nernst Layer ('Depletion' Layer)

Before the potential is stepped from, e.g. 0 to 1.6 V, the electrode will be surrounded by only Tl^+, the concentration of which will be the same throughout the analyte solution.

Conversely, after the voltage is switched, Tl^+ ions will be oxidized to Tl^{3+} according to the reaction, $Tl^+ \rightarrow Tl^{3+} + 2e^-$. Very soon, the electrode will be surrounded with a layer of solution that contains no Tl^+. It is said then to be **depleted** of Tl^+. The volume around the electrode, which now contains no Tl^+, is called the **depletion region**, the **depletion layer** or sometimes, the **Nernst layer**. Its thickness is often given the symbol δ. As the length of time increases following the potential being stepped, i.e. as the extent of electrolysis increases, so the thickness of the Nernst layer (δ) increases.

DQ 6.6

If the electrode is surrounded with a layer that contains no Tl^{+},[†] why is there any current at all at longer times?

Answer

The value of [Tl^{+}] in the solution bulk remains essentially constant since only a tiny proportion of the overall amount of Tl^{+} is oxidized, but at the surface of the electrode we can say, to a good approximation, that [Tl^{+}] $= 0$. Very soon after the potential is stepped, Tl^{+} from the bulk diffuses[‡] *toward the electrode, thereby attempting to 'even out' the* ***concentration gradient****, i.e. to replenish the Tl^{+} that was consumed at the commencement of the step. We need to recognize, however, that these thallium ions will not remain as Tl^{+} for long as they will be oxidized 'immediately' to form Tl^{3+}, i.e. as soon as they impinge on the electrode.*[§] *The end result is that a concentration gradient will soon form after the potential has been stepped.*

The plots shown in Figure 6.3 show the variations of the concentrations of both reduced and oxidized forms of Tl (of reactant Tl^{+} in Figure 6.3(a) and product Tl^{3+} in Figure 6.3(b)). Each concentration is depicted as a function of the distance from the electrode | solution interface where oxidation is effected: these curved traces are often termed ***concentration profiles****. Each of the figures incorporate a series of concentration profiles, drawn as a function of time, to show how the Nernst layer increases in thickness during electrolysis, because the extent of electromodification has increased with time.*

DQ 6.7

Why are we allowed to use concentration instead of activity if equation (6.2) is written in terms of activity?

Answer

In Chapter 3, we looked at the way the activity coefficients can be more or less equalized if there is a swamping electrolyte in solution (see Section 3.4.4, SAQ 3.9 and Figure 3.8). By the nature of the species studied in a chronoamperometric experiment, (a) a swamping electrolyte is added to the solution in order to minimize migration effects, and

---

[†] While we have said that the Nernst layer is 'depleted', in fact it still contains a tiny amount of Tl^{+} (the exact amount can be calculated from the variant of the Nernst equation, equation (6.2)).

[‡] Because diffusion is the sole form of mass transport – see the introductory paragraph to this chapter for a rationale.

[§] While we talk about 'immediate' reaction, it is wise to note that all reactions occur within a *finite* time-scale – even if a very short one.

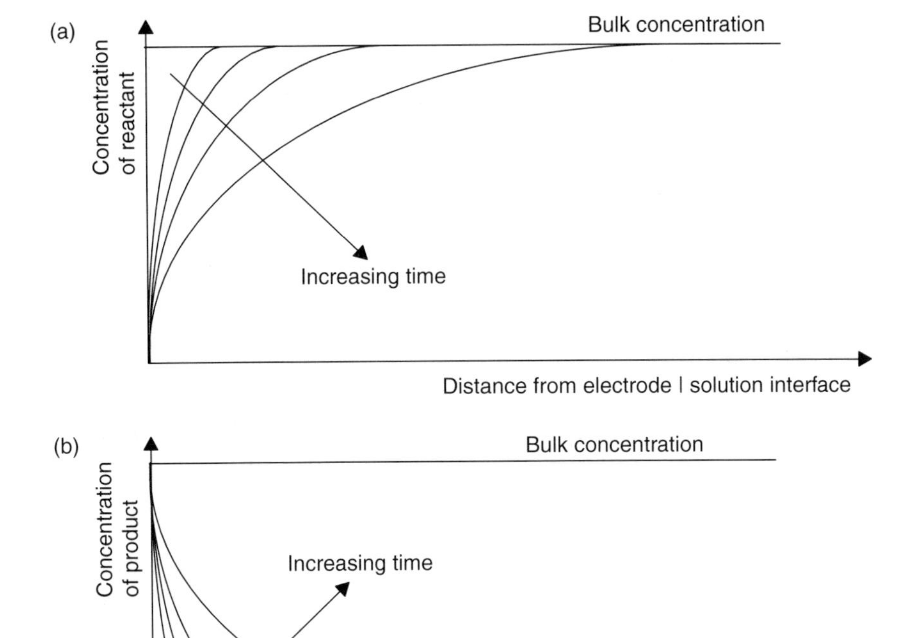

Figure 6.3 Plots of concentration against distance from the electrode | solution interface ('concentration profiles') as a function of time during the chronoamperometry experiment for: (a) the concentration of Tl^{+} (as reactant) remaining in solution; (b) the concentration of Tl^{3+} (as product). Movement of the material through the solution is by diffusion, i.e. a convection-free situation.

(b) both the oxidized and reduced forms of the analyte are soluble in the solvent. As a consequence of these two conditions, we can safely say that the activity coefficient of each redox state is the same. In summary, from the definition of activity (in equation (3.12)), we can employ, concentrations *instead of* activities, *since* $\gamma_{ox}/\gamma_{red} = 1$, *which can then be cancelled out in equation (6.2).*

DQ 6.8

All of the concentration profiles shown in Figure 6.3 relate to the condition that $E_{\text{applied}} \ll E^{\ominus}_{\text{O,R}}$, implying that all of the analyte is consumed at the working electrode. What happens if $E_{\text{applied}} \approx E^{\ominus}_{\text{O,R}}$?

Answer

From equation (6.2), if $E_{applied} \approx E^{\ominus}_{O,R}$ *then* $a(O)/a(R) \approx 1$. *While the concentration profiles might be similar in shape to those in Figure 6.3, the profile will not intercept the* y*-axis at* c_{bulk}, *but at a smaller concentration, in accordance with the Nernst equation.*

Worked Example 6.1. How much of the Tl^+ will be converted to Tl^{3+} if $E_{\text{applied}} = E^{\ominus}_{\text{O,R}}$?

We will start by assuming that the activity coefficients of $Tl^+{}_{(aq)}$ and $Tl^{3+}{}_{(aq)}$ are the same, in which case activities and concentrations may be considered as being interchangable. If $E_{\text{applied}} = E^{\ominus}_{\text{O,R}}$, then from equation (6.2) we see that $c_O/c_R = 1$. In other words, half of the thallium remains as Tl^+ while the other half has been converted to Tl^{3+}. The concentration profiles will follow the traces shown in Figure 6.4.

Worked Example 6.1 introduces us to an important result, namely that below a certain overpotential, not all of the analyte at the surface of the electrode will be electromodified. At or above a crucial value of η, all of the material around the electrode is converted. The associated current, if conversion is total, is then

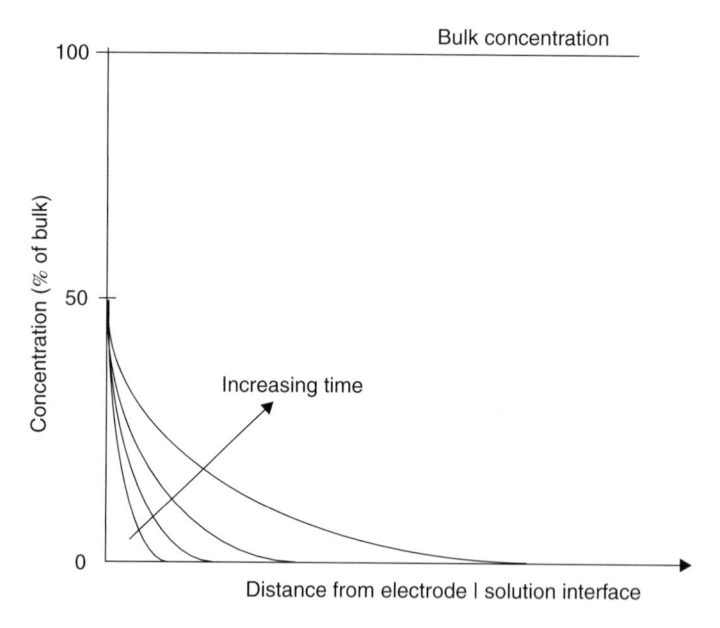

Figure 6.4 Plots of concentration profiles as a function of time during the chronoamperometric oxidation of Tl^+; the applied potential, E_{applied}, was $E^{\ominus}_{Tl^{3+},Tl^+}$.

said to be **limiting**. We shall see in the next section that the limiting current is directly proportional to the concentration of analyte. This proportionality is what allows dynamic voltammetry to be such a powerful analytical tool.

6.2.2 The Cottrell Equation

For a quantitative description of the way that a chronoamperometric current trails off with time, we employ the **Cottrell** equation, as follows:

$$I_{\text{lim}} = nFAc\sqrt{\frac{D}{\pi t}} \tag{6.3}$$

where n is the number of electrons transferred in the reaction, $O + ne^- = R$, F is the Faraday constant, t is the time elapsing after the potential was stepped and D is the diffusion coefficient.

We need to realize that the concentration term† in equation (6.3) relates to the *bulk* concentration of analyte rather than the local concentration: adsorption or electrolysis will both cause differences such that $c_{\text{surface of electrode}} \neq c_{\text{bulk}}$. From now on, we will assume that c here is the **bulk** concentration, unless stated otherwise.‡ All other concentrations will be identified with a suitable subscript.

Note that this equation relies on the current being *limiting*, i.e. the potential must be stepped to an extreme overpotential to ensure that $I_{\text{lim}} \propto c_{\text{analyte}}$. If the potential is not sufficient for the current to be limiting, then it is usually found that I is still proportional to $t^{-1/2}$, but a smaller concentration than c_{analyte} needs to be included within the collection of proportionality constants.

The Cottrell equation, as written here, relates to an electrode in the form of a cylindrical wire. One end of the wire will be embedded in a non-conductive sleeve (e.g. glass), so that only one end of the wire will ever be in contact with the analyte solution. If the wire has a length h and a diameter r, then the surface area A of the wire is given by:

$$A = \pi r^2 + 2\pi rh \tag{6.4}$$

In electroanalysis, the area is conventionally considered to have units of cm^2. If the electrode is fractal (see Section 5.1.2 and Figure 5.5), then the *electrochemical* area, rather than the *geometric* area, is employed as 'A'.

The diffusion coefficient D can be thought of as the velocity of the analyte as it moves from the bulk of the solution towards the electrode just prior to the electron-transfer reaction. Because D is a velocity, larger values of D relate to a faster motion of analyte through the solution, while smaller values relate to slower motion. It is assumed, when deriving equation (6.3), that diffusion is linear

---

† Remember from Chapter 1 that the electrochemist often expresses concentration in units of mol cm^{-3} rather than mol dm^{-3}.

‡ Some texts indicate the bulk concentration as c^{∞} or $[C]^{\infty}$, where the superscript can be taken to mean 'at an infinite distance from the electrode'.

(in one direction only) toward a planar electrode. Luckily, the same equation is obtained for electrodes having most of the commonly employed geometries – not just for cylindrical wire electrodes.

DQ 6.9

The Cottrell equation is derived with the assumption that diffusion is the sole form of mass transport. How can we confirm that it actually is?

Answer

If equation (6.3) holds, then a graph of I_{lim} *against* $t^{-1/2}$ *– a* ***Cottrell plot*** *– should be linear and pass through the origin. By corollary, such Cottrell plots represent a simple test of whether diffusion is indeed the sole form of mass transport: a non-zero intercept indicates that non-faradaic currents contribute to* I_{lim}*. The linear part of the Cottrell plot indicates those times at which diffusion is the sole form of mass transport – usually at the shortest times. Any curvature in a Cottrell plot indicates that migration and/or natural convection are contributing to the overall mass transport, perhaps in varying proportions.*

DQ 6.10

How are chronoamperometric experiments useful to the analyst?

Answer

Actually, chronoamperometry is not a commonly performed electroanalytical technique. Probably its most common application is to determine the electrochemical area of an electrode if the concentration and diffusion coefficient of the analyte are already known.

Additionally, it can often, however, be a good idea to perform chronoamperometric transients over a wide range of times and voltages to ascertain those experimental conditions which do indeed yield a linear Cottrell plot that passes through the origin, i.e. to ascertain those experimental conditions over which diffusion is indeed the sole form of mass transport.

When determining a diffusion coefficient D from a Cottrell plot, it is important to determine the current for as long a time as possible, in order to ensure reliability. With suitable precautions, as much as 10 s may be possible. Certainly, 4 or 5 s should be attempted as a minimum and, in practice, data are best obtained over a range of, say, 1 ms to 5 s.

6.3 Polarography at Hg Electrodes

Before we start this section, it is necessary to note that two conventions exist concerning the way the polarograms or voltammograms are depicted. The **IUPAC**

convention has a normal x-axis, i.e. with the potential running from negative (at the left-hand side) to positive (at the right-hand side). Oxidative currents are positive and reductive currents are negative. A moment's thought shows that the trace of current (as 'y') against potential may progress clockwise or anticlockwise, depending on the values of E_i and E_λ (as defined below).

More popular than the IUPAC convention is the so-called **polarography convention**, which is the reverse of the IUPAC presentation: here, the polarogram starts at a potential of about 0 and progresses toward negative potentials (i.e. toward the right-hand side during scanning). Such a polarograph is thus drawn with the start potential on the left-hand side, with reductive currents being drawn as positive.

DQ 6.11

The polarographic convention seems contrary to common sense. How did it evolve?

Answer

Few oxidative electrode reactions are possible at liquid mercury surfaces (see below), so only negative potentials can be employed. The earliest polarographers found it convenient to scan toward more negative potentials during a 'run'.

A third convention is rapidly growing in popularity. The voltammogram (as below) is always drawn going clockwise from an origin at the left-hand side of the page. This origin represents E_{initial}, whether the potential first progresses toward a positive or a negative potential. We will use this third convention here at all times.

6.3.1 The Polarographic Experiment

Polarography and **linear-sweep voltammetry** (**LSV**) are terms which, today, are often (erroneously) considered to be more or less interchangeable. Strictly speaking, polarography is a voltammetric method in which the working electrode is mercury, while the working electrode employed in an LSV experiment is a solid, such as platinum, gold or glassy carbon (GC). Experimentally, the two techniques are very similar. The composition of the reference electrode is often a saturated calomel electrode (SCE), while the counter electrode (CE) is not particularly important.

Voltammetry was described briefly in the previous chapter, when we first looked at stripping techniques. To recap: during the experiment, the potential is ramped from an initial value, E_i, to a final value, E_f (see Figure 6.5). The potential of the working electrode is **ramped**, with the rate of dE/dt being known as the **sweep rate**, v. The sweep rate is also called the **scan rate**. Note that the value of v is always cited as a *positive* number.

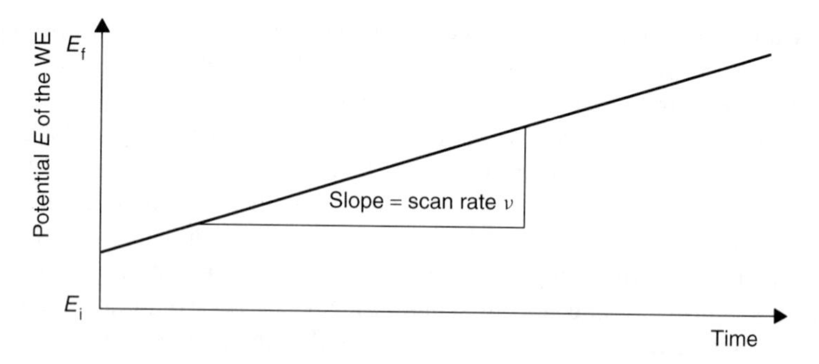

Figure 6.5 Potential is varied at a constant rate of dE/dt during voltammetric techniques such as polarography, linear sweep voltammetry and cyclic voltammetry. The scan rate ν is always cited as a positive number.

The rate of dE/dt is always kept constant during a potential **scan**, although obtaining a series of polarograms as a function of ν can be extremely informative. (The variation of the sweep rate can be regarded as being equivalent to a variation of varying the time-scale of observation of an electrochemical experiment, as will become clear later in Section 6.4.3.)

For both techniques, the analyte (in the concentration range 10^{-2}–10^{-5} mol dm^{-3}) is dissolved in a still solution that also contains supporting electrolyte, so the sole form of mass transport is *diffusion*. Usually, the potential is scanned from a value of E_i at which the analyte is electro-inactive to a final potential E_f at which the current is limiting. The resultant plot of current (as y) as a function of potential (as x) is termed a **polarogram**.

6.3.2 Polarography: the Dropping-Mercury Electrode

Figure 6.6 shows a schematic diagram of the apparatus required as a working electrode for polarography. Such a set-up is almost universally called a **dropping mercury electrode (DME)**, with the mercury drop being immersed in a cell that is essentially the same as that shown in Figure 6.1.

DQ 6.12

In the previous chapter, it was emphasized that the non-faradaic component of current should be minimized. Is non-faradaic current a problem in polarography?

Answer

Liquid mercury is quite easily oxidized at anodic potentials (when immersed in water) so most of the measured current would be due to the reaction $Hg^0 \rightarrow Hg^{2+}$. *Mercury is electrochemically stable for a considerable range of cathodic potentials (for example, from +0.1 to*

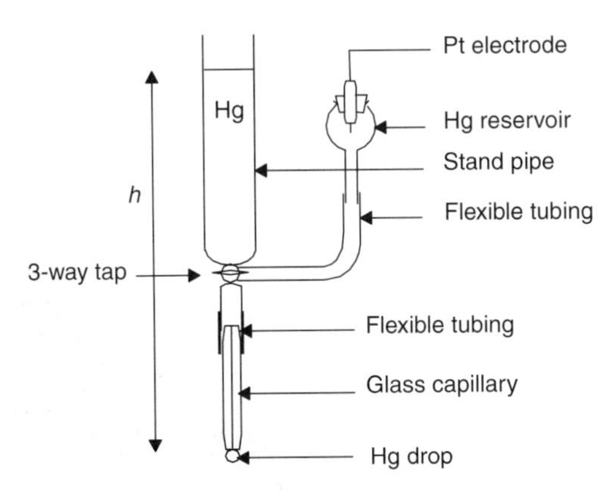

Figure 6.6 Schematic representation of a typical dropping-mercury electrode (DME) for polarography, where the DME acts as a working electrode in a cell such as that shown in Figure 6.1. The platinum electrode at the top right of the diagram is needed to give an electrical connection. The rate of mercury flow is altered by adjusting by changing the height *h*.

−1.6 V in 0.1 mol dm^{-3} KCl solution). We say that mercury has a wide ***cathodic window****. The exact range of the window will depend on the solution composition.*

The second problem we encounter during cathodic electrochemistry is caused by oxygen. Gaseous oxygen from the atmosphere dissolves in water to a concentration of about 10^{-3} mol dm^{-3} (which explains why fish can 'breathe' under water). Dissolved oxygen is electroactive at negative potentials, so the reaction, $O_2 + 2e^- + 2H_2O \rightarrow H_2O_2 + 2OH^-$, occurs readily, with the hydrogen peroxide formed in this way also being electroactive, i.e. $H_2O_2 + 2e^- \rightarrow 2OH^-$. This being a four-electron process, the associated non-faradaic currents can be quite large. Figure 6.7 below shows a polarogram of oxygen in water.

Molecular oxygen must be removed from solution if analyses are to be performed at cathodic potentials. The best means of excluding O_2 is to gently bubble gaseous nitrogen or argon through the solution for about 10 min before the commencement of polarography.† This process of removing oxygen is sometimes called **sparging**. Such sparging should be discontinued *during* the polarographic analysis because bubbling of this sort would cause convection, and we need to maintain diffusion control.

During analysis, it is a good idea to protect the solution from diffusion of atmospheric oxygen by gently blowing inert gas over the *surface* of the solution.

† The inert gas dissolves preferentially, causing oxygen to be released from solution, and thus carried away from the solution – a process sometimes called **entrailment**. Neither argon nor nitrogen are electroactive when dissolved in aqueous solution.

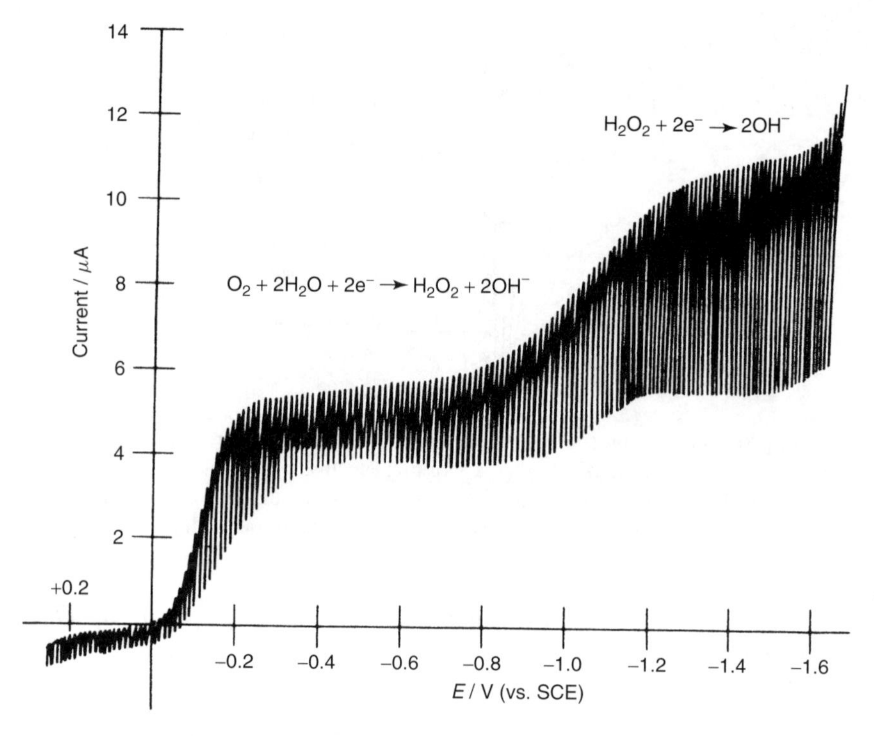

Figure 6.7 Polarogram of air-saturated water, i.e. a polarogram of *oxygen*. The solution also contains KNO_3 as an inert ionic electrolyte (0.1 mol dm^{-3}) and Triton X-100 (a non-ionic surfactant) as a current maximum suppresser (see Section 6.8.1). From Bard, A.J. and Faulkner, L.R., *Electrochemical Methods: Fundamentals and Applications*, © Wiley, 1980. Reprinted by permission of John Wiley & Sons, Inc.

DQ 6.13

Why a mercury *drop*?

Answer

At the heart of the polarographic apparatus is a fine-bore capillary through which mercury flows at a constant rate. Mercury emerges from the end of the capillary as small droplets, which are formed at a constant, controlled rate of between 10–60 drops per minute. During each 'drop cycle', the spherical drop emerges, grows in diameter and then falls.† *The duration of the drop cycle is sometimes given the symbol* τ.

† In some apparatus, the drop falls under the influence of gravity, i.e. when it reaches a critical size of, say, a couple of mm in diameter. In a more sophisticated apparatus, drops of a more carefully defined size are delivered either by forcing the mercury through the capillary, e.g. by assisting the delivery via a flow of nitrogen gas, or by using an electromechanical drop 'dislodger'. If the drop falls by gravity alone, then the frequency can be readily adjusted by variation of the height *h* in Figure 6.6.

DQ 6.14

Furthermore, what are the implications of using a mercury drop in a DME?

Answer

The first implication is that polarography can only be a tool for cathodic *electroanalysis for the reasons given above (although there are a few exceptions to this).*

Secondly, the surface of the drop is constantly being renewed, therefore meaning that its surface is always clean. Problems caused by contamination are more readily controlled than with a conventional solid electrode, thus leading to reproducible current–potential data.

Finally, the current at a DME is directly proportional to the electrochemical area of the electrode. (Being a liquid of high surface tension, the electrochemical area of a mercury drop is the same as its geometric area.) As the mercury drop grows during each cycle (see Figure 6.8), so the surface area also increases during each cycle and hence currents will fluctuate in a cyclical manner.

Figure 6.8 shows a polarogram obtained during the reduction of chromate ion at a DME. The appearance of a polarogram is often described as 'sawtoothed', with the large fluctuations in current being wholly attributable to the periodic changes in size of the mercury drops. At the left-hand side of the polarogram, the analyte is electro-inactive since (from equation (6.2)) the oxidized form will be the only stable form. We see that no current flows. As the potential becomes more cathodic, so reduction of the analyte occurs and a current is generated. The current reaches a **plateau** at higher (more negative) potentials. The magnitude of the current at the plateau is independent of potential, and is called a **diffusion current** (I_d). The latter is a *limiting* current because its magnitude is directly proportional to the bulk concentration of chromate ion (as analyte). The trace is often called a polarographic **wave**, because of its shape.

There are other aspects to consider when performing polarographic analyses with the DME, as summarized in Table 6.1.

The magnitude of the polarographic diffusion current is given by the **Ilkovic** equation:[†]

$$I_d = 708nD^{1/2}m^{2/3}t^{1/6}c_{\text{analyte}} \tag{6.5}$$

where n is the number of electrons per redox reaction, D is the diffusion coefficient of the analyte prior to electromodification, τ is the lifetime of the drop

[†] The derivation of the Ilkovic equation assumes that the drops of mercury are spherical, the rate of flow of mercury is constant, the way analyte diffuses toward the DME obeys the Cottrell equation, and the diffusion current is truly limiting, i.e. that $c_{\text{surface}} = 0$.

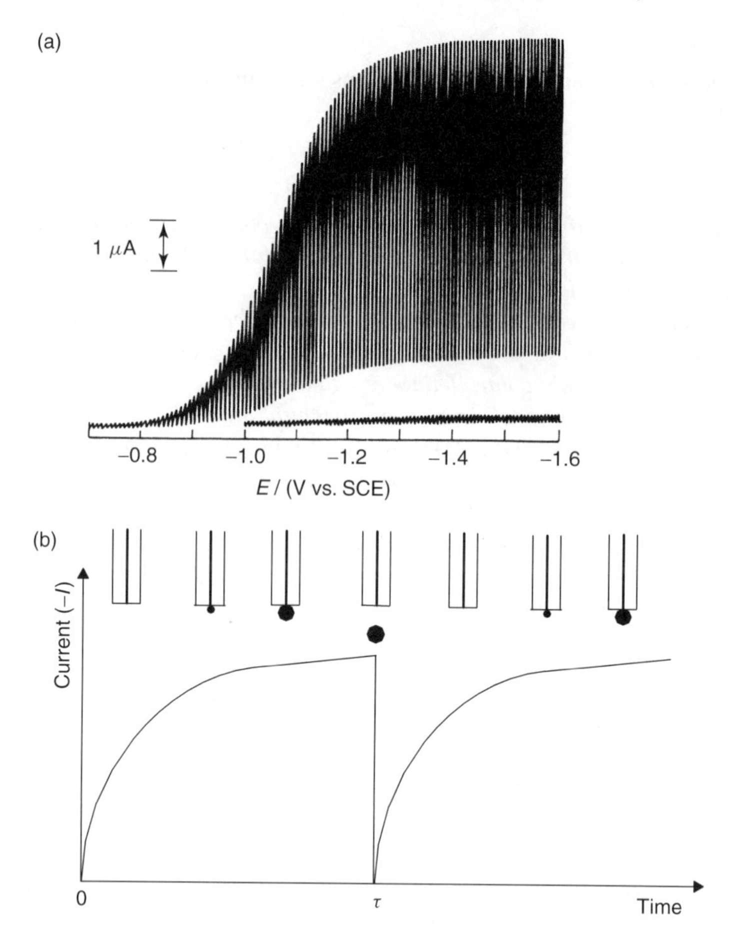

Figure 6.8 (a) Polarogram of the reduction of chromate ion at a dropping-mercury electrode, with $[CrO_4^{2-}] = 1$ mmol dm^{-3} in degassed 0.1 mol dm^{-3} NaOH solution, and using a scan rate ν of 20 mV s^{-1}. The small lower trace represents the residual current obtained at the same DME but in the absence of chromate. (b) The sequence of operation and the way that drop size dictates the current; note that current is negative since reduction is involved. From Bard, A. J. and Faulkner, L. R., *Electrochemical Methods: Fundamentals and Applications*, © Wiley, 1980. Reprinted by permission of John Wiley & Sons, Inc.

between its formation and fall, and m is the flow rate of mercury in kg s^{-1}. Some typical values of D are given in Table 6.2.

For completeness, we should note that I_d is in fact the *average* current since the current fluctuates according to where in the drop cycle we measure it. Because currents have a 'sawtoothed' shape, they are sometimes written as $\overline{I}$ or $\overline{I}_d$.

Table 6.1 The principal advantages and disadvantages of the dropping-mercury electrode (DME) as an analytical tool

Advantages

1. A new electrode surface is presented to the solution every second or so (i.e. with each new drop), which eliminates (or minimizes) the effects of adsorptive contamination
2. Following from the above, the reproducibility of polarograms is very high
3. Most metals can be reduced at a mercury drop to form an amalgam, thus maintaining a relatively pure Hg surface
4. The potential at which water is 'split' is quite negative, thus further broadening the number of metals which can be analysed by polarography (i.e. extending the range of $E_{1/2}$ values)

Disadvantages

1. Anodic analysis is impossible for most cations because mercury is so readily oxidized (although the amounts of some neutral species such as vitamin C can be quantified at a DME)
2. Experimentally, the DME is cumbersome to assemble and operate
3. The cleaning-up of spent mercury is costly
4. Elemental mercury is highly toxic

Table 6.2 Representative diffusion coefficients for various ions in KCl of concentration 0.1 mol dm^{-3}, in water at 25°C. Note the non-IUPAC units for D

Analyte	D(cm^{2}s^{-1})
Zn^{2+}	6.73
Cd^{2+}	7.15
Pb^{2+}	8.67
Methyl viologen[a]	8.60

[a] See Table 4.1 for the structure of this organic dication.

Figure 6.9 is a redrawing of the polarogram shown in Figure 6.8, but with the current depicted as a *smooth* line to represent time-averaged currents, with these averages being taken as the midpoints between each sawtooth's minimum and maximum in the previous figure.

Figure 6.9 also shows the **residual** current, which could be caused by IR drop (see Chapter 3) or by the reduction of species other than analyte. The magnitude of the current, I_d, should be determined with respect to the residual current, rather than from the baseline at $y = 0$. One way of determining the residual current is to measure the polarogram with all experimental variables unaltered except that no analyte is in solution. The small trace at the foot of Figure 6.8(a) was obtained in such a way.

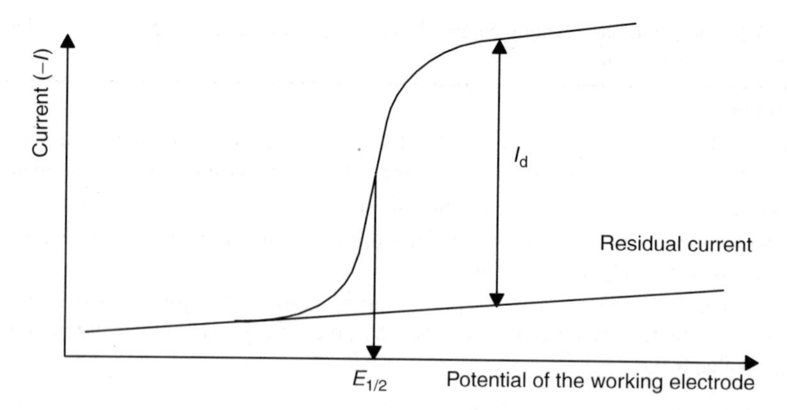

Figure 6.9 Polarogram showing a *time-averaged* current, i.e. redrawn without the 'sawtoothed' effect caused by the cyclic nature of the mercury drops; the half-wave potential, $E_{1/2}$, and the residual current are also indicated. The magnitude of the diffusion current, I_d, is determined with respect to the residual current.

Worked Example 6.2. Cadmium ion is reduced at a DME and the diffusion current measured as 24.3 μA. Previously, a sample of 2.0×10^{-4} mol dm^{-3} Cd^{2+} was analysed at the same DME with the same drop time, and the diffusion current was then found to be 15.2 μA. What is the concentration of the sample? (Assume that the residual current was known, and has been subtracted for both samples.)

From straightforward ratios:

$$I_{d1}c_2 = I_{d2}c_1$$

so:

$$c_2 = \frac{I_{d2}}{I_{d1}} \times c_1$$

and therefore:

$$c_2 = \frac{24.3}{15.2} \times 2.0 \times 10^{-4} \text{ mol dm}^{-3} = 3.2 \times 10^{-4} \text{ mol dm}^{-3}$$

SAQ 6.1

Following the worked example above, what is the concentration of Cd^{2+} if the diffusion current is 9.04 μA ?

While a calculation of the above type is easily performed, a **calibration graph** would be a far superior approach since any slight non-linearity of the response can more readily be accounted for, and residual currents can be detected more easily. Alternatively, a **multiple addition method**, such as a *Gran* plot (see Section 4.3.2), would improve the accuracy of these answers.

6.3.3 *Treatment of Polarographic Data: Obtaining* $E_{1/2}$ *and its Use*

The potential of the mercury-drop working electrode at the point where the current has reached exactly half its maximum value (that is, $1/2 \times I_d$) is called the **half-wave potential** ($E_{1/2}$). The latter occurs at a potential value that is characteristic of the analyte.† A single polarogram is sufficient to both *identify* the analyte (from the value of $E_{1/2}$ – comparing it with values in standard tables) and *quantify* its concentration (from I_d and a known standard).

DQ 6.15

How do we determine $E_{1/2}$ *accurately*?

Answer

It is sometimes possible to determine $E_{1/2}$ *just by inspection, that is, find the potential at which the current is exactly half of the diffusion current* I_d. *In practice, however, such values of* $E_{1/2}$ *are not particularly accurate, so we need an alternative approach. We use the* ***Heyrovsky–Ilkovic*** *equation as follows:*‡

$$E = E_{1/2} + \frac{RT}{nF} ln\left(\frac{I_d - I}{I}\right) \quad (6.6)$$

where all terms are familiar, and where I *is the current at the potential* E.

In order to obtain $E_{1/2}$ *from the above equation, we plot a graph of* $ln[(I_d - I/I)]$ *against* E. *The intercept on the* x*-axis gives then an* accurate *value of* $E_{1/2}$.

Worked Example 6.3. Consider the reduction at a DME of Fe^{2+} to form Fe. What is the *accurate* value of $E_{1/2}$ from the data shown below? Take $I_d = 3.24$ μA and assume a temperature of 20°C.

E/(V vs. SCE)	−0.395	−0.406	−0.415	−0.422	−0.431	−0.445
I/μA	0.48	0.97	1.46	1.94	2.43	2.92

† Strictly speaking, we should note that the value of $E_{1/2}$ also depends strongly on the solvent in which the analyte is dissolved. In fact, comparison between experimental and literature values of $E_{1/2}$ should be treated with caution, also on account of the fact that different electrolyte mixtures may have been employed.

‡ It is very common to see this equation cited in an alternative form, with the last term being written as '$-(RT/nF) ln\ [I/(I_d - I)]$'. Both forms are correct: the difference simply illustrates the laws of logarithms.

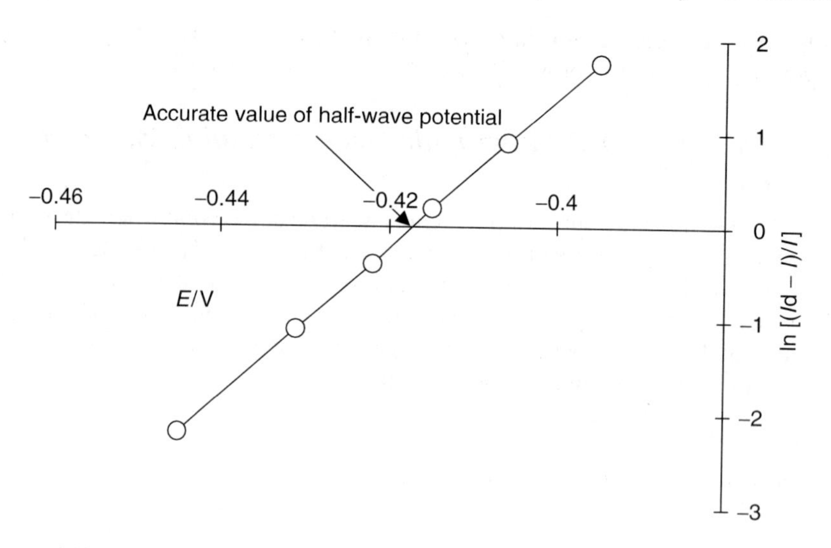

Figure 6.10 Logarithmic analysis of a polarographic wave by using the Heyrovsky–Ilkovic method (via equation (6.6)). The intercept on the x-axis is an *accurate* value of $E_{1/2}$ for the Fe^{2+}, Fe couple in water.

These data refer to the sloping portion of the polarogram. To obtain $E_{1/2}$, we plot a graph of ln $[(I_d - I)/I]$ as 'y' against E as 'x' (as depicted above in Figure 6.10). The intercept on the x-axis is $E_{1/2}$, which is seen to have a value of −0.417 V.

SAQ 6.2

The following data refer to the reduction of a mercurous complex at a DME. Draw a Heyrovsky–Ilkovic plot to determine both $E_{1/2}$ and the number of electrons transferred in the electrode reaction.

E/(V vs. SCE)	0.97	0.98	0.99	1.01	1.02	1.03	1.04	1.05
I/μA	2.134	4.255	7.718	17.10	20.64	22.83	25.01	25.01

DQ 6.16

Does the gradient of a Heyrovsky–Ilkovic graph have any significance?

Answer

From the way that the Heyrovsky–Ilkovic equation (equation (6.6)) is written, the gradient of a graph of ln $[(I_d - I)/I]$ *as 'y' against* E *as 'x' should be* (nF/RT)*. If the gradient of the graph is indeed* (nF/RT)*, then* *we can say that the electrode reaction is likely to be* ***reversible*** *in the electrochemical sense.*[†]

If the gradient is too low, however, then we can imply that the (nF/RT) *term has been modified, i.e. that the reaction at the DME is electrochemically* ***irreversible****, such that a variant of the Heyrovsky–Ilkovic equation now needs to be employed, as follows:*

$$E = E_{1/2} + \frac{RT}{\alpha nF} ln\left(\frac{I_d - I}{I}\right) \qquad (6.7)$$

where α *is a constant (having a value in the range* $0 < \alpha < 1$*). A wholly reversible electron-transfer process has a value of* $\alpha = 1$.

SAQ 6.3

By using a suitable graphical method, show that the data presented in SAQ 6.2 above represent a *reversible* electrode reaction.

Multiple analytes. The polarographic method can monitor several analytes at once, provided that the values of $E_{1/2}$ for each analyte are widely separated. In this context, 'widely separated' means that $\Delta E_{1/2}$ should preferably by no less than about 0.15 V. Figure 6.11 shows a polarogram for a solution containing a mixture of heavy metal cations. This figure clearly shows several different waves. The half-wave potential, $E_{1/2}$, for each wave is characteristic of the couple and, given care during measurement, can be thought of as invariant. The relative height of each wave is in proportion to the concentration of analyte.

Note that only the first of the three diffusion currents is determined with respect to the residual current, with the subsequent diffusion currents being determined as shown on the figure. We will also need a separate calibration curve for each analyte.

6.3.4 Determination of Equilibrium Constants: Shifts in $E_{1/2}$ *on Complexation*

We will consider in this section the case where a metal cation, M^{n+}, readily forms a complex with q number of ligands L to form a complex of final stoichiometry ML_q.

The usual situation we encounter is that where the ligands are electro-*inactive* but the cation is electroactive, both before and after complexation. However, the potential at which reduction occurs will shift following complexation: reduction of the uncomplexed cation is characterized by a half-wave potential, $E_{1/2(\text{free})}$,

[†] As a 'rule of thumb', reversibility (in the electrochemical sense) implies that the electron-transfer reaction is sufficiently swift for the current to obey equation (6.6) 'instantly' and that no chemical processes accompany the electron-transfer reaction – see Section 6.3.4.

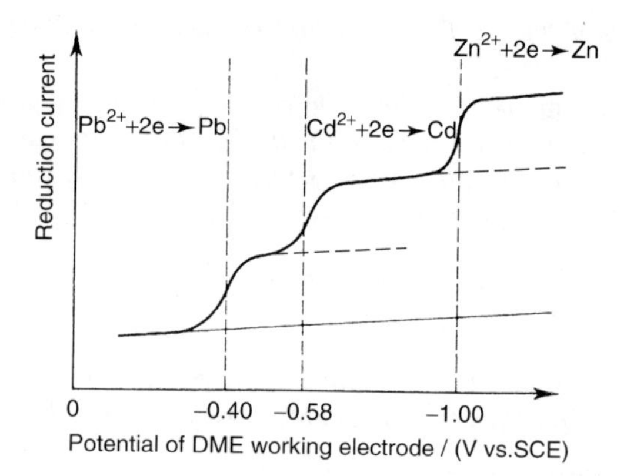

Figure 6.11 Polarogram of a solution containing three analytes, showing three different 'waves'. The half-wave potential, $E_{1/2}$, for each is characteristic of the respective analyte couples, while the wave heights reflect the relative concentrations of each ion. The trace has been smoothed to remove the 'sawtoothed' effects seen in Figures 6.7 and 6.8. The solution also contained KCl (0.1 mol dm^{-3}) as a swamping ionic electrolyte, and Triton X-100 (a non-ionic surfactant) as a current maximum suppressor.

while the potential at which reduction of the complexed cation occurs is characterized by $E_{1/2(\text{complex})}$.[†]

DQ 6.17

Why is $E_{1/2}$ shifted by complexation?

Answer

A fraction of the electrons on the metal cation help form the new bonds created during the complexation process. With a different electron density on the central cation, the energy needed to reduce the metal will be different before and after the complexation reaction.

The extent of complexation is given by the formation constant K, as follows:

$$K = \frac{a(\text{ML}_q)}{a(\text{M})a(\text{L})^q}$$

A relatively good gauge of the magnitude of K may be obtained by determining the magnitude of the shift in $E_{1/2}$ that accompanies complexation.

It can readily be shown that:

$$E_{1/2(\text{complex})} - E_{1/2(\text{free})} = -\frac{0.0591}{n} V \log K - \frac{0.0591}{n} V \log c_{\text{L}}^q \qquad (6.8)$$

[†] We need to recognize that both free and complexed cations are **solvated**, albeit to differing extents.

where we have assumed that activity can be replaced with concentration, and the ligand concentration is c_L.

Worked Example 6.4. The half-wave potential, $E_{1/2}$, for the reduction reaction, '$Cu^{2+}{}_{(aq)} + 2e^- \rightarrow Cu_{(Hg)}$', is 0.02 V (vs. SCE) if copper ion is uncomplexed, but $E_{1/2} = -0.22$ V if ammonia solution (at a concentration of 1 mol dm^{-3}) is added to the reaction solution. What is $K_{\text{complexation}}$?

As always, we must first discern or find out (from standard tables) the reaction stoichiometry involved. In this case, addition of ammonia to cupric ion effects the well-known reaction, $Cu^{2+} + 2NH_3 \rightarrow [Cu(NH_3)_2]^{2+}$.

Next, we calculate the shift in $E_{1/2}$ to be -0.24 V. Inserting the appropriate values into equation (6.8), we obtain:

$$-0.24 \text{ V} = -\frac{0.0591}{2}\text{V} \log K - \frac{0.0591}{2}\text{V} \log 1^2$$
$$= -0.02955\text{V} \log K - 0$$

which gives, after dividing throughout by V:

$$8.12 = \log K$$

and therefore:

$$K = 1.3 \times 10^8$$

SAQ 6.4

Consider the complexation reaction between Fe^{2+} and diethylenetriamine (det) to form $[Fe(det)]^{2+}$: the equilibrium constant K is 10^6 at 20°C in 0.1 mol dm^{-3} KNO_3. If the electrode reaction is the reduction, $Fe^{2+} + 2e^- \rightarrow Fe$, calculate the shift in $E_{1/2}$ expected if the concentration of det is 0.01 mol dm^{-3}.

In fact, superior results are obtained by plotting a graph of the shift in half-wave potential ($\Delta E_{1/2} = E_{1/2\ (\text{complex})} - E_{1/2\ (\text{free})}$) against $\log c_L$. Such a plot should give a straight line with an intercept that is related to K.

This graphical method is generally a good means of obtaining values of K to an accuracy of about ± 5–10%. The accuracy is maximized if each value of $E_{1/2}$ is obtained by using the Heyrovsky–Ilkovic procedure (via equation (6.6) as seen earlier). Furthermore, the values of $E_{1/2}$ (both complexed and free) are more accurate if well separated.

Obtaining plots by using equation (6.8) here is also a good means of obtaining the value of q, the number of ligands incorporated with M^{n+} in the complex. We recall from the laws of logarithms that $\log a^b = b \log a$, so the last term in equation (6.8) may be rewritten as $q \times (0.0591/n \log c_L)$, meaning that the value of q can be readily obtained via the gradient.

SAQ 6.5

From the data given below, determine the equilibrium constant K and the preferred stoichiometry of the complex formed between Co(II) and ethylenediaminetetraacetic acid (EDTA) in water. Take $E_{1/2}$ for the di-reduction of Co^{2+} at Hg to be −0.13 V.

c_{EDTA}/mol dm^{-3}	0.010	0.020	0.030	0.040	0.050	0.060	0.070	0.080	0.090	0.100
$E_{1/2}$/V	−0.553	−0.562	−0.567	−0.571	−0.574	−0.576	−0.578	−0.580	−0.581	−0.583

6.4 Linear-Sweep and Cyclic Voltammetry at Solid Electrodes

Having discussed polarography, we shall now look at voltammetry. The principal difference between polarography and voltammetry is the nature of the working electrode (WE). In polarography, a dropping-mercury electrode (DME) is used as the WE, while the WE employed during voltammetry is usually solid. The hanging mercury-drop electrode (HMDE), discussed in the previous chapter, is an exception to this latter generalization, and is sometimes used in voltammetry.

Like polarography, voltammetry is a microanalysis technique, so only a small proportion of the solution is ever modified by the processes occurring at the electrode.

The potential of the working electrode is ramped at a scan rate of ν. The resultant trace of current against potential is termed a **voltammogram**. In **linear-sweep voltammetry** (LSV), the potential of the working electrode is ramped from an initial potential E_i to a final potential E_f (cf. Figure 6.2). Figure 6.12 shows a linear-sweep voltammogram for the reduction of a solution-phase analyte, depicted as a function of scan rate. Note that the x-axis is drawn as a function of *overpotential* (equation (6.1)), and that the peak occurs just after $\eta = 0$.

During **cyclic voltammetry**, the potential is similarly ramped from an initial potential E_i but, at the end of its linear sweep, the direction of the potential scan is reversed, usually stopping at the initial potential E_i (or it may commence an additional cycle). The potential at which the reverse occurs is known as the **switch potential** (E_λ). Almost universally, the scan rate between E_i and E_λ is the same as that between E_λ and E_i. Values of the scan rates $\nu_{forward}$ and $\nu_{backward}$ are always written as *positive* numbers.

Figure 6.13 shows a voltammogram for a simple solution-phase couple: such a plot is known as a **cyclic voltammogram** (CV). The adjective *cyclic* arises from the closed loop drawn within the plot. The shape of the CV shown in Figure 6.13 is typical for a couple that is wholly reversible in the thermodynamic sense: other simple diagnostic tests for electro-reversibility are listed in Table 6.3.

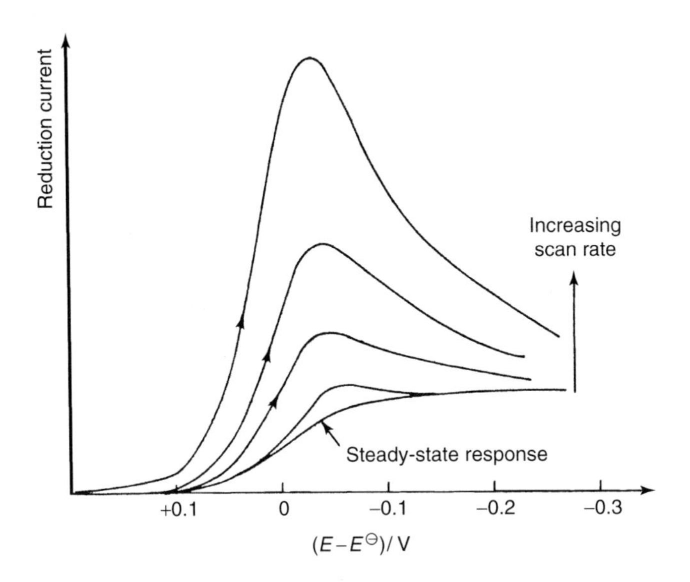

Figure 6.12 Linear-sweep voltammogram for the reduction reaction, $O + ne^- \rightarrow R$, at a solid electrode, shown as a function of the scan rate ν. The solution was under diffusion control, which was achieved by adding inert electrolyte and maintaining a still solution during potential ramping. Note that the x-axis has been normalized to $E^{\ominus}$, that is, the x-axis represents an overpotential. Reproduced from Greef, R., Peat, R., Peter, L.M., Pletcher, D. and Robinson, J., *Instrumental Methods in Electrochemistry*, Ellis Horwood, Chichester, 1990, with permission of Professor D. Pletcher, Department of Chemistry, University of Southampton, Southampton, UK.

There is seen to be a peak formed in both the forward and reverse sides of the CV. These peaks are of similar shape and, if fully **reversible**, their magnitudes will be identical. Oxidation has occurred during the forward part of the CV, with oxidation taking place during the reverse part. If the scan was 'going negative' from E_i, then reduction would occur during the forward part of the scan, and oxidation during the reverse.

As with other forms of voltammetry, the magnitude of the current is proportional to concentration, so the equality between $I_{p(\text{forward})}$ and $I_{p(\text{back})}$ (where 'p' represents 'peak') implies a quantitative retrieval of electromodified material (which follows from Faraday's laws). At the completion of the CV, when the potential has returned to E_i, there is still a slight current, indicating that a small amount of material (juxtaposed to the electrode) has still not been reduced. This slight residual current would dissipate to zero if we were to force the potential of the electrode to more negative potentials beyond E_i.

DQ 6.18

What is meant by 'reversible' in the context of the above CV?

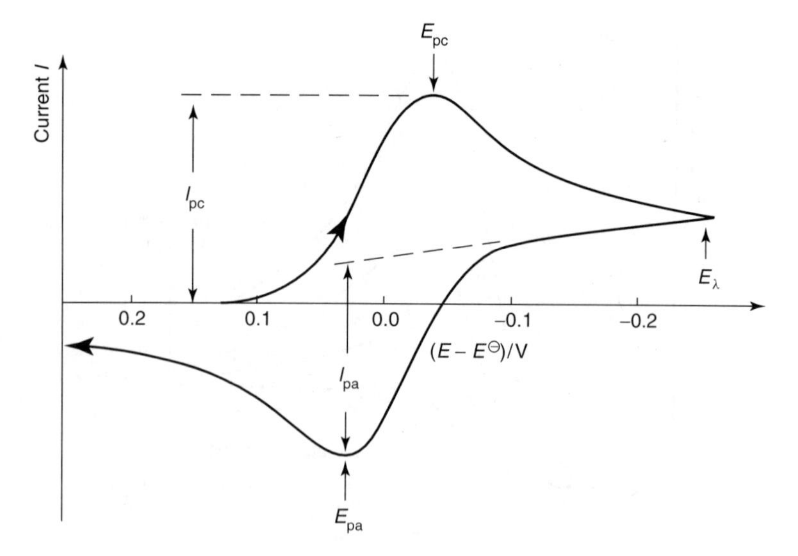

Figure 6.13 Schematic cyclic voltammogram for the reduction reaction at a solid electrode. As in Figure 6.12, the solution was under diffusion control, which was achieved by adding inert electrolyte and maintaining a still solution during potential ramping. The initial solution contained only the oxidized form of the analyte couple, so the upper (cathodic) peak represents the reaction, $O + ne^- \rightarrow R$, while the lower (anodic) peak represents the electrode reaction, $R \rightarrow O + ne^-$. Note also that the x-axis represents overpotential, so the peaks are centred about $E^{\ominus}$.

Table 6.3 Diagnostic tests for the electrochemical reversibility of a redox couple, carried out by using cyclic voltammetry[a]

1. $I_{pc} = I_{pa}$
2. The peak potentials, E_{pc} and E_{pa}, are independent of the scan rate ν
3. $E^{0'}$ is positioned midway between E_{pc} and E_{pa}, so $E^{0'} = (E_{pa} + E_{pc})/2$
4. I_p is proportional to $\nu^{1/2}$
5. The separation between E_{pc} and E_{pa} is 59 mV/n for an n-electron couple

[a] pc, peak cathodic; pa, peak anodic.

Answer

As a 'bottom-line' definition, the CV will look like that shown in Figure 6.13 only if (i) the ratio of activities of the oxidized and reduced forms of the redox couple satisfies the Nernst equation for the potential of the working electrode, and (ii) the rate of electron transfer, k_{et}*, is extremely fast. In practice, we mean 'fast' such that the rates of analyte diffusing to and from the electrode are limiting. A couple fulfilling all of the criteria given in Table 6.3 will probably be reversible. The couple is almost certainly not reversible if one or more of these criteria are not fulfilled.*

DQ 6.19

What do the potentials of the CV peaks tell us?

Answer

We will call the potential of the peak E_p*. The potential of the cathodic peak is then* E_{pc}*, while the potential of the anodic peak is* E_{pa}*.*

In polarography, we obtained the half-wave potential $E_{1/2}$ *by analysing the shapes of the polarographic wave.* $E_{1/2}$ *is a useful characteristic of the analyte in the same way as* $E^{\ominus}$*. In cyclic voltammetry, the position of* both ***peaks*** *(both forward and back in Figure 6.13; cathodic and anodic, respectively, in this example) gives us thermodynamic information. Provided that the couple is fully reversible, in the thermodynamic sense defined in Table 6.3, the two peaks are positioned on either side of the **formal electrode potential*** $E^{0'}$ *of the analyte redox couple, as follows:*

$$\frac{E_{pa} + E_{pa}}{2} = E^{0'} \tag{6.9}$$

*The formal electrode potential (also called the **formal potential** or the **formal redox potential**) is conceptually similar to the standard electrode potential,* $E^{\ominus}$.

DQ 6.20

So how does the formal electrode potential relate to the standard electrode potential?

Answer

In the Nernst equation (equation (3.8)), the electrode potential $E_{O,R}$ *and the standard electrode potential* $E^{\ominus}_{O,R}$ *are related, one to the other, in terms of the* activity a. *We can write a variant of the Nernst equation in which the concentrations* c *are used instead, as follows:*

$$E_{O,R} = E^{0'}_{O,R} + \frac{RT}{nF} \ln \left(\frac{c_O}{c_R} \right) \tag{6.10}$$

From the definition of activity and concentration given earlier in equation (3.12), we can say that the formal and standard electrode potentials are related as follows:

$$E^{0'}_{O,R} = E^{\ominus}_{O,R} + \frac{RT}{nF} ln \left(\frac{\gamma_O}{\gamma_R}\right) \quad (6.11)$$

SAQ 6.6

Derive equation (6.11) from equations (6.10), (3.8) and (3.12).

SAQ 6.7

When will the standard and formal electrode potentials be the same?

SAQ 6.8

Following on from SAQ 6.7 – experimentally, how can this equality be brought about?

The difference between $E^{\ominus}$ and $E^{0'}$ is often ignored because such differences are so slight. We will not refer to the formal potential again in this text.

DQ 6.21

Polarography is able to determine multiple analytes (cf. Figure 6.11). Can cyclic voltammetry be used in the same way?

Answer

Yes – several analytes can indeed be followed with cyclic voltammetry, with one pair of peaks (anodic plus cathodic) per couple. The couples can represent separate analyses, or can represent one analyte capable of displaying multiple redox states. Figure 6.14 shows a CV of vanadium pentoxide, with the CV comprising two separate redox couples. The anodic peaks are labelled as 'pa' and the cathodic peaks as 'pc'. Additional subscripts, (1) and (2), show how the two sets of forward and reverse peaks relate, one to another.

Occasionally, when the voltammetry solution is non-aqueous, it is difficult to find a reference electrode of known potential. If this is the case, it is useful to add a tiny amount of ferrocene, $Fe(cp)_2$, to the voltammetry solution. The $Fe(cp)_2^{+\bullet}$, $Fe(cp)_2$ couple is wholly reversible in almost every solvent system except water, so the CV will contain all of the peaks of the analyte of interest, plus a small pair of peaks due to the ferrocene couple. The potentials of the peaks of interest can then be cited with respect to the $E^{\ominus}$ for the ferrocene couple in the solvent system in question (cf. adding tetramethylsilane (TMS) to an NMR sample).

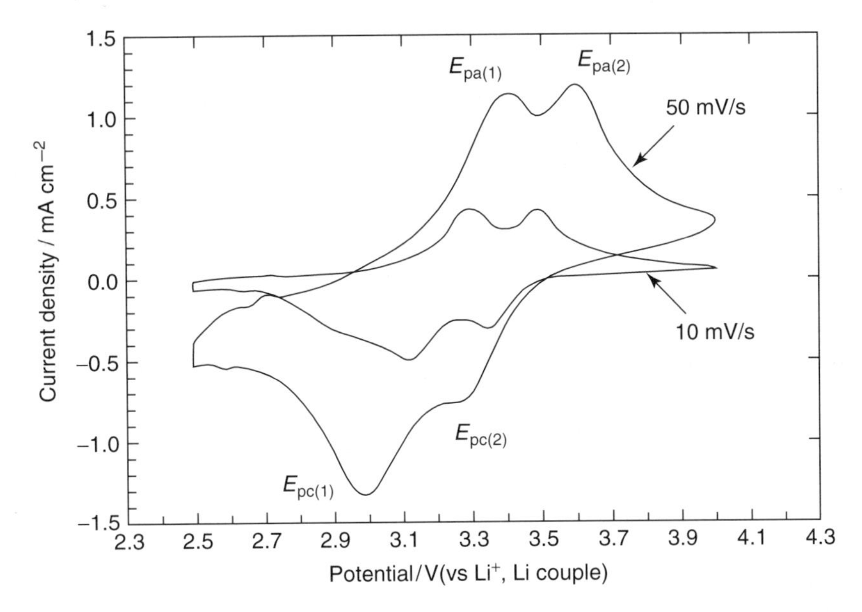

Figure 6.14 Cyclic voltammogram obtained for a multiple-electron-transfer system, where a thin film of sputtered V_2O_5 on a platinum working electrode has been immersed in an electrolyte solution of propylene carbonate containing $LiClO_4$ (1.0 mol dm^{-3}). From Cogan, S. F., Nguyen, N. M., Perrotti, S. J. and Rauh, R. D., 'Electroctromism in sputtered vanadium pentoxide', *SPIE*, **1016**, 57–62 (1989). Reproduced by permission of the International Society for Optical Engineering (SPIE).

DQ 6.22

Is cyclic voltammetry more useful than linear-sweep voltammetry?

Answer

Cyclic voltammetry is probably the most commonly encountered technique for studying dynamic *electrochemistry. It is useful for discerning kinetics, rates and mechanisms, in addition to thermodynamic parameters which are usually obtained at equilibrium.*

The forward half of the CV is identical to a linear-sweep voltammogram. The back half of the CV represents the reverse electron-transfer processes occurring at the working electrode: if the peak on the forward limb of the CV represents the oxidation reaction, $Fe^{2+} \rightarrow Fe^{3+} + e^-$, then the reverse limb represents the reduction reaction, $Fe^{3+} + e^- \rightarrow Fe^{2+}$. We can gain much information if the peak current of the reverse limb is smaller than the peak current during the forward part of the cycle (see next section). Such information cannot be obtained in a LSV experiment because no reverse limb is traversed.

One of the simplest ways of indicating the **time-scale** τ of the cyclic voltammetry experiment is to calculate the time required for one voltammetry cycle.† During a single cycle, the potential traverses the potential $E_{\mathrm{i}} \rightarrow E_{\lambda}$ and then $E_{\lambda} \rightarrow E_{\mathrm{i}}$, so:

$$\tau = \frac{2(E_{\mathrm{i}} - E_{\lambda})}{\nu} \tag{6.12}$$

From equation (6.12), it is clear that the time-scale is best varied by varying the scan rate ν. We will look at the way in which the peak currents vary with the time-scale in Section 6.4.4.

Worked Example 6.5. If $E_{\mathrm{i}} = 0$ V and $E_{\lambda} = 1.0$ V, what is τ when the scan rate is 20 mV s^{-1}?

From equation (6.12):

$$\tau = \frac{2 \times 1.0\ \mathrm{V}}{0.02\ \mathrm{V\ s^{-1}}} = 100\ \mathrm{s}$$

SAQ 6.9

A CV starts at 0 V, goes to −1.1 V, and returns to 0 V. Calculate the time-scale of the CV if the scan rate ν is 100 mV s^{-1}.

We see by comparing Worked Example 6.5 and SAQ 6.9 that the simplest way of varying the time-scale of a cyclic voltammogram is to vary the scan rate ν: a shorter time-scale is achieved with a faster scan rate.

6.4.1 The Randles–Sevčik Equation

The magnitude of the peak current, I_{p}, in a cyclic voltammogram is a function of the temperature, bulk concentration, c_{analyte}, electrode area, A, the number of electrons transferred, n, the diffusion coefficient, D, and the speed at which the potential is scanned, ν, according to the **Randles-Sevčik** equation, as follows:

$$I_{\mathrm{p}} = 0.4463 nFA \left(\frac{nF}{RT}\right)^{1/2} D^{1/2} \nu^{1/2} c_{\mathrm{analyte}} \tag{6.13}$$

Other terms have their usual meanings.‡ The second name is pronounced ‘Sev-chick’.

† There are different ways of gauging the time-scale. Merely for convenience, we have chosen the definition ‘τ = time for one cycle’ – see later for more meaningful definitions of τ.

‡ Remember that the electrochemist uses the so-called electrochemical units, so A has units of cm^2, D has units of $\mathrm{cm}^2\ \mathrm{s}^{-1}$ and c has units of mol cm^{-3}. Other units are consistent with the SI system (see Chapter 1), so remember to convert ν from mV s^{-1} to V s^{-1} before carrying out any calculations using equation (6.13).

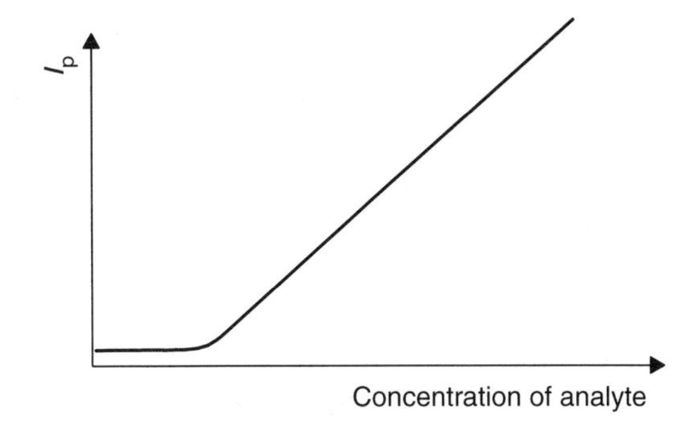

Figure 6.15 Plot of peak current (I_p) in a voltammogram (either linear-sweep or cyclic) against analyte concentration. The linear portion obeys the Randles–Sevčik equation, while the horizontal plateau at low $c_{analyte}$ values is usually caused by non-faradaic components of I_p, such as double-layer charging.

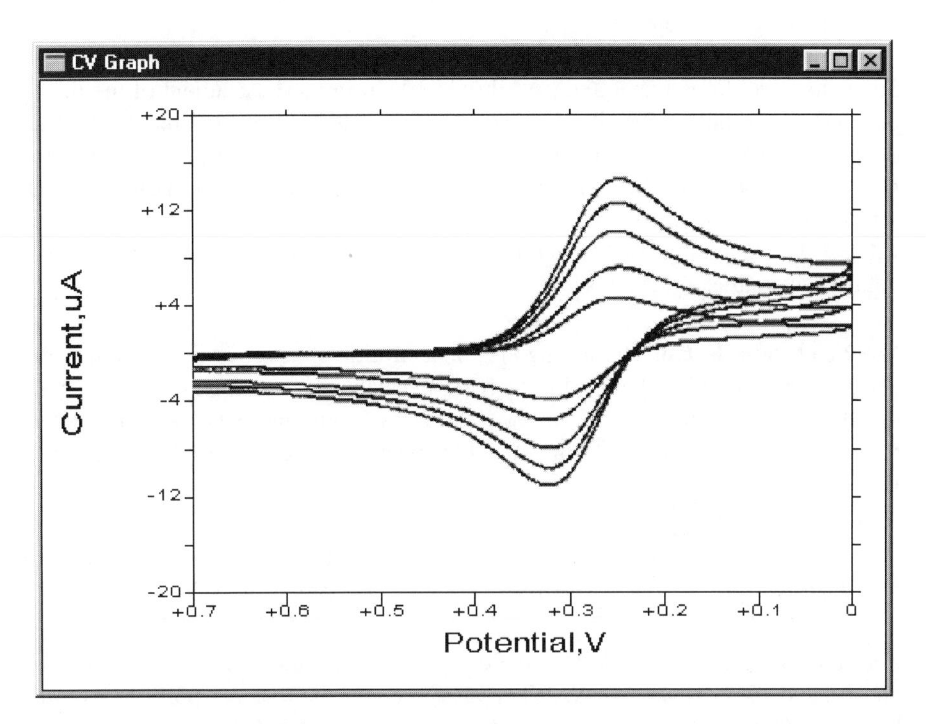

Figure 6.16 Cyclic voltammograms as a function of scan rate. This figure comprises traces simulated by the DigiSim® program for a reversible one-electron couple, with the fastest scan rate being shown outermost. Reprinted with permission from *Current Separations*, Vol. 15, pp. 25–30, copyright Bioanalytical Systems, Inc., 1996.

If the electrolyte composition is constant in terms of temperature, solvent, swamping electrolyte, etc., then we can use the Randles–Sevčik equation to determine the concentration of analyte by the construction of a suitable calibration curve. Figure 6.15 represents such a curve. Note that all values of I_p were determined at a constant scan rate ν. Such curves should yield accurate values of concentration (to within a few %) in the range c. 1 mol dm^{-3} down to as low as 10^{-5} mol dm^{-3} at the lower extreme. Below the lower concentration limit, the non-faradiac current resulting from charging of the electrode double-layer becomes comparable in magnitude to I_p. The extreme low-concentration portion of the graph therefore appears as a horizontal plateau, implying that I_p becomes independent of concentration. In fact, drawing a graph such as that shown in Figure 6.15 can yield an approximate value of $I_{\text{double-layer charging}}$ by taking the I value for the horizontal portion.

Alternatively, the CV can be drawn as a function of the scan rate ν (see Figure 6.16). Note here how the magnitude of the peak current increases with the increasing scan rate.

A plot of I_p as 'y' against $\nu^{1/2}$ as 'x' is often called a **Randles–Sevčik plot**. Such plots should be linear and pass through the origin. The gradient of the linear portion of a Randles–Sevčik plot can be employed to determine the concentrations of analyte if the diffusion coefficient D is known accurately. Unfortunately, D is not independent of c_{analyte} and is often quite unknown. Note, however, that a Randles–Sevčik plot is one of the best ways of determining an experimental value of D if a literature value is unavailable, provided that the electrode reaction if fully reversible.

Worked Example 6.6. Determine the concentration of aqueous Fe^{2+} ion in nitric acid (of concentration 1 mol dm^{-3}) from the data given below, which refer to the electrode reaction $Fe^{2+}{}_{(aq)} \rightarrow Fe^{3+}{}_{(aq)} + e^-$. From other studies, D is known to have a value of 5×10^{-5} cm^2 s^{-1}. The electrochemical area of the electrode is 1.3 cm^2.

ν/mV s^{-1}	5	10	20	40	80	160
I_p/mA	16.4	18.8	21.6	25.6	31.2	33.2

Figure 6.17 shows the Randles–Sevčik plot of current I_p against $\sqrt{\nu}$ constructed with these data. The gradient of the linear portion of the plot is 8.68×10^{-2} A (V $s^{-1})^{-1/2}$. Using equation (6.13), this gradient is $0.4463nFA(nF/RT)^{1/2}D^{1/2}[Fe^{2+}]$, which, when written in terms of electrochemical units, becomes $2.47 \times 10^3 \times [Fe^{2+}]$ cm^3 mol^{-1} A (V $s^{-1})^{-1/2}$. Accordingly:

$$2.47 \times 10^3 \times [Fe^{2+}] \text{ cm}^3 \text{ mol}^{-1} \text{ A V}^{-1/2} \text{ s}^{1/2} = 8.68 \times 10^{-2} \text{ A (V s}^{-1})^{-1/2}$$

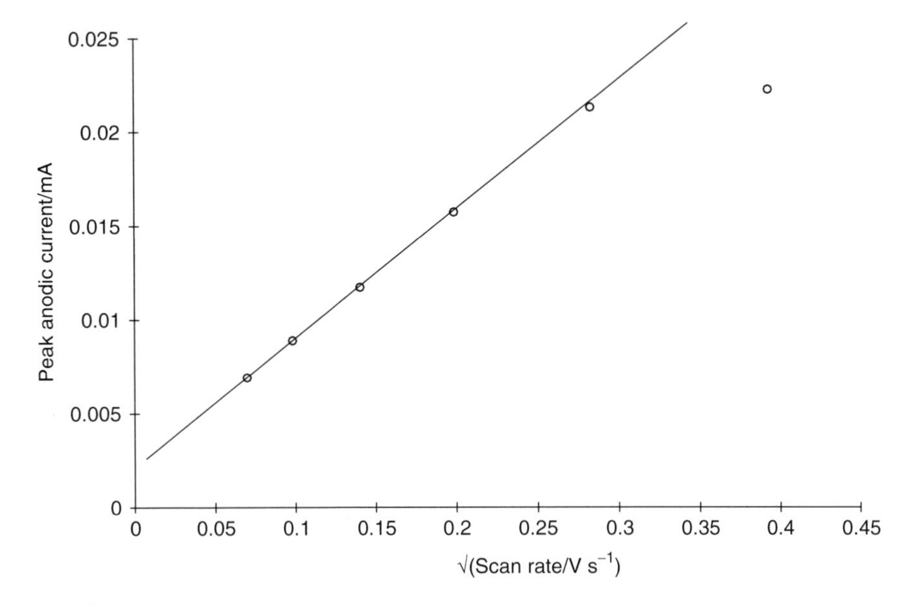

Figure 6.17 A Randles–Sevčik plot of I_p against $\nu^{1/2}$. Data refer to the oxidation of aqueous ferrous ion at a stationary platinum wire electrode. The non-linearity at the higher scan rates represents the demand for flux at the working electrode being too great since ν is too fast, while the non-zero intercept is caused by non-faradaic currents contributing toward the overall I_p.

which, after cancelling out the units, gives:

$$[Fe^{2+}] = 3.51 \times 10^{-5} \text{ mol cm}^{-3}$$

or, in more familar notation:

$$[Fe^{2+}] = 0.0351 \text{ mol dm}^{-3}.$$

DQ 6.23

Why does the plot in Figure 6.17 deviate from the Randles–Sevčik equation?

Answer

Figure 6.17 broadly follows the Randles–Sevčik equation insofar as most of the plot of I_p *against* $\sqrt{\nu}$ *is linear. The figure does, however, have a non-zero intercept, and is not linear at the highest scan rates.*

The simplest reason for a non-zero intercept is non-faradaic current, the causes of which are discussed in Section 6.8.2.

Figure 6.17 also deviates from linearity at the fastest scan rates (typically when $\nu > 200-500$ *mV* s^{-1}*) since the peak currents* I_p *are* *somewhat smaller than expected. Under-sized currents of this type are caused by v being too fast. Diffusion is not a particularly efficient means of mass transport.*[†] *The deviation in Figure 6.17 is one means of telling us that insufficient analyte is moving through solution to the electrode, which explains why equation (6.13) breaks down. In fact, we find that eventually* I_p *reaches a plateau because the flux of analyte at the electrode attains its maximum value.*

Sometimes, there is no linear portion to a Randles–Sevčik graph, and the data yield a curved plot. The derivation of equation (6.13) assumes that diffusion is the sole means of mass transport. We also assume that all diffusion occurs in one dimension only, i.e. perpendicular to the electrode, with analyte arriving at the electrode | solution interface from the bulk of the solution. We say here that there is ***semi-infinite linear diffusion****.*

In this case, the first reason for deviation from linearity is that the exponent on the scan rate is not 1/2, but some other fraction. If the exponent is greater than 1/2, then we assume that an additional means of mass transport supplements the diffusion – with either convection or migration also being involved. Conversely, if the exponent is much less than 1/2, then we usually assume that the analyte is adsorbed *at the electrode, implying that there is little mass transport of analyte at all.*

6.4.2 The Effect of Slow Electron Transfer: Semiconducting Electrodes

Up until now, all of the electron-transfer reactions we have considered were said to be so fast as to be 'instantaneous' relative to the rate of mass transport of analyte to and from the working electrode. Since the only mode of mass transport is diffusion (although see next chapter), and diffusion is intrinsically slow, the assumption that the rate of electron transfer, k_{et}, is very fast is usually a safe one. Occasionally, however, k_{et} is sufficiently slow that we observe two rates, both that of mass transport and of electron transfer.

DQ 6.24

When is k_{et} likely to be sufficiently slow that its rate is visible?

Answer

In the overwhelming majority of analyses, an electrochemist will choose an electrode constructed from platinum, gold, silver, mercury or some

---

[†] A simple proof that mass transport by convection is more efficient than movement by diffusion alone can be afforded by considering a cup of tea: we *stir* sugar into a cup of tea rather than let the sugar move by diffusion alone.

other material having a metallic conductivity. Detailed reasons for this choice were discussed in DQ 2.3 on the magnitude of (faradaic) currents.

*Occasionally, the electroanalyst employs electrodes made from amorphous silicon (*a-*Si), metal oxides or some other semiconductor. Unfortunately, when following the effects of shining light on an electrode – the topic of photoelectrochemistry – the most interesting effects are generally obtained with semiconductors such as* a-*Si.*

Later, in Chapter 8 (Section 8.1.2), we will look at in situ *spectroelectrochemistry – the simultaneous monitoring of electrochemical processes with UV–visible spectroscopy. One of the best electrodes for this purpose is a thin film of In(III) oxide doped with Sn(IV) oxide. Although this mixture of oxides has a relatively good electronic conductivity σ, the magnitude of σ is never high.*

DQ 6.25

If a poor conductor is inevitable, how is the slow rate of k_{et} seen in the CV?

Answer

Figure 6.18 shows a simulated CV in an attempt to illustrate the effects of slow electron transfer. At very slow scan rates, while the speed of electron movement through the electrode is slow, the rate of charge uptake at the electrode | solution interface is comparable (currents I *are small), so the effect of a slow* k_{et} *will be undisclosed.*

As the scan rate v *increases, an additional kinetic 'bottleneck' is introduced. Because the electrons move slowly through the electrode, there is a perceptible time lag between the potential at the voltage source (generally a potentiostat) and that at the electrode | solution interface. Since all potentials experience this 'lag', the peaks are shifted to more extreme potentials, thus giving the CV a 'stretched' appearance.*

Because the CV is stretched, the separation between the peaks ΔE_p *(anodic and cathodic) increases from its theoretical values of 59/*n *mV, which characterizes a fully-reversible electron-transfer reaction (see Table 6.3). The magnitude of the overshoot depends on the time lag, and therefore as the scan rate increases, so the separation between the peaks increases, thus causing the CV to look even more stretched.*

One simple, although by no means foolproof test of whether the distortion is caused by slow electron transfer is to plot ΔE_p *(as 'y') against the (scan rate* $v)^{1/2}$ *(as 'x'). The graph will be linear, and have an intercept of 59/*n *mV if slow motion of the electrons through the electrode is the main cause of the stretch.*

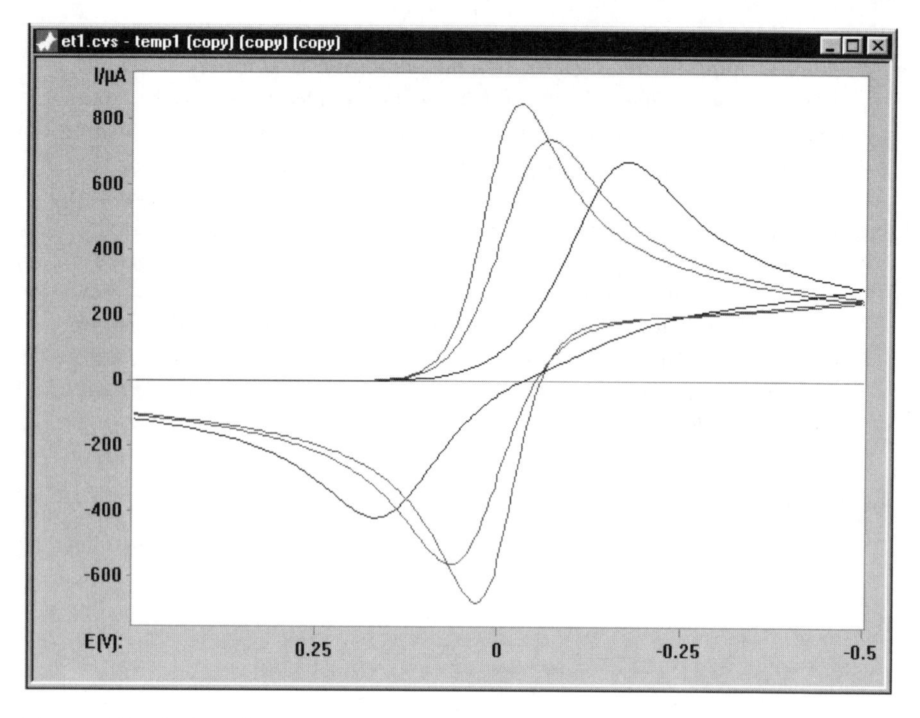

Figure 6.18 Cyclic voltammograms as a function of scan rate to show the effects of a slow rate of electron transfer as caused by poor electronic conductivity through the working electrode reaction. This figure comprises traces simulated by the DigiSim® program. The fastest scan rate ν is shown outermost; notice how the CV looks stretched as ν increases. Figure courtesy of Dr Adrian Bott, Copyright Bioanalytical Systems, Inc., 2000.

SAQ 6.10

Plot such a graph for the redox couples used in Figure 6.14, and ascertain the intercept potentials, and hence a value of *n* for each couple.

6.4.3 Quantification of Diffusion Phenomena

Occasionally, we want to know the value of a diffusion coefficient D. The Cottrell equation (equation (6.3)) is one simple means of obtaining D, i.e. by plotting $I_{\text{lim}(t)}$ against $t^{-1/2}$ and obtaining a value for the gradient. ($I_{\text{lim}(t)}$ here means the limiting current as a function of time.)

The Randles–Sevčik equation usually gives a more accurate value since all measurements are performed under pseudo-steady-state conditions. In practice, a Randles–Sevčik plot of I_{p} against $\nu^{1/2}$ is drawn for a redox couple of known concentration $c_{analyte}$, and the gradient of the graph is then determined. Knowing $c_{analyte}$ and the value of the gradient, a value of D may be obtained to within 2–5%.

SAQ 6.11

An aqueous solution of herbicide is known to contain methyl viologen (MV^{2+}, see Table 4.1) as the active ingredient, at a concentration of 2×10^{-4} mol dm^{-3}. Prior to voltammetric analysis, inert electrolyte was added to the solution and dissolved oxygen removed by sparging with N_2. A Randles–Sevčik graph of I_p against $\nu^{1/2}$ was drawn with I_p data for the reduction reaction, $MV^{2+} + e^- \rightarrow MV^{+\bullet}$. The linear part of the graph had a gradient of 7.09×10^{-5} A (V s^{-1})$^{-1/2}$ at 25°C, and the electrode area A was 0.45 cm^2. Calculate the diffusion coefficient D of MV^{2+} in water.

6.4.4 Mechanistic Data from Voltammetry

The power of cyclic voltammetry is seen when we look at the way it can distinguish between different types of electrochemical processes.

Reinmuth notation. In the electrochemical world, the sequence of electrode and/or chemical reactions that occur are described by a simple shorthand code. Simple electron-transfer reactions are called 'E' reactions. In the same shorthand system, a multiple electron-transfer reaction such as $Fe^{3+} \rightarrow Fe^{2+} \rightarrow Fe$ is an 'EE reaction', i.e. the product of an electron-transfer process itself undergoes a second electron-transfer process. (Note that the two electron-transfer processes might occur at the same time, in which case it is merely an 'E' reaction.) The vanadium pentoxide system illustrated in Figure 6.14 is another example of an EE system.

Chemical reactions are designated as 'C', so if the product of electron transfer undergoes a *homogeneous* chemical reaction we say that it is an 'EC' reaction. The 'C' terms are often given a superscript or subscript to show why type of chemical reaction occurs, e.g. disproportionation, dimerization or catalytic. Table 6.4 lists many of the commonly encountered Reinmuth terms.

Remember Reinmuth notation is read from left to right, so an EC reaction occurs with a chemical reaction *following* an *initial* electrode (electron-transfer) reaction.

SAQ 6.12

Consider an electron-transfer reaction that forms an intermediate which can undergo a homogeneous reaction, and where the product of the chemical

Table 6.4 Reinmuth notation for use within mechanistic electrochemistry, where the terms shown can be combined, and the sequence of the steps should be read from left to right (after Reinmuth, W. H., *J. Am. Chem. Soc.*, **79**, 6538–6364 (1957))[a]

Term	Reaction
E	Single ET reaction
EE	Two consecutive ET reactions (the product of an ET reaction undergoes a second ET reaction)
EEE	Three consecutive ET reactions
CE	ET preceded by a chemical reaction
EC	ET followed by a chemical reaction
CEC	ET that is both preceded and followed by a chemical reaction
ECE	ET followed by a chemical reaction, the product of which undergoes another ET reaction
C_{disp}	Chemical reaction – disproportionation
C_{dim}	Chemical reaction – dimerization
C′	Chemical reaction – catalytic

[a]ET, reversible electron transfer.

reaction can itself undergo a subsequent electron-transfer reaction. What is the Reinmuth notation for this overall process?

The necessity for the somewhat long-winded wording used in SAQ 6.12 explains why Reinmuth first proposed such a shorthand notation (in 1957).

Reversibility. The first aspect we analyse with cyclic voltammetry is 'electrochemical reversibility'. Table 6.3 above lists the simplest voltammetrically determined tests of reversibility. A system that fulfills each of these criteria is probably electro-reversible, while a system that does not fulfill one or more of the criteria is certainly not fully electro-reversible. The CV shown in Figure 6.13 is that of a fully electro-reversible couple in a single electron-transfer ('E') reaction.

Following chemical reactions. We will now consider **following chemical reactions**, which are also called **coupled chemical reactions** or **EC reactions**. We will illustrate such a situation with the bromine–bromide couple.

We commence the CV cycle with a solution containing only potassium bromide (i.e. the reduced form of the couple). The value of $E^{\ominus}_{Br_2,Br^-}$ is 1.09 V (see Appendix 2), so the bromine–bromide couple is fully reversible, and therefore the two peaks – anodic and cathodic – are positioned at an equal distance either side of 1.09 V (see Table 6.3). The forward peak represents the electrode reaction, $2Br^- \rightarrow Br_2 + 2e^-$, while the reverse peak represents the reverse electrode reaction, $Br_2 + 2e^- \rightarrow 2Br^-$. We recall that the CV peak height is proportional to concentration according to equation (6.13). The two peaks in the CV of bromine–bromide have the same magnitude, which again shows complete reversibility. Without a *following chemical reaction*, the CV of the bromine–bromide couple is just that of a simple E reaction.

We will now consider the differences in the CV caused by the chemical reaction. We are assuming that the reaction is homogeneous, that is, all reactants and products are wholly soluble. Elemental bromine is relatively stable in aqueous solution, although it will behave as an oxidizing agent. A typical oxidation process represents an addition-type reaction across an unsaturated bond, such as that found in an alkene. We therefore consider the CV of a similar voltammetry solution, i.e. one that contains not only KBr but also a small amount of alkene. In this context, the simplest alkene to consider is allyl alcohol ($CH_2{=}CH{-}CH_2{-}OH$), where the alcohol group is required in order to ensure solubility of the alkene in water. In all other respects, the allyl group behaves as a typical alkenic double bond, so elemental bromine adds rapidly to form the corresponding dibromo compound.

DQ 6.26

How can cyclic voltammetry detect that an EC reaction is occurring?

Answer

Before we start, we recall (from equation (6.13)) that the magnitude of the CV current peak is proportional to concentration. Allyl alcohol and brominated allyl alcohol are not electroactive at all, so the CV only monitors the relative amount of bromine and bromide. The forward peak of the CV has a magnitude I_{pa} *(where 'pa' = 'peak, anodic'), and* $I_{pa} \propto$ *[bromide], because bromide is converted at the peak to form bromine.*

The return, reverse peak is smaller than the forward peak if the CV of the Br_2, Br^- *couple is run with allyl alcohol in solution;* $I_{pc} \propto$ *[bromine], where 'pc' = 'peak, cathodic'. The fact that* I_{pc} *is smaller than* I_{pa} *implies that some fraction of the electrogenerated bromine has been consumed by the chemical reaction between* Br_2 *and allyl alcohol: the CV thus contains evidence of an EC reaction. Figure 6.19 below shows the cyclic voltammogram for the EC reaction between bromine and allyl alcohol, while Table 6.5 lists a series of simple diagnostic tests for an EC reaction.*

DQ 6.27

Can the CV tells us anything about the *rate* of the bromination reaction?

Answer

When looking at the return CV peak in Figure 6.19, we see that it decreases in magnitude as the scan rate slows down. This is a 'time-scale'

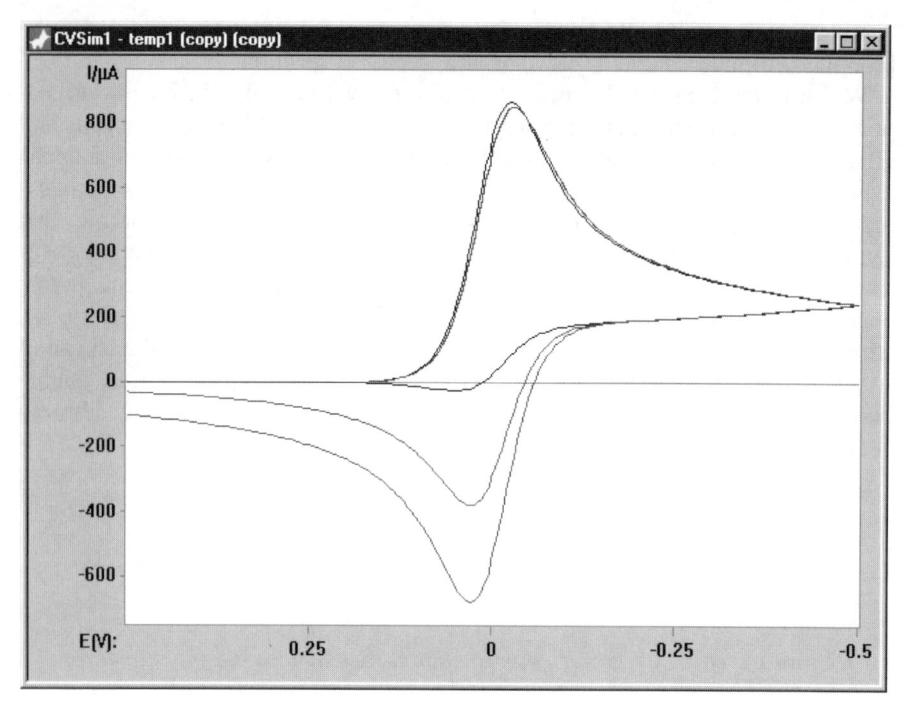

Figure 6.19 Cyclic voltammograms as a function of scan rate to show the effects of an EC reaction. This figure comprises traces simulated by the DigiSim® program. The fastest scan rate is shown outermost; note how the reverse peak is essentially absent at slow scan rates. Reprinted with permission from *Current Separations*, Vol. 18, pp. 9–16, copyright Bioanalytical Systems, Inc., 1999.

Table 6.5 Simple diagnostic tests for a coupled chemical (EC) reaction, carried out by using cyclic voltammetry (after Nicholson, R. S. and Shain, I. *Anal. Chem.*, **36**, 706–723 (1994), and Nadjo, L. and Saveant, J. M., *J. Electroanal. Chem.*, **48**, 113–145 (1973))

1. The ratio of forward and back peaks will deviate from unity: $I_{p(back)}/I_{p(forward)} < 1$
2. The current function $I_p/\nu^{1/2}[C]$ decreases with ν, but only slightly (<5%), i.e. the steepness of the peak increases
3. E_p shifts away from $E^{\ominus}$
4. E_p shifts with the scan rate ν: at slow scan rates (where the chemical reaction is dominating), $\partial E_p/\partial \log \nu = 30/\nu$ mV for a first-order coupled chemical reaction

effect. We looked at the time-scale of a cyclic voltammogram just before Section 6.4.1. while a simple definition of τ was given in equation (6.12).

As an analyst, if we were to perform a series of simple experiments to study the kinetics of bromine consumption, we would start the experiment (at a time we call $t = 0$*) and then monitor the amount of bromine remaining as a function of time* after $t = 0$. *Probably the simplest way of monitoring this process would be to remove aliquots of reaction solution after various lengths of time, and titrate each of these, e.g. with thiosulfate, to determine the amount of bromine remaining in each sample. In effect, we say here that 'chemical kinetics' is the study of the proportion of the matter that is initially present as a function of the reaction time-scale* τ.

Determination of the kinetic parameters by using cyclic voltammetry is conceptually very similar to this: $\tau = 0$ *is taken to be the time at the formation of the intermediate (here* Br_2*), i.e. at the forward current peak* I_{pa}, *and the time when it is monitored at* $t = \tau$, *i.e. at the current peak for the reverse electrode process,* I_{pc}. *The time-scale of the reaction,* τ, *is given by the following equation:*

$$\tau = \frac{(E_{pa} - E_{\lambda}) + (E_{pc} - E_{\lambda})}{\nu} \qquad (6.14)$$

so the simplest means of varying the time-scale of observation, τ, *is to change the scan rate* ν, *with a faster* ν *equating to a shorter* τ.

SAQ 6.13

In a cyclic voltammogram, E_{pa} = occurs at 0.560 V while E_{pc} occurs at 0.501 V. If $\nu = 50$ mV s^{-1}, what is τ when $E_{\lambda} = 1.0$ V?

The proportion of the originally generated intermediate (Br_2) that 'survives' during a time τ is readily gauged by the ratio of the forward and back peak currents and by saying that $I_{pc}/I_{pa} = [Br_{2(retrieved)}]/[Br_{2(formed)}]$.

When the scan rate is fast – say 500 mV s^{-1} – the CV is obtained very quickly, so any bromine generated at the electrode will have too little time to react to any significant extent before it is electro-reduced back to bromide. The ratio, I_{pc}/I_{pa}, will be close to unity at this scan rate.

Conversely, if ν is slow (e.g. 10 mV s^{-1}) then the electrogenerated bromine stays in solution for quite some time before being 're-reduced' to bromide (so τ is long). The scope for chemical reaction is seen to be greater at 10 mV s^{-1} than at 500 mV s^{-1}. We see therefore that the I_{pc}/I_{pa} ratio will be less than 1 because some of the bromine has been removed from solution by the reaction with allyl alcohol.

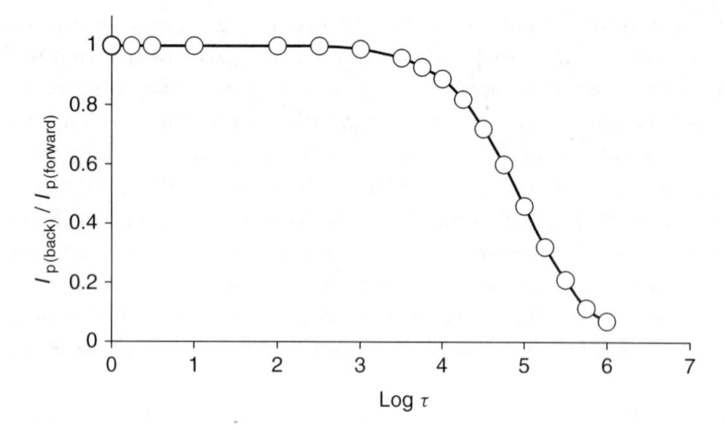

Figure 6.20 Experimental plot of the ratio of peak currents, $I_{p(back)}/I_{p(forward)}$, against $\log(\tau)$ for an EC reaction, where τ is the time-scale of the CV.

Above a critical scan rate, I_{pc}/I_{pa} always has a value of 1 because ν is so fast that there is no time for a chemical reaction to occur at all, and the electrogenerated bromine is retrieved quantitatively.† Below another critical scan rate, I_{pc}/I_{pa} is always 0 because ν is so slow that *all* of the bromine has reacted.

Between these two critical scan rates, I_{pc}/I_{pa} varies between 1 at the fast extreme to 0 at the slower extreme. Figure 6.20 show a plot of I_{pc}/I_{pa} against the logarithm of the time-scale τ. From this figure, we can again see that fractions of the electrogenerated bromine are retrieved as a function of the scan rate.

DQ 6.28

How is the rate constant of reaction k determined from a CV?

Answer

A similar graph to that shown in Figure 6.20 can be computed theoretically, with an identical y*-axis but with the* x*-axis scale being slightly different, i.e. as* $\log(k\tau)$ *(see Figure 6.21), where* k *is the rate constant of the reaction (here, the reaction must be* homogeneous *and assumed to be* first-order*). It can be seen that the shape of the plot in Figure 6.21 is the same as that shown in Figure 6.20.*

In practice, we draw a graph like Figure 6.20 with our data for the reaction, Br_2 + *allyl alcohol, with* I_{pc}/I_{pa} *as a function of log t. We then superimpose our graph on top of the computed graph (cf. Figure 6.21) – for this reason, it is a good idea to draw our graph of* I_{pc}/I_{pa} *against* $\log t$ *on tracing paper. The two graphs have the same*

† We could say that the 're-reduction efficiency' was 100%.

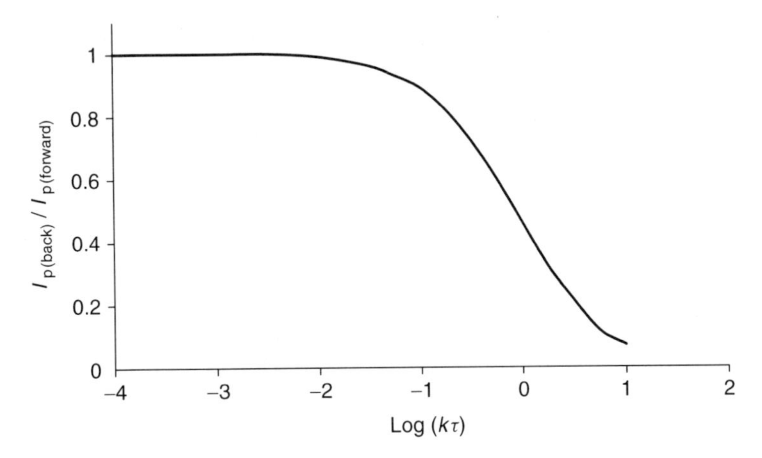

Figure 6.21 Computed plot of the ratio of peak currents, $I_{p(back)}/I_{p(forward)}$, against log $(k\tau)$ for an EC reaction, where τ is the time-scale of the CV and k is the rate constant of the first-order homogeneous reaction. Notice how the plot has a similar shape to that shown in Figure 6.20, but the x-axis is offset by the amount log k.

shape but, because of the slightly different axes, there is an 'offset' between the two x-*axis scales. This offset yields* k.

Worked Example 6.7. When a graph of I_{pc}/I_{pa} against log τ is superimposed on a graph of I_{pc}/I_{pa} against log $(k\tau)$, there is an offset on the x-axis of 6.2. How do we calculate the rate constant k from these graphs? The concentration of allyl alcohol is 0.12 mol dm^{-3}.

By saying that an offset exists, we are saying that log $(k\tau)$ is different from log τ. The numerical difference is 6.2. We can state this in another way, as follows:

$$\log (k\tau) = \log \tau + 6.2$$

So, after rearranging according to the laws of logarithms, we obtain:

$$\log k + \log \tau = \log \tau + 6.2$$

The log τ term is common to both sides, and can be removed. Therefore:

$$\log k = 6.2$$

$$k = 10^{6.2}$$

which gives:

$$k = 1.58 \times 10^{6}\ \text{s}^{-1}$$

Notice that k here relates to a *first-order* reaction. In this case, k is a pseudo-rate constant, implying that the allyl alcohol was in excess. The true rate constant is obtained straightforwardly as $k_2 = k_{pseudo}$/[allyl alcohol]. This gives:

$$k_2 = \frac{1.58 \times 10^6\ s^{-1}}{0.12\ mol\ dm^{-3}} = 1.3 \times 10^7\ (mol\ dm^{-3})^{-1}\ s^{-1}$$

for the reaction between Br_2 and allyl alcohol in water at 25°C.

SAQ 6.14

From the following data, what is k_{pseudo} for the homogeneous reaction between electrogenerated bromine and styrene in DMF solution? Assume that the potential differences, $(E_{p(forward)} - E_\lambda)$ and $(E_{p(back)} - E_\lambda)$ add up to 1.0 V.

ν/mV s^{-1}	0.01	0.1	1	10	100	200	1000
$I_{p\ back}/I_{p}$ forward	0.04	0.18	0.45	0.82	0.97	0.99	1.0

(Hint – draw a graph of $I_{p(back)}/I_{p(forward)}$ against log (τ). You will need the definition of τ from equation (6.14). Superimpose this graph over Figure 6.21 (tracing paper is a good idea) and note the offset, which is log k.

6.5 Improving Sensitivity: Pulse Methods

The analytical sensitivity of 'classical' polarographic or voltammetric methods is usually quite good at about 5×10^{-5} mol dm^{-3}. At the lowest concentrations of analyte, however, the currents caused by double-layer effects or other non-faradaic sources causes the accuracy to be unacceptably low. Pulse methods were first developed in the 1950s to improve the sensitivity of the polarographic measurements made by pharmaceutical companies. At present, two pulse methods dominate the analytical field, i.e.'normal pulse' and 'differential pulse'. Square-wave methods are also growing steadily in popularity.

During polarography, the cycle of drop production, growth and replenishment causes the current to oscillate, thus producing the 'sawtoothed' effect seen in Figures 6.7 and 6.8. The currents are largest at the *end* of each drop's lifetime, so if the current is sampled only at the end of each drop cycle, then the signal-to-noise ratio would be considerably improved since the faradaic currents would be greater. This idea is the basis of **sampled DC polarography**: in a typical sampled DC polarography experiment, the current is measured only during the last 15% of each drop cycle. By this means, the sensitivity of polarography can be increased slightly to about 10^{-5} mol dm^{-3}.

6.5.1 Normal Pulse Voltammetry

A greater improvement in sensitivity is obtained if the drop remains unpolarized during the majority of its lifetime. In practice, instead of a simple potential ramp

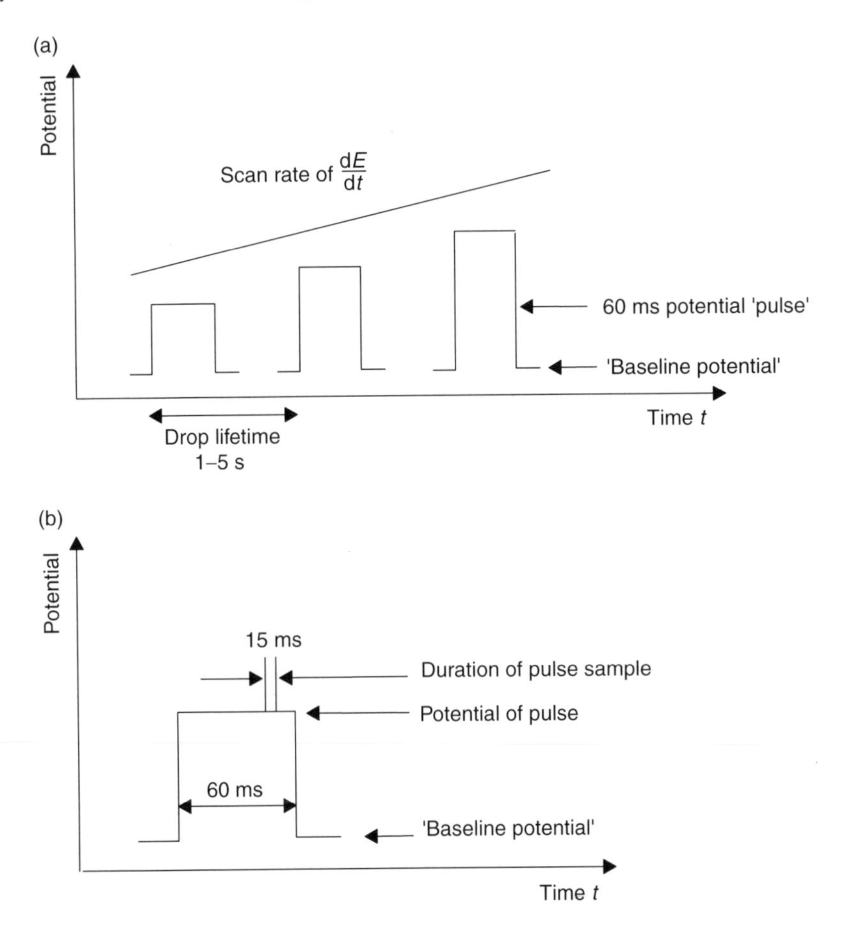

Figure 6.22 (a) The time-dependence of potential applied to a DME during normal pulse polarography: the baseline potential remains constant during the experiment, but the heights of the pulses increase at a rate of dE/dt. (b) Enlarged area showing the sampling of the current during a period of a few ms near the end of each pulse.

being applied to the working electrode (cf. Figure 6.5), a succession of square-wave pulses are applied to the dropping-mercury electrode (DME) according to the schematic diagram in Figure 6.22, with one pulse per drop. The drops have a relatively short lifetime (say 1.5 s). For the majority of the time, the potential of the drop is such that no electrolysis can occur. We call this potential the base potential, E_b. The electrolysis time is clearly shorter with such a pulsing procedure than during continuous polarization, and so the depletion region around the electrode is never allowed to become extensive. If the current is sampled only during the last few milliseconds of each pulse, then the advantages of sampled DC polarography (above) can also be gained.

The separation between each drop is commonly about 1.5 s. The height of each pulse is slightly higher than the preceding one such that a scan rate of dE/dt is superimposed on the peaks of the pulse.

DQ 6.29

Why does pulsing enhance the quality of polarographic analyses?

Answer

One of the main problems encountered during normal polarography is that changes occur to the **capacitance** C *of the double-layer around the drop – we call this 'charging' – which thus induces non-faradaic charging currents (see Section 5.1.2). The magnitude of* C *varies with time, drop size and potential. If the current sampling is extremely short and performed near the end of the pulse, then it can also be assumed that the drop does not increase in size during sampling. Furthermore,* dE/dt *does not really change during the lifetime of the pulse and the area of the drop remains the same at the onset of each pulse because the current is sampled from the same point in the drop cycle. In consequence, the capacitance changes are small and have decayed to a small*

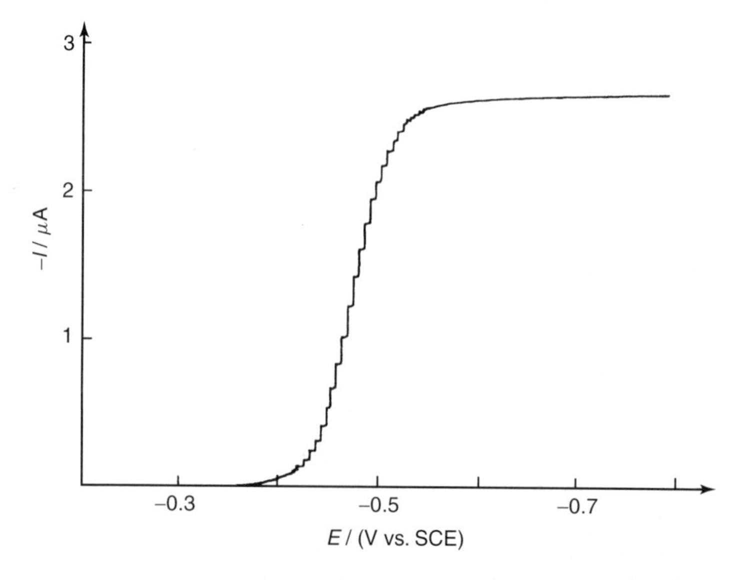

Figure 6.23 Normal pulse polarogram of the reduction of Pb^{2+} (10^{-4} mol dm^{-3}) at a DME. The ionic electrolyte was KNO_3 (0.1 mol dm^{-3}). Reproduced from Greef, R., Peat, R., Peter, L. M., Pletcher, D. and Robinson, J., *Instrumental Methods in Electrochemistry*, Ellis Horwood, Chichester, 1990, with permission of Professor D. Pletcher Department of Chemistry, University of Southampton, Southampton, UK.

and steady value by the end of each pulse, reaching c. 1% of the capacitance seen during normal polarography. It is valid to assume (to a very good approximation) that the current measured is wholly faradaic.

In practice, a polarogram obtained via pulsing at a DME generates a shape that is very similar to that of a 'normal' polarogram, with the main difference being that the 'sawtoothed' effect caused by current oscillations is lost, and the current now comprises a series of short, flat segments. The polarograph thus has a 'cleaner' appearance (see Figure 6.23 below). In practice, we also find that the height of the polarographic wave (which is the analytic signal to compute with) is significantly higher than that obtained with a 'normal' polarographic wave. In consequence, much smaller concentrations can still yield a measureable current, and the concentration limit with pulse polarography is enhanced to about 10^{-7} to 10^{-8} mol dm^{-3}.

6.5.2 Differential Pulse Voltammetry

In many respects, **differential pulse voltammetry** is more similar to classical polarography than to the normal pulse methods (see above). A linear potential ramp of $\mathrm{d}E/\mathrm{d}t$ is applied to the working electrode (see Figure 6.24). However, in common with 'normal' pulse voltammetry, a succession of pulses are also applied to the working electrode. (The WE is often a DME, and then we refer to 'differential pulse *polarography*'.)

The difference between the baseline potential and the peak is ΔE, and is a constant. In normal pulse voltammetry, no electrolysis occurs at times between each pulse because the baseline potential E_b is constant, and is chosen to be sufficiently anodic for currents to be zero.

Conversely, since peaks are superimposed on a ramped baseline, the peaks obtained during **differential pulse voltammetry** are not square since the plateau of the peak increases at the constant rate of $\mathrm{d}E/\mathrm{d}t$ that the baseline follows.

DQ 6.30

Why is the above technique termed *differential* pulse?

Answer

In normal pulse voltammetry, the current is sampled for a short period just before the drop is dislodged. The current monitored is assumed to be constant with time. In the differential pulse method, the current is monitored twice per drop: the first sample is taken just before the rise in potential when the pulse starts, while the second is taken at the end of the current pulse just before it decreases back to the baseline. The difference between these two currents is ΔI_{pulse} The differential pulse voltammogram is then a plot of current difference against potential. In

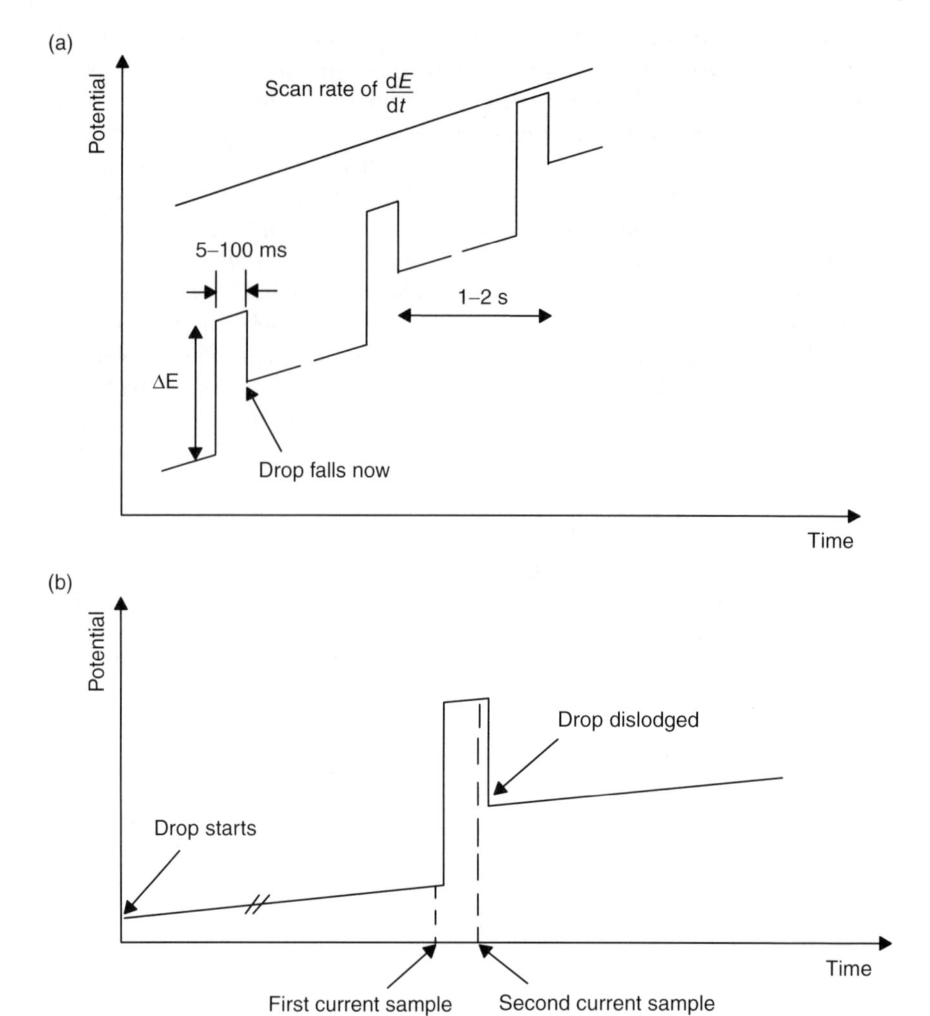

Figure 6.24 (a) The time-dependence of potential applied to a DME during differential pulse polarography; the height of each pulse is the same at ΔE. While the baseline between each pulse is independent of time, the potential of the baseline increases at a mean rate of dE/dt. (b) Enlarged area, showing the current being sampled twice per pulse, once just before the pulse and then before the drop is dislodged at the end of each pulse.

> *effect, the difference method here is a means of measuring the* differential *of current, i.e. the* rate of change *of current passage.*

The difference between the faradaic components of the two current samples is usually negligible so, in practice, the current difference ΔI_{pulse} is only significant

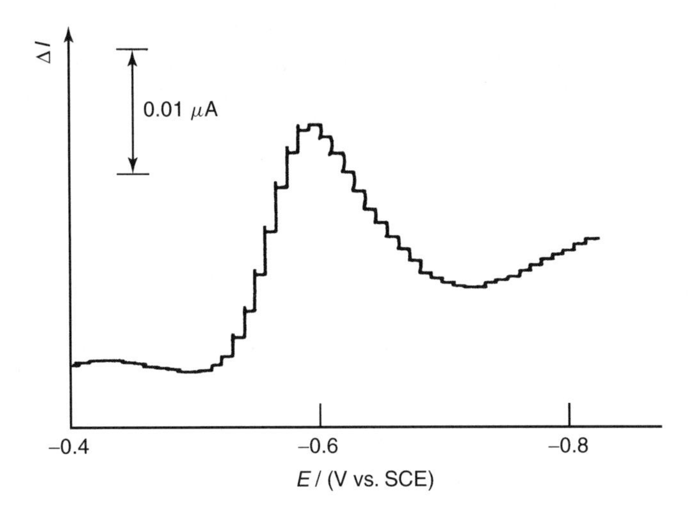

Figure 6.25 Differential pulse polarogram of 10^{-6} mol dm^{-3} Cd^{2+} in hydrochloric acid (0.01 mol dm^{-3}) at a DME. The separation in potential, ΔE, during the pulse was −50 mV. Note that the y-axis here is *not* current but *differential* current. Reproduced from Greef, R., Peat, R., Peter, L.M., Pletcher, D. and Robinson, J., *Instrumental Methods in Electrochemistry*, Ellis Horwood, Chichester, 1990, with permission of Professor D. Pletcher, Department of Chemistry, University of Southampton, Southampton, UK.

when redox activity occurs. In summary, ΔI_{pulse} is zero unless there is reduction of analyte at the working electrode. The magnitude of ΔI_{pulse} increases more rapidly when the potential is near to $E_{1/2}$ than at other potentials, so the differential pulse voltammogram contains a peak at $E_{1/2}$ (e.g. see Figure 6.25), whereas other methods such as normal pulsed polarography comprise a *wave*. (As with normal pulse voltammetry, the peak is stepped in appearance rather than being perfectly smooth.) The potential of the peak can help identify the cation in solution, in a similar manner to normal polarography.

The height of the current peak[†] ΔI_{p} is proportional to analyte concentration and ΔE according to the Osteryoung–Parry equation as follows:

$$\Delta I_{\text{p}} = \frac{n^2F^2A}{4RT}\left(\frac{D}{\pi t}\right)^{1/2} c_{\text{analyte}}\Delta E \tag{6.15}$$

where all terms have their usual meaning, and t is the time between pulses. The rooted term in brackets is seen to be very similar to one seen above in the Cottrell equation (equation (6.3)), implying that diffusion of analyte to the electrode is an important determinant of the magnitude of ΔI_{p}.

[†] This differential peak height is sometimes written as ∂I_{p}, although note that it is also written as just I_{p}.

Larger values of ΔE are preferred since increasing the magnitude of the current peak will also increase the accuracy of the technique, although we need to be aware that such larger values will also decrease the ability of the technique to analyse more than one analyte at a time. We note, however, that it is easier to determine the magnitude of a peak than of a wave, particularly if the wave is superimposed on a sloping baseline.

DQ 6.31

What are the analytical implications of these differential methods?

Answer

Since the technique is differential by nature, it is the area *under a peak which is proportional to concentration, so the Osteryoung–Parry equation is merely an approximation. This explains why many workers prefer to work with peak area rather than say that peak height is proportional to concentration (equation (6.15)). In fact, there is usually a trade-off between several different experimental factors, which are listed in Table 6.6 below.*

Experimentally, an analyst will run several standards (at constant ΔE*) to calibrate the analysis, and will then determine the amounts of analyte in solution. A 'standard additions' method such as a Gran plot will further enhance the accuracy of measurement (e.g. see Section 4.3.2).*

Despite these possible drawbacks, differential pulse voltammetry is one of today's most popular electroanalytical tools. Its principal advantages over normal pulse voltammetry are twofold: (i) many analytes can be sampled with a single voltammogram since the analytical peaks for each analyte are quite well resolved, and (ii) by working with a differential current, and hence obtaining a voltammetric peak, *the analytical sensitivity can be improved to about* 5×10^{-8} *to* 10^{-8} *mol dm*$^{-3}$*. This sensitivity is clearly superior to normal pulse voltammetry.*

6.5.3 Square-Wave Pulse Voltammetry

An alternative and more recent electroanalytical tool is **square-wave voltammetry** (which is probably now employed more often than normal or differential pulse voltammetry). In this technique, a potential waveform (see Figure 6.26) is applied to the working electrode. Pairs of current measurements are then made (depicted on the figure as t_1 and t_2); these measurements are made for each wave period ('cycle'), which is why they are drawn as times after t_0 (when the cycle started). The current associated with the forward part of the pulse is called $I_{forward}$, while the current associated with the reverse part is $I_{reverse}$. A square-wave voltammogram is then just a graph of the difference between these two

Table 6.6 Experimental considerations when obtaining data by using differential pulse voltammetry

1. The accuracy of the measurement is increased by using a *differential* method
2. Accuracy is also enhanced by increasing the magnitude of ΔE
3. Unfortunately, larger values of ΔE cause analyses of multiple-analyte mixtures to be more difficult since peak resolution is decreased
4. Peak resolution is enhanced by use of a slower scan rate because the lines of the voltammogram are shorter (i.e. the trace appears to be less 'jerky')
5. In addition, superior accuracy is obtained by integrating the peaks to obtain *areas*, although measuring a peak *height* is far simpler. Provided that the peak is symmetrical, then peak height is proportional to peak area, and I_p can still be used
6. Differential pulse voltammetry is particularly susceptible to adsorption of species on the electrode, which can have drastic implications for peak shape. If adsorption is suspected, then peak area should be used rather than peak height

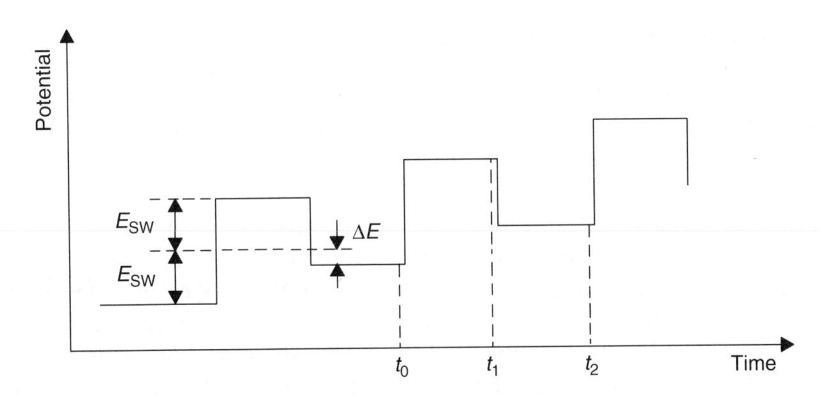

Figure 6.26 The time–potential profile for square-wave voltammetry.

currents as a function of the applied potential. A square-wave voltammogram for the reduction of ferric oxalate (a reversible system) is shown in Figure 6.27. The height of the peak is directly proportional to analyte concentration, so $I_{difference}$ is the analytical signal measured.

The value of E_{SW} (see Figure 6.26) is typically about 50 mV, while ΔE is commonly about 5 mV. A third variable is the scan rate ν, which can be as fast as 1 V s^{-1} (see below). Finally, the value of τ is often about 20–40 ms, corresponding to a frequency of about 500–250 Hz if $\nu = 1$ V s^{-1}.

Square-wave voltammetry affords several advantages to the electroanalyst. First, we note that $I_{difference}$ is larger than either $I_{forward}$ or $I_{reverse}$, so the height of the voltammetric peak is usually quite easy to read, thereby increasing the accuracy. A detection limit of about 10^{-8} mol dm^{-3} is attainable under optimized conditions.

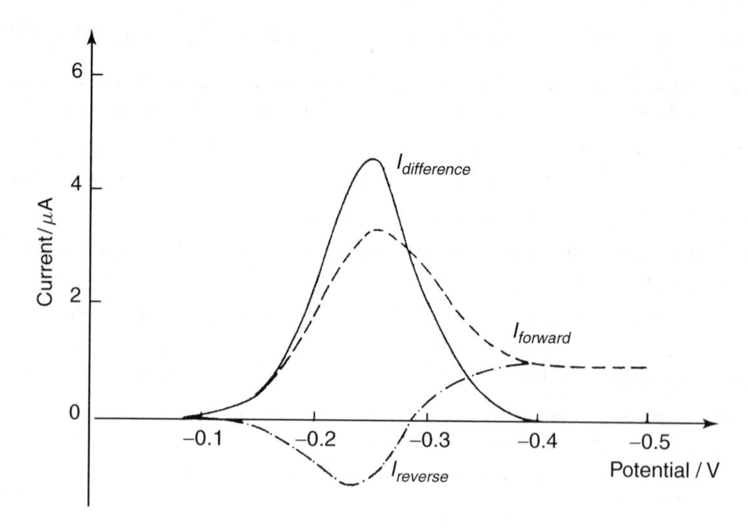

Figure 6.27 Square-wave voltammogram obtained for the electro-reduction of a ferric oxalate complex (5×10^{-4} mol dm^{-3}) in aqueous oxalate buffer; $\tau = 33.3$ ms, $E_{SW} = 30$ mV and $\Delta E = 5$ mV. Reprinted with permission from Turner, J. A., Christie, J. H., Vukovic, M. and Osteryoung, R. A., *Anal. Chem.*, **49**, 1899–1903 (1977). Copyright (1977) American Chemical Society.

The second major advantage of the square-wave procedure is the way that capacitive contributions to the overall current are minimized, and so the scan rate can be increased dramatically – a scan rate ν of 1 V s^{-1} is easily achievable.

Thirdly, oxygen need not be excluded from the analyte solution: provided the voltammetric peak is more cathodic than that for the reduction of oxygen, then the magnitude of both $I_{forward}$ or $I_{reverse}$ will incorporate an equal current due to the reduction of O_2. Since $I_{forward\ (O_2)}$ and $I_{reverse\ (O_2)}$ are equal, they cancel out and $I_{difference}$ is seen to be independent of the oxygen concentration in the analyte solution.[†]

DQ 6.32

Can such concentration limits be attained in practice with these pulse methods?

Answer

The lower detection limits cited above for these pulsed voltammetric and polarographic methods are summarized below in Table 6.7. Whenever we cite a concentration limit, *we are saying in effect that this is the very lowest that it is possible to achieve. It should be emphasized, however,*

[†] Oxygen still needs to be excluded if the analyte is oxygen-sensitive, i.e. to prevent EC-type reactions occurring.

Table 6.7 Lower electroanalytical detection limits for various polarographic and voltammetric methods

Technique	Lower detection limit ($mol\ dm^{-3}$)
Normal polarography	5×10^{-3}
Cyclic voltammetry	1×10^{-5}
Sampled DC polarography	1×10^{-5}
Normal pulse polarography	$10^{-7}-10^{-8}$
Differential pulse polarography	$10^{-8}-5 \times 10^{-8}$
Square-wave polarography	1×10^{-8}
Anodic stripping voltammetry	$10^{-10}-10^{-11}$

that in practice the limits cited are merely extremes, i.e. the concentrations of many (probably most) cations cannot be determined accurately at such low levels.

6.6 Stripping Voltammetry

In the previous chapter, we encountered a form of coulometry known as 'stripping'. We can combine both stripping and voltammetry in the powerful technique of **stripping voltammetry**.[†] As we have seen, the potential of the working electrode is ramped during a voltammetric or polarographic experiment. The resultant current represents the rate at which electroactive analyte reaches the surface of the electrode, that is, current $I \propto$ flux j.

The magnitude of a voltammetric current peak I_p is proportional to the concentration of analyte. In consequence, we can readily construct a calibration graph of I_p against [analyte] from known standards. We then perform our analyses by measuring I_p, and read off the concentration of analyte. This is generally quite a good method.

The most commonly encountered form of stripping is **anodic stripping voltammetry** (ASV). As with normal stripping, a hanging mercury-drop electrode (HMDE) is the working electrode. The HMDE is cathodized to exhaust the solution but, rather than *stepping* the potential to strip the drop (as in the previous chapter), we *ramp* the potential during ASV at a scan rate ν of between 2 to 100 mV s^{-1}. The resultant peak current I_p has a magnitude that is proportional to the number of moles of analyte within the amalgam of the drop. (For more accurate work, we would integrate the area under the peak.) Figure 6.28 shows an ASV trace of copper that had been pre-concentrated on the surface of a HMDE, and then oxidatively removed.

[†] The technique here should strictly be termed 'linear-sweep stripping voltammetry' in order to differentiate it from the related methods of pulse and differential stripping voltammetry, which are not discussed in this present text.

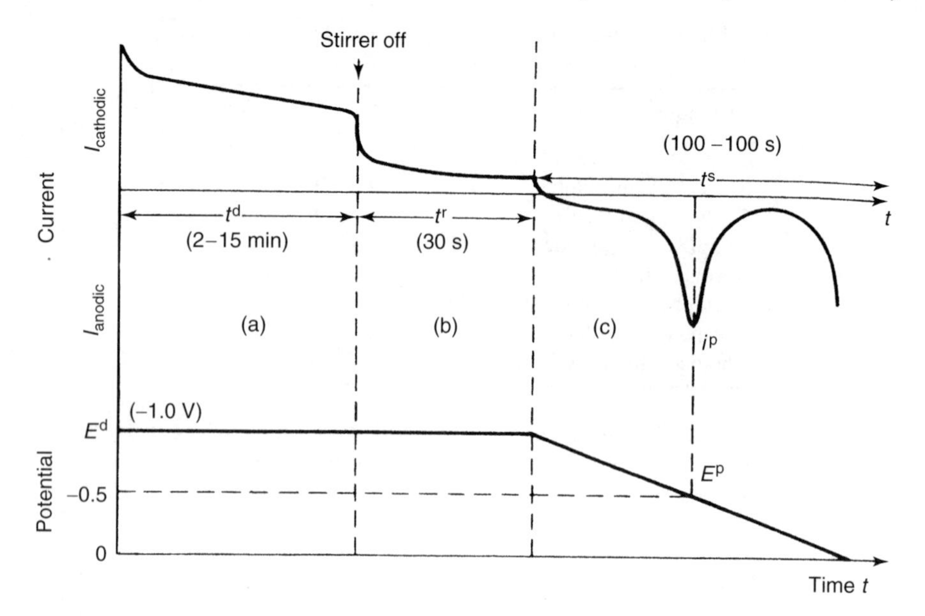

Figure 6.28 Stripping voltammogram obtained for the determination of Cu (II) in aqueous solution. The bottom of the figure also indicates the way that the potential applied to the hanging mercury-drop electrode (HMDE) varies during the ASV procedure. All values shown are intended to be typical rather than prescriptive: (a) pre-concentration of the stirred solution; (b) a short 'rest' period (without stirring); (c) anodic *stripping* during a potential ramp of 10–100 mV s^{-1}, again without stirring. Note the peculiar shape of the voltammogram which arises because I_{anodic} is negative (by definition), while $I_{cathodic}$ is positive. From Bard, A. J. and Faulkner, L. R., *Electrochemical Methods: Fundamentals and Applications*, © Wiley, 1980. Reprinted by permission of John Wiley & Sons, Inc.

Cathodic stripping voltammetry (CSV) is more rarely required than ASV, but is useful for the determination of halide(s), sulfide or other anions. Occasionally, cations of intermediate valency, such as Mn^{2+} or Pb^{2+}, can be analysed by using CSV, e.g. by reducing cations within a layer of metal oxide on an electrode made of carbon.

DQ 6.33

Why is there a sloping baseline shown in the top trace in Figure 6.28?

Answer

The sloping baseline shows that, in addition to the removal of copper, a second, additional process which involves charge flow is also occurring. In fact, the sloping baseline seen in Figure 6.28 is again illustrating the capacitive charging of the double-layer.

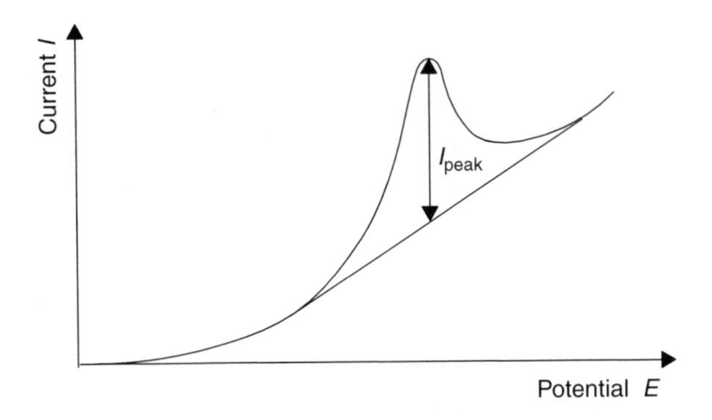

Figure 6.29 Schematic representation of the adjustment which is made to compensate for the sloping baseline in stripping voltammetry, itself due to charging of the electric double-layer at the WE.

Part of Figure 6.28 has been redrawn in Figure 6.29 to show schematically how we can readily compensate for the existence of a capacitive background current by extrapolating the current from more negative potentials. In practice, we measure the peak current by drawing a vertical line from the tip of the peak to where it strikes the extraplant.

Consideration of stirring. In stripping voltammetry, it is normal to employ a stationary electrode and a solution which is gently stirred. An alternative method is to have a still solution and an electrode that is rotated. If the solution is stirred, then the rate of stirring should be reproducible and controlled. Exhaustive electrolysis can be performed without stirring but the time required for deposition is likely to be quite long.

In fact, in most modern analyses, the only time an unstirred solution is employed is when coulometry is performed in combination with a differential pulse voltammetry procedure (see Section 6.5).

Consideration of deposition time. Sometimes, it is not possible to completely exhaust the solution, for example if there is insufficient time for such a procedure. If complete depletion of an analyte solution is necessary and time is short, then a small portion of the liquid will be taken and this will then be depleted.

Such a pre-concentration by stripping (as a means of exhaustive electrolysis) can achieve a lower concentration limit of about 10^{-10}, or even 10^{-11}, mol dm^{-3} if performed carefully. It is more common, however, to remove only a small portion of the analyte from solution – say, about 2%.

DQ 6.34

If only a portion of the solution is removed (c. 2%), then how do we know what fraction has been removed, i.e. how is the concentration determined?

Answer

The choice of 'about 2%' is made to ensure that the composition of the analyte solution remains effectively constant. A constant-composition solution is useful in inhibiting side reactions since, commonly, the reaction(s) at the mercury drop vary with solution composition. In particular, the electrode reaction alters as depletion of the solution approaches completeness, i.e. as all cations of analyte are removed from solution.

A complete *removal of analyte from solution is not necessary if the conditions during pre-concentration are held constant; so, by careful choice of calibration and fixed deposition times, the measured voltammetric response (i.e. the ASV peak height) is still a direct measure of analyte concentration.*

Derived methods. A mercury-film electrode (MFE) is superior to an HMDE because stirring of the solution can be performed much more vigorously, thereby enhancing the efficiency of the mass transport of analyte to the electrode. Faster stirring is allowed because there is no longer the chance that the mercury drop will be displaced and thus lost by the solution's movement. The usual substrate employed for the mercury film is graphite.†

Alternatively, a electrode of bare graphite can be employed, which is particularly useful for the determination of mercury ions (for which, mercury itself is wholly unsuitable). While graphite will clearly not form an amalgam, it does have the major advantage that a very wide range of potentials are available before electrochemical 'splitting' of any water-containing electrolyte will occur; metallic platinum or gold do not have particularly large cathodic ranges in aqueous solutions.

Nevertheless, the quantitative sensitivity when using ASV is of the order of 10^{-10}, or even 10^{-11}, mol dm^{-3} of analyte. This improvement over other coulometric methods arises solely from the pre-concentration by a factor of up to 1000 times, thereby decreasing the relative magnitude of charging currents.

6.7 The Glucose Sensor: a Worked Example of Voltammetric Analysis

The glucose sensor is one of the best known everyday applications of the electroanalytical ideas we have met in this chapter. Such sensors provide a fast and reliable determination of the glucose concentration in blood. People with diabetes frequently require such assays, and pathology laboratories in hospitals and clinics also need these data, so analyses of this type need to be quick, cheap and above all, accurate and reliable.

---

† The metallic mercury is deposited electrochemically from an aqueous solution of Hg(II).

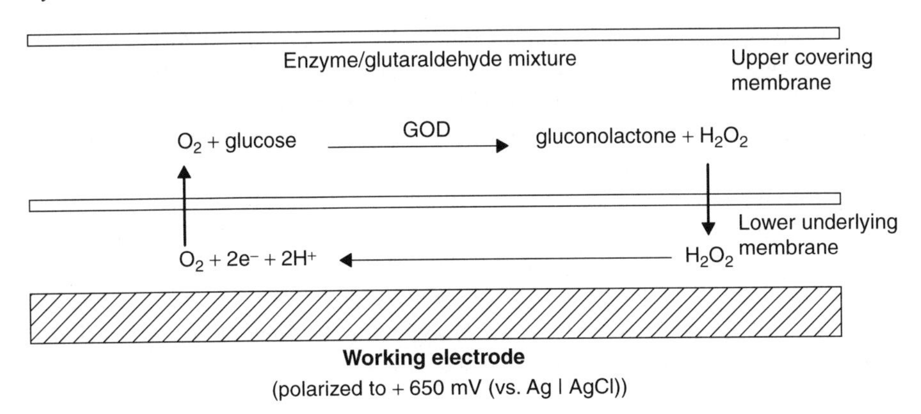

Figure 6.30 Schematic representation of a glucose sensor operating by diffusion across a perm-selective membrane (as represented by the vertical arrows); 'GOD' is glucose oxidase.

There are two broad classes of glucose sensor commonly encountered: the first employs a mediator (see Chapter 5), while the second employs a perm-selective membrane.

Sensors for hospital use: glucose sensor operating via a membrane. Figure 6.30 shows a cutaway drawing of a glucose sensor based on a membrane-type operation. An outer biocompatible membrane forms the outer surface of the sensor and represents the interface between the sensor and the analyte sample. A sample is placed on the biocompatible membrane, and material passes through into the working interior of the sensor where a current I is generated, according to the following equations:

$$\text{glucose} + O_2 \xrightarrow{\text{glucose oxidase}} \text{gluconic acid} + H_2O_2 \tag{6.16}$$

$$H_2O_2 \xrightarrow[\text{(vs. SSCE)}]{+650\text{ mV}} 2H^+ + 2e^- + O_2 \tag{6.17}$$

The current I from this second reaction is proportional to c_{glucose} over limited concentration ranges; generally, the concentration of glucose in the sample is determined from a calibration curve.

Below the outer membrane is a filter, usually composed of an anionic polymer, e.g. based on salicylate. Its precise composition and dimensions (thickness, pore size, amount and type of plasticizer, fillers, etc.) are optimized in order to 'tailor' the diffusion rates of material crossing the filter from the analyte solution toward the working electrode of the sensor. Ideally, some uncharged molecules, such as H_2O_2, will traverse the filter so fast that, in effect, the filter is 'invisible' to them. Conversely, we want other types of ion to traverse the filter so slowly that they can be said to be *blocked*. In effect, we want a filter with variable permeability – we term these **perm-selective membranes**.

The filter blocks contaminant anions such as ascorbate or urate, but uncharged molecules such as glucose are relatively unhindered as they cross into the sensor interior. Unfortunately, neutral (unionized) paracetamol will also cross the membrane, so diabetics should not use this analgesic at all. Once inside the sensor, such molecules are oxidized, e.g. with electrogenerated hydrogen peroxide (neutral H_2O_2 is formed at about 650 mV with respect to an internal silver | silver chloride reference electrode). The current generated is related to c_{glucose}, with the values being obtained from a calibration graph.

Sensors for home use: glucose sensor operating with a mediator. In the second type of sensor, the amount of glucose is monitored amperometrically via the current. At the heart of the assay is enzymatic oxidation of glucose by glucose oxidase (GOD). The enzyme itself is not electroactive, so a mediator will oxidize it. The mediator of choice is ferrocene ($Fe(cp)_2$), which can readily oxidize to form a stable radical cation ($Fe(cp)_2^{+\bullet}$) (see Figure 6.31).

Because the oxidation of an analyte is effected with a mediator, the potential of the working electrode need not be as anodic as the potential needed for the first type of sensor, implying that anions such as ascorbate (see above) will remain unnoticed as *electro-inactive* contaminants. By remaining unoxidized, the magnitude of the non-faradaic current is kept to a minimum.

For patients at home, these sensors have been marketed (by Medisense) to look like a typical credit card; a sample of blood is placed on an uncovered portion on the flat of one side, and a readout obtained when the sensor is plugged into a portable display unit. This type of sensor can yield an accurate determination of [glucose] in the concentration range 0–25 mmol dm^{-3} within about 20–30 s.

Alternatively, mediator-based sensors have been marketed as 'Glucose Pens' – *The Medisense Pen* – in which the pen's 'nib' is in fact attached to the external part of the sensor. With this sensor, the finger of a patient is pricked and a drop of blood is then placed on the sensor tip. These electroanalytical

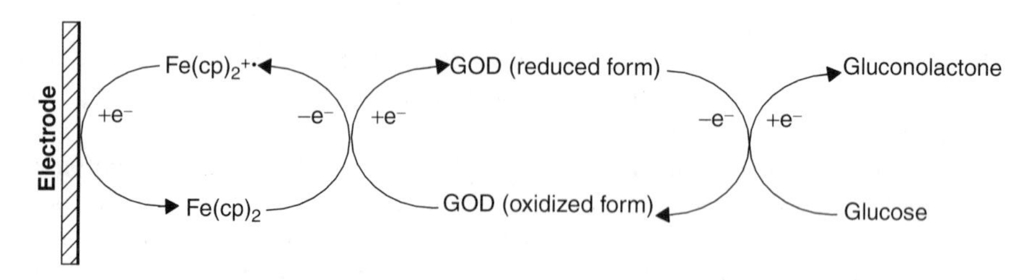

Figure 6.31 Schematic representation of a glucose sensor operating with ferrocene to mediate the redox chemistry of glucose oxidase (GOD).

devices can now be made so cheaply that they can be disposed of following each analysis.

6.8 Causes and Treatment of Errors

6.8.1 Polarographic 'Peaks'

We saw above that the polarographic current rises from zero to a current *plateau*. The plateau may be horizontal, or it might be gently sloping upwards: we called this rise a *residual current*. Occasionally, there is also a current *peak* superimposed on the wave (see Figure 6.32). Such peaks are of two types, i.e. maxima of the first kind and maxima of the second kind. Both are caused by enhanced rates of mass transport at the Hg | solution interface, as described in the following.

A current **maximum of the first kind** has the form of a sharp, straight line which starts to form just before the main polarographic wave (curve 'a' in Figure 6.32). Such a maximum can be considerably larger than the wave itself, although it will usually drop suddenly back to the 'normal' wave. Maxima of the first kind are caused by convective effects, as electrolyte flows past the surface of the mercury drop, resulting from surface tension differences at various points on the surface of the drop.

A current **maximum of the second kind** has the form of a relatively small, rounded peak at the start of the plateau (curve 'b' in Figure 6.32). These peaks are observed in solutions of high ionic strength, and tend to be more common if the rate of mercury flow is fast. While their cause is still debated, it is likely

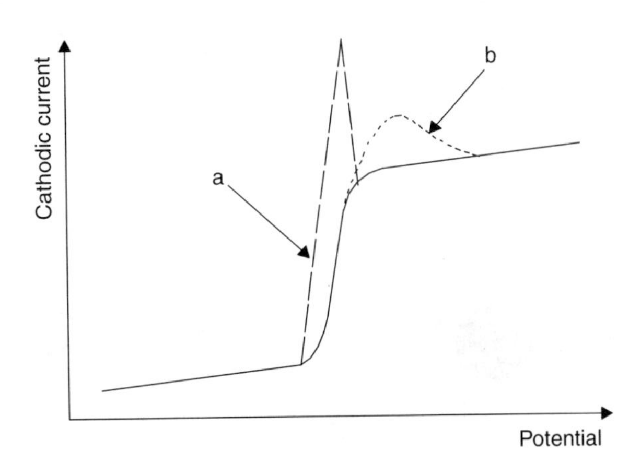

Figure 6.32 Illustration of polarographic current maxima, where the continuous line is the correct, undistorted polarograph (drawn as the mean current, i.e. without the 'sawtoothed' effect of drop replenishment), and the dashed and dotted lines represent current maxima of the first and second kinds, respectively.

that they are a feature of electrolyte movement around the drop as it forms, i.e. also a convective effect.

The magnitudes of these maxima can be decreased almost completely by adding a surfactant to the analyte solution – we call such an additive a **current maximum suppresser** (or, sometimes, a **depolarizer**). The usual suppresser employed is Triton X-100 (a non-ionic surfactant (detergent)), which is added to the solutions at concentrations of no more than 0.002 mol dm^{-3}.

6.8.2 IR *Drop: the Luggin Capillary*

We first met the concept of IR drop in Chapter 3 (Section 3.6.4). IR drop is also a cause of error in dynamic electroanalysis. The analyte solution has an electronic resistance of R, and a potential E is applied between the reference and working electrodes, so an additional current I is induced. The magnitude of such an ohmic current I is E/R, according to Ohm's law (equation (2.2)).

All of the electroanalytical techniques described in this present chapter have made use of the general relationship, 'faradaic current $\propto$ analyte concentration', according to Faraday's laws. It is therefore important that such non-faradaic currents be minimized. First, the resistance of the solution can be minimized by adding an inert electrolyte to the solution in swamping concentration. (Adding a swamping electrolyte also decreases the extent of mass transport by migration.)

The other means of decreasing R is to minimize the distance d between the working and reference electrodes, R being directly proportional to d.

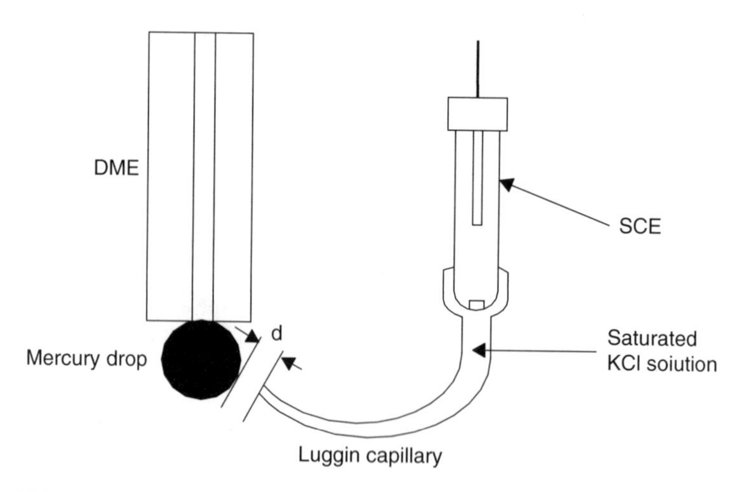

Figure 6.33 Schematic representation of a Luggin capillary used for minimizing IR drop. Calculations of uncompensated solution resistance require a knowledge of the distance d between the reference tip of the capillary and the working electrode (depicted here as a dropping-mercury electrode (DME)).

DQ 6.35

Won't a reference electrode (RE) impede the diffusion pathways if placed directly in front of the working electrode?

Answer

Placing the RE directly *in front of the working electrode (WE) does indeed impede the diffusion of analyte toward the latter. Conversely, if the RE is placed very close to the WE, but* beside *it rather than in front of it, then* d *can be decreased without disrupting the flux.*

An alternative, and usually superior, method involves using a **Luggin capillary** (see Figure 6.33). At heart, a Luggin capillary is a reference electrode connected to a tube filled with KCl solution at high concentration, so the resistance of the internal solution is tiny. In addition, the capillary is slender in order to minimize disruption of the flux at the electrode surface, and its tip is narrow to ensure that none of the internal KCl solution seeps into the analyte solution.

In practice, the tip of the capillary is placed as close to the working electrode as possible. There are seen to be two portions of solution between the WE and RE, i.e. analyte solution between the electrode and the capillary tip and KCl solution within the capillary itself. The resistance of the internal KCl solution is negligible, so only the short length of analyte solution is deemed to have a resistance – and only a few mm separate the tip and the surface of the WE.

IR drop cannot be excluded, but the use of a Luggin capillary can significantly decrease it.

Summary

Voltammetry and polarography are dynamic electroanalytical techniques, that is, current flows. A three-electrode cell is needed to allow accurate and simultaneous determination of current and potential. The electrode of interest is the working electrode, with the other two being the reference and counter electrodes.

Voltammetry and polarography are performed under diffusion control, which is ensured by keeping the solution still, and using an excess of inert electrolyte. The latter also has the effect of equalizing all activity coefficients in solution, so values of *concentration*, rather than activity, may be derived during measurements.

Solutions for polarographic analysis should contain a surfactant in low concentration to act as a current maximum suppresser, and all traces of oxygen should be removed.

The current at a dropping-mercury electrode (DME) is a function of potential, as described by the Heyrovsky–Ilkovic equation (equation (6.6)). At less extreme overpotentials, the current rises from zero to a maximum. The potential at which the current has reached half its maximum value is termed the half-wave potential, $E_{1/2}$, the value of which is characteristic of the analyte.

At extreme overpotentials, the current is independent of potential. This maximum current is said to be limiting, that is, current $\propto c_{bulk}$. It is termed the *diffusion* current, I_d. The dependence of I_d on concentration, drop speed, etc., is described by the Ilkovic equation (equation (6.5)), although calibration graphs or standard addition methods (Gran plots) are preferred for more accurate analyses.

In voltammetry, the electrode is a solid conductor. The surface of the electrode is not refreshed constantly as it is for a DME, so voltammograms do not have a 'sawtoothed' shape, but are smooth. Rather than a current *plateau*, I_d, voltammograms contain a *peak* current, I_p, with the magnitude of the peak being directly proportional to the bulk concentration of analyte, according to the Randles–Sevčik equation (equation (6.13)).

Cyclic voltammetry is a powerful tool for following mechanisms since varying the scan rate ν is equivalent to varying the time-scale of observation, τ. In order to obtain the rate constant k of the homogeneous 'C' reaction, the CV is obtained as a function of the scan rate, with k of the reaction then being determined from working curves calculated from theoretical principles.

Cyclic voltammetry is also a powerful method for quantitative determinations since the height of the CV peak, I_p, is proportional to the bulk concentration of the analyte. The electroanalytical sensitivity limit is about 10^{-5} mol dm^{-3} when using normal cyclic voltammetry.

Alternative voltammetric methods that improve the sensitivity of voltammetry as an electroanalytical tool are normal pulse voltammetry (with a lower detection limit of 10^{-8} mol dm^{-3}), differential pulse voltammetry (with a detection limit of 10^{-7}–10^{-8} mol dm^{-3}) and square-wave pulse voltammetry (with a detection limit which is perhaps as low as 10^{-8} mol dm^{-3}).

The voltammetric sensitivity can be improved further by analyte pre-concentration in conjunction with stripping analyses (cf. Chapter 5). Anodic stripping voltammetry (ASV) (Section 6.5) is the best known of the stripping techniques, and is capable of detecting concentrations as low as 10^{-11} mol dm^{-3}. Differential pulse voltammetry, when applied to stripping, can further improve the accuracy of electroanalytical measurement and, in principle, further improve the sensitivity of the technique.

The most serious causes of error are (i) wave and peak distortion caused by excessively fast scan rates, which are themselves caused by diffusion being an inefficient mass transport mechanism, (ii) current maxima caused by convective effects as the mercury drop forms and then grows, and (iii) IR drop, i.e. the resistance of the solution being non-zero. Other causes of error can be minimized by careful experimental design.

Chapter 7

Analysis by Dynamic Measurement, B: Systems under Convection Control

Learning Objectives

- To appreciate that convection-based systems can only be usefully employed if the movement of the analyte solution is reproducible over the face of the electrode.
- To learn that laminar flow is reproducible, while turbulent flow cannot be modelled, and thus the latter leads to irreproducible results.
- To find that the limiting current at a rotated disc electrode (RDE) is directly proportional to the concentration of analyte, according to the Levich equation.
- To appreciate that there are two commonly employed ways of expressing a frequency of rotation, i.e. both angular and linear (ω and f, respectively), and that the Levich equation is formulated in terms of angular frequency.
- To appreciate that, experimentally, the best way to perform analyses at the rotated disc electrode (the most popular hydrodynamic electrode) is at a constant rotational frequency and with the face of the disc well below the surface of the liquid.
- To understand why the counter electrode should be larger than the working electrode, and to learn where to place electrodes in the voltammetry solution in order to avoid eddy currents.
- To appreciate that rotated electrodes are best employed in tandem with calibration curves which have been constructed for known analyte concentrations, and obtained at rotation speeds previously shown to allow only laminar flow.

- To appreciate that deviation from the Levich equation is likely to stem from non-limiting currents (the overpotential η is not extreme enough), breakdown of mass transport (η is too extreme) and turbulent flow.
- To realize that with *stationary* electrodes, the relationship between the limiting current, I_{lim}, and the rate of solution flow, V_f, will depend on whether a flow cell or a channel electrode is employed.
- To realize that many analysts employ an empirical approach by saying $I_{lim} \propto c_{analyte}$ at fixed flow rates.
- To find that calibration graphs are a superior way of maintaining laminar flow, e.g. graphs of limiting current I_{lim} vs. $c_{analyte}$ as a function of flow rate or rotation speed.
- To learn what a wall-jet electrode is, and understand why it can be employed for electroanalyses despite operating with a turbulent flow.
- To learn that the rotated ring-disc electrode (RRDE) is one of the most powerful analytical tools for following the kinetics of fast homogeneous reactions.
- To appreciate that during RRDE analyses, one reagent is in the solution while the other is electrogenerated at the disc electrode, and that the proportion of the electro generated reagent that remains after reaction is monitored at the ring electrode, and is a function of ω and k_2.
- To learn that current is exponentially related to the applied potential, according to the Tafel and Butler–Volmer equations.
- To understand why a Tafel plot of log $|I|$ against η has its specific shape, and learn that the central slopes of such plots allow kinetic data to be obtained.
- To learn how to calculate the rate constant of electron transfer, k_{et}, from a Tafel plot of log $|i_0|$ (as 'y') against overpotential η (as 'x').
- To learn that when the rate of electron transfer is slow, a useful approach is to construct Koutecky–Levich plots of 1/(current) against $1/\omega$, taking the intercepts of the lines as a function of potential, and then drawing the Tafel plot from these mass-transport-limiting values of the current.

7.1 Introduction to Convective Systems

In the previous chapter, we discussed *dynamic* electroanalytical techniques such as polarography and voltammetry. Each technique in that chapter was similar insofar as the principal mode of mass transport was diffusion. Mass transport by migration was minimized by adding an inert ionic salt to the electroanalysis sample and convection was wholly eliminated by keeping the solution still ('quiescent').[†]

[†] Some small extent of natural convection is always inevitable as caused, e.g. by warming of the solution around the current-bearing electrodes. The extent of this convection can be assumed to be negligible by comparison with

In this present chapter, we revisit dynamic electroanalytical techniques, that is, ones where the concentrations of analyte change at the electrode in accompaniment with current flow, but will now look at those methods in which **convection** is the principal mode of mass transport.

As in the previous chapter, mass transport by migration can be assumed to be absent (or at least minimized) by the prior addition of an inert ionic salt to the electroanalysis solution.

Mass transport by diffusion can never be totally eliminated if there are differences in concentration throughout the solution (e.g. as caused by current flow), but convection is such an efficient form of mass transport, when compared with either migration or diffusion, that it is safe to assume that diffusion is quite negligible in comparison.

In the electroanalytical context discussed here, 'convection' will be defined as 'mass transport occurring with the movement of solution relative to the face of an electrode'. Within this corpus, there are two extremes, as follows:

(i) movement of an electrode through an otherwise still solution;
(ii) movement of solution past a stationary electrode.

We will consider both of these.

Both (i) and (ii) rely on convection. Although different in matters of experimental detail, it is now common to see the techniques comprising these two classes being grouped together as employing **hydrodynamic electrodes**. The word 'hydrodynamic' means, literally, movement of water, since the terms deriving from the 'Greek' for motion and water are *dynamos* and *hydra*, respectively. The term 'hydrodynamic' is usually taken to mean any convection-based system. Nevertheless, water is usually the principal choice of solvent in electroanalytical measurements, so all of the systems we will discuss below can safely be considered to be hydrodynamic in the very strictest sense.

7.2 The Rotated Disc Electrode

7.2.1 Discussion of the Experiment

The **rotated disc electrode** (RDE) is one of the most commonly employed hydrodynamic electrodes. Figure 7.1 shows a schematic representation of a typical RDE. The electrode itself is a flat, circular disc of metal, graphite or an other conductor, and has a radius of r; its area A, therefore, is straightforwardly πr^2. The disc is embedded centrally into one flat end of a cylinder of an insulatory material such as Teflon or epoxy resin. Behind the face of the electrode is an

---

forced convection except at the highest currents. Convection cannot be ignored if gas is generated at the electrode, thereby causing movement of solution during movement to the solution surface.

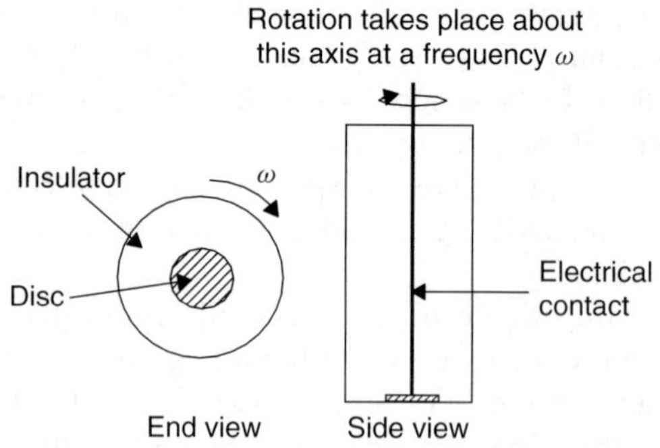

Rotation takes place about this axis at a frequency ω

Insulator

ω

Disc

Electrical contact

End view

Side view

Figure 7.1 Schematic representation of a rotated disc electrode (RDE) of radius r.

electrical connection leading ultimately to the potentiostat which controls the potential of the electrode (cf. Section 6.1).

During measurement, the RDE is immersed in the analyte solution. The insulator extends beyond the edge of the flat disc electrode for a further 5–8 mm or so. This additional distance between the electrode and the edge of the cylinder shaft is needed in order to minimize the effects of eddy currents (see Section 7.5 below).

DQ 7.1

How is the rotated disc electrode employed during analyses?

Answer

The basis of any RDE analysis continues to be the three-electrode voltammetry cell we first met in Figure 6.1, with the RDE as the working electrode (WE). A large-area platinum electrode is a good choice of counter electrode (CE) for most analyses, and a saturated calomel electrode is usually a suitable choice of reference electrode (RE). Both the RE and CE are kept stationary. Whatever the reference, it should be placed within a Luggin capillary (see Section 6.8.2) if at all possible. As with the systems discussed in the previous chapter, all traces of oxygen should be removed from solution prior to the analysis, e.g. by sparging with de-oxygenated nitrogen for a few minutes.

The RDE is immersed in the solution of analyte, with the face of the disc immersed – by at least 10 mm – below the surface of the liquid, again to minimize eddy currents. If possible, the disc should also be 30 or 40 mm above the floor of the electrochemical cell in order to maintain a good and reproducible flow of solution over the face of the disc.

The disc is spun at a fixed, known frequency about its central axis, and the current through the disc is then measured as a function of the potential applied (expressed as an overpotential η) and rotation speed ω.

DQ 7.2

Why was a *large-area* counter electrode mentioned in the previous discussion question?

Answer

Convection is a highly efficient form of mass transport, so the flux of analyte at the surface of the RDE can be very high. In addition, the area of the RDE is often greater than the area of, e.g. the mercury drop at the centre of a polarographic determination, again causing the currents through the working electrode to be higher than those encountered in the previous chapter – in fact, I *can be as large as a tenth of an ampere.*

We also recall that the working and counter electrodes will pass the same current, albeit with one as an oxidative current while the other is reductive. By corollary, if I_{RDE} *is large, then* $I_{counter\ electrode}$ *will also be large.*

We need to remember from earlier (equation (1.1)) that the current density i *relates to the amount of current passing through an electrode per unit area, i.e.* $i = I/A$. *In effect, equation (1.1) implies that if the RDE is large and the counter electrode is small, then the current density* i *through the counter electrode will be greater than that through the RDE working electrode. Moreover, it is the WE that we are interested in, so we cannot allow its performance to be inhibited if we want a reliable response.*

For any electrode, there is an intrinsic maximum current density it can pass – this follows simply because of the electrical conductivity of the metal from which the electrode is made and the maximum flux of analyte that can impinge on the electrode surface per unit time, itself a function of the rotation speed and solution viscosity and density. We see that it is quite possible for a small counter electrode to reach its maximum current density, and thus limit the overall current flowing through the working electrode, all because the counter electrode is smaller than the working electrode. In summary, we must always remember that I_{WE} *cannot exceed* I_{CE}*: they cannot be different in magnitude, only in terms of sign.*

In order to prevent such a limitation, it is advisable for the counter electrode to be as large as practicable, and certainly larger than the working electrode.[†] *With a large CE, the magnitude of the working-electrode current reflects only the flux of analyte solution, thereby obviating the kinetic limitations due to the counter electrode.*

[†] This experimental consideration should also be borne in mind when working with solutions under diffusion control, although the maximum current density is rarely reached since currents are so much smaller in the absence of convection.

In consequence, it is common practice with an RDE to have a CE made of platinum mesh, placed concentrically around the disc electrode. Care should be taken that the CE is not so close that it could cause additional eddy currents (see Section 7.5 below) near the face of the RDE.

SAQ 7.1

If the maximum current density through a working electrode is 0.1 A cm^{-2} and the current is 12 mA, what is the minimum area that the counter electrode can have?

SAQ 7.2

Following on from SAQ 7.1, if we want a current through the counter electrode of 12 mA and the area of the working electrode is 0.2 cm^2, will the maximum current density be surpassed?

DQ 7.3

Why are hydrodynamic electrodes employed at all if there are possible problems with the maximum current densities?

Answer

To reiterate, convective currents are generally much greater than those measured through electrodes of a similar area immersed in still solutions, thus resulting in convective currents generally being more accurate. In addition, the minimum sensitivity of an electroanalytical measurement during a convective analysis can often be greatly improved when compared to that obtained when rates of diffusion dictate the mass transport.

Secondly, because the solution is deliberately *stirred during a convective measurement, the analytical data are generally more reproducible than those obtained from diffusion-controlled measurements,*[†] *thereby improving the precision.*

DQ 7.4

How reproducible *is* the solution flow over the face of an RDE?

Answer

The schematic diagram in Figure 7.2(a) indicates the direction of solution flow over the face of the RDE during stirring, while the end view in

[†] The usual problem encountered with diffusion-controlled measurements is inadvertent movement of the solution, i.e. unwanted convection, which is unlikely to be laminar, as defined below.

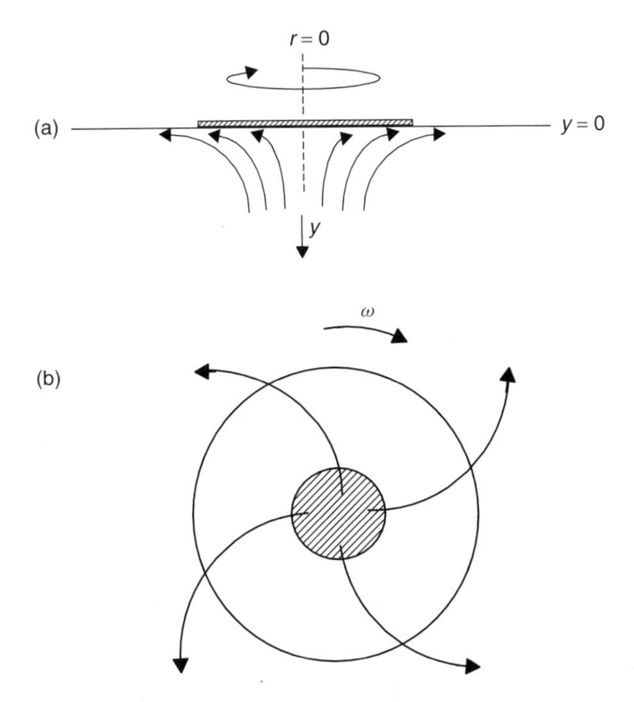

Figure 7.2 (a) Schematic representation of the fluid flow from the solution bulk and toward the face of an RDE ($y = 0$), where the disc is spun about $r = 0$ at a frequency ω. (b) End view showing the radial movement of solution over the face of the disc during rotation.

Figure 7.2(b) shows the radial movement of solution over the face of the disc during rotation; these lines of flow are sometimes called ***streamlines****.*

In order to explain the shape and relative directions of these streamlines, consider an infinitesimally small volume of solution sitting on the face of the RDE. As the electrode rotates, so centrifugal forces draw this volume increment away from its position on the disc face, out towards the rim of the cylinder and hence into the solution bulk. In fact, all of the solution on the face of the electrode is flung outward. Children will appreciate the 'flinging out' of solution in this way since they experience something very similar while sitting on a playground roundabout that is rotating very fast!

Movement of solution away from the disc in this way would cause a vacuum to form at the disc centre, and so more solution is drawn in, from the solution bulk towards the electrode face. In effect, the rotation of the disc induces a sort of 'pump' action, with solution continually being drawn from the bulk, travelling toward the surface of the disc electrode, over the disc and hence back into the solution bulk. It is this 'pump action' that we follow as the flux of analyte solution at the RDE.

The flow of solution past the central disc is remarkably reproducible provided that:

(i) the face of the RDE is well positioned in the solution (as discussed above);
(ii) the RDE spins concentrically about its central axis, i.e. it does not 'wobble';
(iii) the counter electrode is large and is not placed so close to the rim of the RDE that it disrupts the solution flow.

7.2.2 The Levich Equation

As analysts, we wish to know the relationship between the current at the RDE and the concentration of analyte. The problem was first solved by the Russian electrochemist **Veniamin Levich** in the 1940s, and so we will use the equation which now bears his name, as follows:

$$I_{\text{lim}} = 0.620nFAD^{2/3}\upsilon^{-1/6}\omega^{1/2}c_{\text{analyte}} \tag{7.1}$$

where the terms encountered in previous chapters retain their usual meanings and ω is the frequency with which the RDE is rotated about its axis. In this equation, υ is the **kinematic viscosity** of the solution (which can be obtained from suitable tables if we need it), and is defined as the ratio of the solution viscosity η_s and the density ρ, according to the following:

$$\upsilon = \frac{\eta_s}{\rho} \tag{7.2}$$

In order to maintain the constant of 0.620 in the Levich equation, the kinematic viscosity is expressed in units of $cm^3\ s^{-1}$.

SAQ 7.3

Calculate the kinematic viscosity υ of propylene carbonate (PC) at 25°C, given that $\eta_s = 2.82 \times 10^{-3}$ N s^{-1} and $\rho = 1.179$ g cm^{-3}.

Note from equation (7.1) that the Levich equation was derived in terms of electrochemical units, so we recall that c_{analyte} is expressed in mol cm^{-3}, A in cm^2 and D in $cm^2\ s^{-1}$. If we prefer other units then we must alter the constant of 0.620.

SAQ 7.4

An RDE of active area 1.2 cm^2 is spun at a frequency ω of 30 rad s^{-1}. Calculate I_{lim} from equation (7.1), given that $\upsilon = 0.01$ $cm^3\ s^{-1}$, $D = 2 \times 10^{-6}$ $cm^2\ s^{-1}$, and $c_{\text{analyte}} = 10$ mmol dm^{-3}, by taking $n = 1$. (Remember to convert the concentration into electrochemical units.)

Worked Example 7.1. A sample of soil, thought to be contaminated with a herbicide, was collected and all traces of the residual herbicide were removed by digesting in hot water. A sample of known concentration (10^{-3} mol dm^{-3} = 10^{-6} mol cm^{-3}) gave a limiting current, I_{lim}, of 3.12 mA. What is the concentration of the herbicide in the soil sample if the limiting current is found to be 0.442 mA?

Assuming that A, ω, D and υ remain unchanged, we can rewrite the Levich equation (equation (7.1)) as follows:

$$I_{lim} = kc_{analyte} \tag{7.3}$$

where k is merely a combination of the above constants.

Step 1. Calibration to determine k. From equation (7.2):

$$k = \frac{I_{lim}}{c_{analyte}} = 3.12 \times 10^3 \text{ mA mol}^{-1} \text{ cm}^3$$

Step 2. Determination of the analyte concentration. From equation (7.3):

$$c_{analyte} = \frac{I_{lim}}{k} = \frac{0.442 \text{ mA}}{3.12 \times 10^3 \text{ mA mol}^{-1} \text{ cm}^3} = 0.142 \times 10^{-3} \text{ mol cm}^{-3}$$
$$= 0.142 \text{ mol dm}^{-3}$$

SAQ 7.5

Another sample of soil was analysed at the same RDE, under the same stirring conditions, but in this case a current of 1.76 mA was recorded. What was the concentration of herbicide in the analyte sample?

It should be noted that, for speed of calculation, we could have taken ratios, i.e. by saying that $c_1/c_2 = I_{lim(1)}/I_{lim(2)}$, in our calculation.

SAQ 7.6

Show that you can obtain the same result for SAQ 7.4 by using ratios.

We should appreciate that a superior (although much more time-consuming) approach would be to start by constructing a calibration curve of I_{lim} (as 'y') against $c_{analyte}$ (as 'x') with known standards. The construction of a calibration curve is always a superior approach: the methodology in the worked example above assumed that currents were wholly faradaic, which might not have been true. If the non-faradaic current was of a significant magnitude, then the calibration curve will not pass through the origin.

For accurate work, we need to appreciate the following:

(1) For concentrated solutions, the kinematic viscosity υ may differ from values in (standard) tables, so we may need to determine its value for ourselves by using a viscometer (which is an easy process). Note, however, that the exponent of '$-1/6$' on υ in equation (7.1) means that most errors are likely to be extremely small.

(2) There are two common ways of expressing a frequency, namely the **angular frequency** ω in radians per second, and the **linear frequency** f (with the SI unit of hertz (Hz)) in cycles per second. Note that f is often cited as cycles per second (c s^{-1}), which is another way of saying Hz.

The relationship between f and ω is as follows:

$$2\pi f = \omega \tag{7.4}$$

We should bear in mind that the Levich equation (equation (7.1)) was derived in terms of *angular* frequency, with the constant of 0.620 in this equation presupposing its continued use.

Worked Example 7.2. What is the angular frequency of an RDE rotated at 750 rpm?

The somewhat old-fashioned unit of rpm stands for 'revolutions per minute', so f for 750 rpm is:

$$\frac{750 \text{ rpm}}{60 \text{ seconds per minute}} = 12.5 \text{ Hz}$$

From equation (7.4):

$$\begin{aligned}\omega &= 2 \times \pi \times f \\ &= 2 \times \pi \times 12.5 \text{ Hz} \\ &= 278.5 \text{ rad s}^{-1}\end{aligned}$$

SAQ 7.7

What is the linear frequency, *in rpm*, of an RDE spun at an angular frequency ω of 6 rad s^{-1}?

SAQ 7.8

Convert a rotation speed f of 12.0 Hz into an angular frequency ω expressed in radians per second.

(3) Thirdly, we need to appreciate how the current term in the Levich equation represents a *faradaic* current, and hence the stipulation that we remove all dissolved oxygen from the solution before our analyses commence. Furthermore, the current is a *limiting* one, so we will commonly perform a few sample experiments before the analysis (usually at fixed frequency) by slowly increasing the potential until a limiting current is reached.

Ramping the potential in this way means that we obtain a *voltammogram*. Figure 7.3 shows a series of voltammograms at an RDE, drawn as a function of the rotation speed.

We should note from this figure that a voltammogram obtained at an RDE looks wholly different from a voltammogram (of the same analyte) obtained under diffusion control (cf. Figure 6.13). We should also appreciate that the voltammetric currents obtained at an RDE are much larger than the corresponding currents obtained in a quiet solution: this reflects the relative efficiencies of the different modes of mass transport (see Section 2.2 and the discussion above).

From the traces in Figure 7.3, it can be seen that a *limiting* oxidative current can only be obtained if $E_{RDE} > 0.4$ (V vs. SCE). The maximum current is a function of the applied potential, which means that we cannot employ the Levich equation if $E_{RDE} < 0.4$ (V vs. SCE).

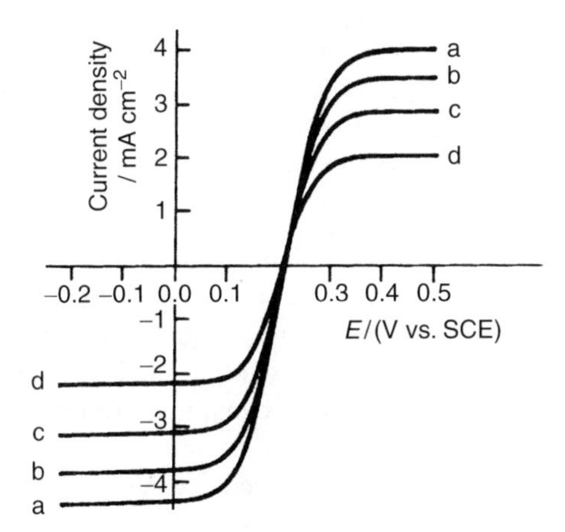

Figure 7.3 Voltammograms at a gold RDE, of current density i as a function of potential E (vs. SCE) and rotation speed f, obtained for a solution of ferrocyanide and ferricyanide (both at a concentration of 10 mmol dm^{-3}) in KCl (0.5 mol dm^{-3}): (a) 20; (b) 15; (c) 10; (d) 5 Hz.

In order to prevent such invalidation, we must produce voltammograms such as those shown in Figure 7.3 (each at constant f) and then determine which potential ranges allow the reliable use of the Levich equation at our RDE for each rotation speed.

Finally, when we know which potentials yield a truly limiting current, we can then determine I_{lim} as a function of concentration (at fixed f) with a series of samples of known concentration and use these data to construct a calibration graph, as mentioned above.

DQ 7.5

Equation (7.1) suggests that larger values of I_{lim} may be obtained by rotating the RDE at ever faster speeds. Is it valid to suppose that a faster ω will always lead to an improved signal-to-noise ratio?

Answer

We might want to increase ω to generate larger limiting currents in order to increase the precision. Improving the precision by this means is valid at low to medium rotation rates (that is, up to about 100 cycles per second) because the motion of solution over the face of the electrode is smooth and reproducible. We say that the flow is ***laminar****. (The word 'laminar' comes from the Latin root* lami*, meaning 'thin layer' or 'plate'.) We can then see how 'laminar flow' implies that solution readily flows over itself in a smooth and reproducible way.*

Above a certain rotation speed, the solution flow suddenly becomes ***turbulent****, and flow is irreproducible, because* ***eddy currents*** *and* ***vortices*** *form around the edges of the electrode.*[†] *Such eddies cannot be modelled, implying that the Levich equation breaks down, and becomes unusable. In other words, at such high rotation speeds, we find that I_{lim} is* not *a function of concentration and rotation speed, thus causing the RDE to be an inappropriate tool for electroanalytical measurements.*

An everyday example will readily prove the above point: a busy stream running over a rocky river bed is an example of turbulent flow, because the direction and magnitude of the flowing water is in a state of continual change (that is, on a microscopic level). This is why our stream makes a 'gurgling' sound, rather than a continuous note. Conversely, water flowing through a smooth pipe is likely to display laminar flow. Figure 7.4 shows schematic representations of both laminar and turbulent flow.

In order to ascertain which rotation speeds correspond to laminar and which to turbulent flow, we construct a **Levich plot** of I_{lim} (as 'y') against $\omega^{1/2}$ (as 'x') (see Figure 7.5). Laminar flow is indicated by the linear part of the plot at

[†] Care: these 'currents' relate to movement of solution rather than electrochemical currents caused by movement of charge.

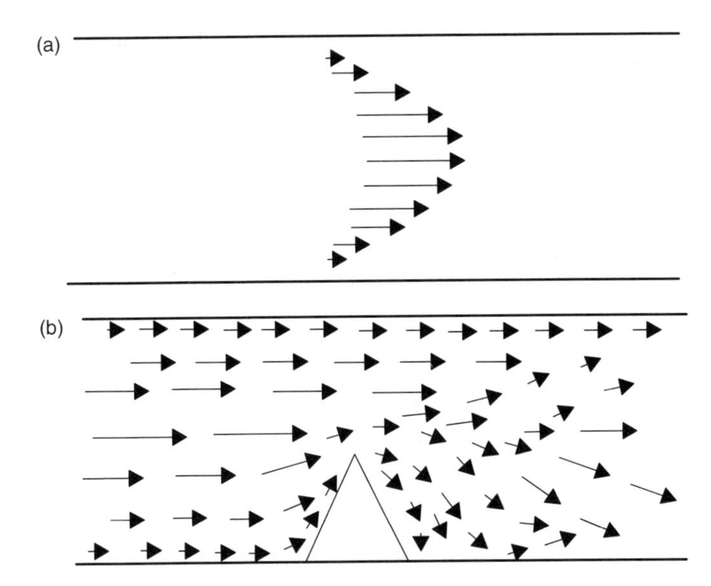

Figure 7.4 Representations of hydrodynamic flow, showing (a) laminar flow through a smooth pipe and (b) turbulent flow, e.g. as caused by an obstruction to movement in the pipe. The length of each arrow represents the velocity of the increment of solution. Notice in (a) how the flow front is curved (known as 'Poiseuille flow'), and in (b) how a solution can have both laminar and turbulent portions, with the greater pressure of solution flow adjacent to the obstruction.

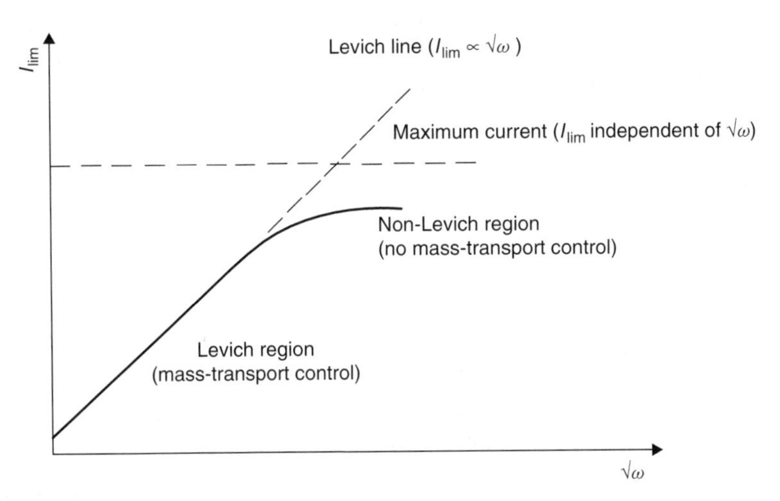

Figure 7.5 A Levich plot of I_{lim} against $\omega^{1/2}$ showing that mass transport at the highest rotation speeds can limit the magnitude of I_{lim}. Note that if the deviations from the Levich line at higher rotation speeds are caused by turbulent flow, then the I_{lim} data are reproducible only if obtained at rotation speeds represented by the linear (laminar) part of the graph.

lower rotation frequencies, while those frequencies which cause turbulent flow are shown by the non-linear and irreproducible portion at higher ω values.

SAQ 7.9

Construct a Levich plot using the data below in order to ascertain the maximum speed that the RDE can safely be rotated at before the onset of turbulent flow.

ω/Hz	20	40	60	80	100	120	140
I_{lim}/mA	9.8 ± 0.1	13.0 ± 0.1	17.0 ± 0.1	19.7 ± 0.1	22.0 ± 0.2	23.0 ± 0.4	23.4 ± 1.0

Looking at the data given in SAQ 7.9, we should have guessed that the flow was not wholly laminar at 100 Hz because the error in the measurement starts to increase, i.e. there is more scatter and 'noise'. With an RDE in front of us, we would have seen a vortex forming around the electrode, and a pen recording the current on an $X-Y$ plotter would be seen to judder and jump up and down in an unpredictable way.

In order to be sure of good quality analytical data, the analyst is recommended to construct a Levich plot such as that shown in Figure 7.5 before starting a series of analyses: analysts will avoid those rotation speeds represented on the graph as turbulent flow. In the jargon of the subject, the analyst may say that a 'laminar regime' is maintained.

The reason why the currents are smaller than those expected from the Levich equation is because turbulent flow results in the entrapment of air within the vortex around the electrode. In effect, the 'active' area (the area in contact with solution) of the electrode decreases in a random way.

DQ 7.6

Do Levich plots deviate from linearity for reasons other than poor flow characteristics?

Answer

Yes. There are two other reasons why a Levich plot is non-linear at higher rotation speeds, as follows:

(i) The rate of electron transfer is slow. We will return to this aspect when we look at slow electron transfer in Section 7.4.3.

(ii) The speed of rotation is so fast that we cannot bring sufficient analyte to the electrode per unit time. We say that 'mass transport has broken down'. In other words, such a Levich plot will look just like that shown in Figure 7.5, although it will be reproducible; *if* turbulent flow *has caused the deviation, then the trace would be* irreproducible.

DQ 7.7

Why does the Levich equation contain a *diffusion* coefficient D if the RDE is a system under *convective* control?

Answer

We said at the start of this chapter that convection is a much more efficient form of mass transport than diffusion, to the extent that diffusion can be ignored. For mass transport through the solution bulk this is unreservedly true, so movement of the analyte-containing solution to the electrode | solution interface is *controlled by convective flow.*

There is a complication, though: a thin (c. 10^{-3} cm) layer of solution exists between the electrode and the bulk solution that is relatively immobile. This forms because of the inherent viscous drag of the solution as it moves over the solid electrode. We call this thin film of immobile liquid the ***diffusion layer****, where the latter has a thickness δ. The thickness of the layer depends on the rotation speed according to the following:*

$$\delta = \frac{1.61\, \upsilon^{1/6} \mathrm{D}^{1/3}}{\mathrm{f}^{1/2}} \tag{7.5}$$

where the factor of 1.61 presupposes a length in cm. Note that the frequency is symbolized as 'f', so we should employ a linear *frequency (that is, in Hz).*

Since analyte cannot move across this layer – from the bulk solution to the electrode – by convection alone,[†] *it must diffuse, with the 'speed' of such diffusion being characterized by the diffusion coefficient* D.

SAQ 7.10

Ferrocene is dissolved in propylene carbonate solution (together with a suitable supporting electrolyte). The solution has a kinematic viscosity of $\upsilon = 0.239\ \mathrm{cm^3\ s^{-1}}$, and the diffusion coefficient of ferrocene is $3 \times 10^{-6}\ \mathrm{cm^2\ s^{-1}}$. Calculate the thickness of the diffusion layer at a frequency of 30 Hz.

SAQ 7.11

Using the data given in SAQ 7.10, how fast must the RDE rotate for the value of δ to decrease to 10^{-3} cm in thickness? In addition, do you think that this rotation speed is attainable in practice?

[†] For this reason, an RDE is very occasionally, although incorrectly, described as a 'diffusive system'.

7.3 Flow Cells, Channel Electrodes and Wall-Jet Electrodes

7.3.1 Flow Cells and Channel Electrodes

In Section 7.2, we looked at electroanalytical systems where the electrode rotates while the bulk of the solution remained still. In this present section, we will reverse this experimental concept by considering the case where it is the solution which flows – this time past a *stationary* electrode. Here, we shall be looking at **flow cells** and **channel electrodes**. The principal mode of mass transport in both cases is convection, since the solution moves *relative* to the electrode.

There is no absolute distinction in the literature between flow cells and channel electrodes. We shall say here that a flow cell contains a tubular electrode (often termed an **annulus**), while a channel electrode system contains a flat (or occasionally curved) electrode. Figure 7.6 shows a typical flow cell with an annular electrode. In contrast, the channel electrode illustrated in Figure 7.7 is flat and embedded inside a rectangular cavity.

For both flow cells and channel electrodes, we assume that the electrode is solid and absolutely immobile. Its surface is flush with the surrounding insulator in which it is embedded, thereby inhibiting the incidence of turbulent flow. In addition, the electrode is polished, again to prevent turbulence.

There are a great many equations available for describing the relationship between current and the flow of analyte solution past a stationary electrode. The

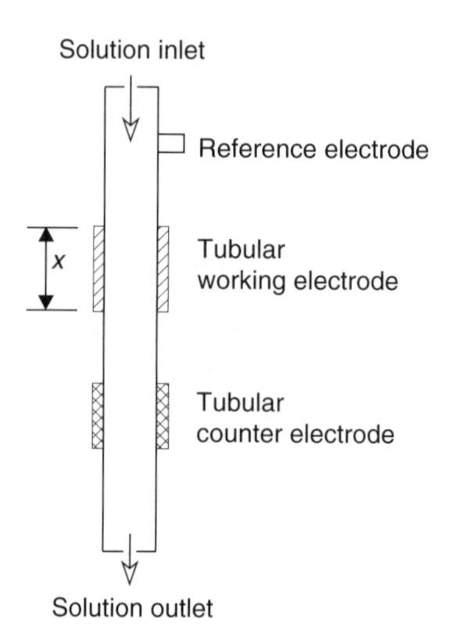

Figure 7.6 Schematic representation of a typical flow cell used for electroanalytical measurements. Note the way in which the counter electrode (CE) is positioned downsteam, i.e. the products from the CE flow *away* from the working electrode.

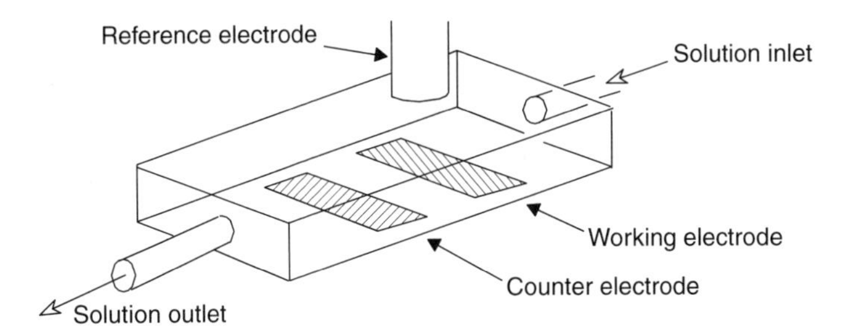

Figure 7.7 Schematic representation of a typical channel electrode system used for electroanalytical measurements. As with the flow cell show in Figure 7.6, note how the counter electrode (CE) is positioned downstream in order to stop the products from the CE flowing over the working electrode (WE); the reference electrode is positioned over the WE.

limiting current, I_{lim}, at each electrode depends on the kinematic viscosity of the solution, υ, the diffusion coefficient, D, and the concentration of analyte, c_{analyte}, in a manner akin to the Levich equation. Note, however, that I_{lim} also depends crucially on the rate of solution flow, V_{f}, over the face of the electrode.

Cell geometry. As with other voltammetric measurements, a minimum of three electrodes are required if reliable results are to be obtained. Ideally, the reference electrode is placed upstream of the working electrode, while the counter electrode is placed downstream.

DQ 7.8

Why does the positioning of the electrodes matter?

Answer

As we have seen many times before, if we want current to flow through the working electrode (WE), then we must also contend with a current of the same magnitude flowing through a counter electrode (CE). Accordingly, the CE will soon be surrounded by the products of electrolysis, which cannot be allowed to flow over the WE since they might contaminate it or 'interfere' with I_{lim} *in other ways. In order to ensure this criterion, we place the CE downstream, so products are washed* away *from the WE. In some cells, the CE is also placed within its own cell compartment, or be slightly recessed from the main flow duct.*

Current is not allowed to flow through the reference electrode (RE), so an RE will not yield any products to contaminate the WE. We place the RE upstream for convenience, although it could equally well be placed downstream.

Unfortunately, the problem of flow cells is more complicated than that of rotating disc electrodes, because we must also appreciate that I_{lim} depends on the cell geometry, i.e. whether the electrode is tubular, curved or flat, and if not tubular, whether its length is longer or shorter than its width, etc. The current is not proportional to the electrode area, implying that its surface is not uniformly accessible.

DQ 7.9

Why do we say that 'the surface is not uniformly accessible'?

Answer

By the time that the solution initially containing analyte reaches the far edge of the electrode, it has been depleted of its analyte content. Conversely, solution will contain analyte when it first flows over the front of the electrode (the so-called ***leading edge****). The concentration of analyte thus varies during the time of solution flow, which explains the mathematical complexity of the flow equations used to describe this situation.*

Furthermore, if the electrode is placed near the tube supplying the solution of analyte, then we also need to know if the tube is itself circular or flat. In addition, does the tube constrict or get wider as the electrode is reached? As a consequence of questions such as these, there are several equations that can be used, with each derived from first principles and each intended to describe $I_{\rm lim}$ in terms of the minutiae of detail concerning the cell geometry.

As a result of such factors, the number and detail of such equations can look intimidating. Since the derivation of any of these hydrodynamic equations is beyond the scope of this introductory text, we will not look at any of these in detail, but shall simplify the situation somewhat by noting that all have a common general form. The following equation will suffice as a workable example:

$$I_{\rm lim} = kc_{\rm analyte}\ V_{\rm f}^{y} \tag{7.6}$$

where $V_{\rm f}$ is the rate of solution flow (expressed as a volume per unit time, typically in $\rm cm^3\ s^{-1}$) and y is an exponent, the value of which depends on the cell design. The value of y is 1/3 for both flow cells and simple channel electrodes. The proportionality constant k incorporates all variables needed to describe the flow conditions through the cell, such as the dimensions of the cell, flow tubes and electrode, together with the diffusion coefficient D.

For the simplest **flow cells** in which the electrode is a simple annulus, the limiting current has the following form:

$$I_{\rm lim} = 5.43nFc_{\rm analyte}\ D^{2/3}\ x^{2/3}\ V_{\rm f}^{1/3} \tag{7.7}$$

Note that $I_{\rm lim}$ is independent of the solution viscosity υ, and x here is the length of the annulus (see Figure 7.6). Equation (7.7) is seen to be entirely consistent with the generalized flow law described by equation (7.6), since all variables are kept constant, thus allowing the conclusion that $I_{\text{lim}} \propto c_{\text{analyte}}$.

SAQ 7.12

Consider the analyte solution mentioned above in SAQ 7.4. For the same analyte, calculate the limiting current as analyte solution flows through a cell at a rate of $15.0 \text{ cm}^3 \text{ s}^{-1}$ (take $x = 1.20$ cm).

Worked Example 7.3. A redox-active dye is eluting from an HPLC column. Since the analyte is redox-active, the HPLC detector is unusual in that it consists of a small annulus of silver, mounted within a short Teflon tube. Eluent from the column contains analyte, trickling at a constant rate, V_{f}, through the cell and over the electrode while the current is monitored. It is assumed that the silver ring only 'sees' the redox-active dye, i.e. the current is wholly faradaic.

After half an hour of elution, the limiting current is 0.45 mA. Calculate the concentration of dye; during a previous experiment, the limiting current was 2.7 mA when a solution of concentration $10.0 \text{ mmol dm}^{-3}$ flowed through the cell.

Methodology. We will use equation (7.6). Even if we don't know the flow rate V_{f}, we *do* know a value of I_{lim} for a solution of known concentration, so we can simplify equation (7.6) by assuming that V_{f} and y stay constant.† We next write $I_{\text{lim}} = k' c_{\text{analyte}}$, where k' now incorporates the V_{f}^y term, and thus we can simplify equation (7.6) even further.

We first determine k' from the limiting current, I_{lim}, measured at the known concentration, and thereby calibrate the flow characteristics of the cell. Only then can we determine the unknown concentration of the dye.

Step 1. Calibration to determine the proportionality constant k'.

$$k' = \frac{I_{\text{lim}}}{c_{\text{analyte}}} = 0.27 \text{ A mol}^{-1} \text{ dm}^3$$

Step 2. Determination of the analyte concentration. (I_{lim} as a function of V_{f}).

$$c_{\text{analyte}} = \frac{I_{\text{lim}}}{k'} = \frac{0.45 \text{ mA}}{0.27 \text{ A mol}^{-1} \text{ dm}^3} = 1.67 \text{ mmol dm}^{-3}$$

SAQ 7.4 and Worked Example 7.3 show how calculations of this type can be long-winded, boring and cumbersome. This explains why many analysts prefer to work with ratios of current.

† The exponent y will remain constant provided that we retain laminar flow and do not change the geometry of the cell.

An even better approach when employing the above simple relationship is to construct a calibration curve of I_{lim} (as 'y') against c_{analyte} (as 'x') with a series of known standards. (Note that we must ensure that the flow rate V_{f} remains constant, for example, by the use of a pump or small constant-head tank.)

In order to ensure that the solution flow is indeed laminar past a tubular electrode, we choose a solution of known and fixed concentration, and measure I_{lim} as a function of V_{f}. A plot of $\log(I_{\text{lim}})$ as 'y' against $\log(V_{\text{f}})$ as 'x' should be linear with a gradient of 1/3. Deviation from this line at fast flow rates indicates those flow rates that should be avoided. The analyst now knows the range of flow rates to keep within. If the solution flow is not completely laminar, then the flow rate is amended until it is – 'amend' in this context almost always means 'decrease ω'.

SAQ 7.13

Show why the above logarithmic analysis is valid.

SAQ 7.14

Continuing with the elution of the same dye from the HPLC column described in Worked Example 7.3, after a further 10 min, I_{lim} had decreased to 0.28 mA. What is the new concentration of dye in the eluent?

A calibration curve would be a safer method of determining c_{analyte}, since compensation for non-faradaic currents can then be made.

For a **channel electrode**, the relationship between limiting current, flow rate and cell geometry is given by the following:

$$I_{\text{lim}} = 1.165nFD^{2/3}\left(\frac{V_{\text{f}}}{h^2/d}\right)^{1/3} w\,X^{2/3}\,c_{\text{analyte}} \tag{7.8}$$

where X is the *length* of the electrode (measured along the same direction as solution flow) and w is the *width* of the electrode (measured perpendicular to the direction of solution flow); h and d are the internal height and width of the channel, respectively. All other terms have their usual meanings. We note that in common with equation (7.7) for a flow cell, the limiting current is independent of solution viscosity.

DQ 7.10

What are the advantages of using a flow cell or a channel electrode?

Answer

Flow cells or channel electrodes are ideal for the continuous monitoring of analyte, for example from an HPLC column or some other form of extraction system, since eluent flows over the electrode all of the time. The limiting current is directly proportional to the concentration of analyte, so a PC or lap-top computer can be interfaced to the current detector for monitoring I_{lim} *as a function of time.*

It would not be possible to employ a normal RDE cell in this manner. If we only had an RDE, we would need to empty the RDE cell and rinse it out between each measurement. The RDE is only capable of performing 'batch-mode' analyses. Furthermore, a few seconds are required at the start of each analysis to reach equilibrium as the solution adjusts to being stirred – there needs to be an equilibration time of c. 4 s between the start of stirring and attaining a constant rotation speed and to allow the transition from turbulent to laminar flow. In consequence, while a flow cell can be thought of as a steady-state monitoring tool, the RDE cannot be regarded as acting in this way.

DQ 7.11

So, what then are the advantages of analysing with an RDE?

Answer

Having given a few worthwhile advantages of performing an electroanalysis by using flow cells or channel electrodes, there are however, considerable experimental problems with such systems. First, maintenance can be difficult since the cell requires dismantling and then reassembly each time it is cleaned. Dismantling will be necessary if contamination of the working electrode is suspected, e.g. as caused by adsorption of analyte or secondary materials from the analyte solution.†

To add insult to injury, a new calibration curve will perhaps need to be constructed after each time the cell is dismantled if the internal geometry of the cell is particularly sensitive (cf. the in situ *EPR cells (discussed in the next chapter) can be very thin).*

From the discussion above, it is clear that the RDE has one major advantage over flow cells in being much easier to maintain. Conversely, flow cells are preferable if continuous monitoring is needed, e.g. from extraction systems or HPLC columns.

7.3.2 The Wall-Jet Electrode

The wall-jet electrode has the general configuration shown in Figure 7.8(a). When using this system, the current is measured while a fine jet or spray of analyte solution is squirted under relatively high pressure towards the centre of the working

† Adsorption is particularly prevalent with organic samples, such as those containing thiols, amines, phenols or aromatic moieties in general, so the flow cell might need to be cleaned after analysing the soil sample considered in Worked Example 7.1.

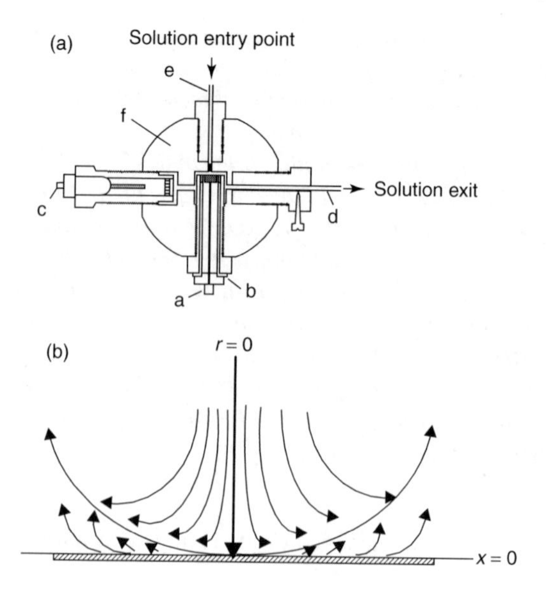

Figure 7.8 Schematic representation of a typical wall-jet electrode used for electroanalytical measurements: (a) contact to Pt disc electrode (the shaded portion at the centre of the figure); (b) contact to ring electrode; (c) AgCl | Ag reference electrode; (d) Pt tube counter electrode; (e) cell inlet; (f) cell body (made of an insulator such as Teflon). (b) A typical pattern of solution flow over the face of a wall-jet electrode, showing why 'splash back' does not occur. Part (a) reproduced from Brett, C. M. A. and Brett, A. M. O., *Electroanalysis*, 1998, © 1998, by permission of Oxford University Press.

electrode. Solution then flows away radially from the nozzle via the solution entry point (e). In order to prevent 'splash back', the diameter of the jet nozzle from which the spray emerges should be narrow (say, no more than 10% of the diameter of the working electrode). Furthermore, the distance that the analyte solution must travel between exit from the jet nozzle and before impinging on the working electrode is critical, i.e. it must not be either too short or too long.

The well-jet electrode can have many different configurations, but it is usually flat and circular. Figure 7.8(b) shows the pattern of solution flow during operation, from which it is easy to see why back splashes are carried away by the flow before they can reach the working electrode.

DQ 7.12

Surely, with such a spray-type operation, it will be extremely difficult to attain laminar flow in the wall-jet electrode?

Answer

The wall-jet electrode is unusual as the solution flow is turbulent at all times. Since this turbulence occurs in terms of fine droplets of analyte solution, and while the turbulence is undoubted, there is, however, an overall coherence in the solution dynamics. Although not laminar, the flow can *be described mathematically.*

The limiting current at a wall-jet electrode is a function of the radius of the circular electrode, r, *the rate of flow of solution,* V_f, *the diameter* a *of the jet supplying the solution of analyte, the diffusion coefficient* D *of the analyte and the bulk concentration of analyte,* $c_{analyte}$. I_{lim} *also depends, in a complicated way, on the distance between the electrode and the nozzle, which we will denote here as* k.

The limiting current at the wall-jet electrode is given by the following equation:

$$I_{lim} = 1.59\,\mathrm{knF}c_{analyte}\,D^{2/3}\,\upsilon^{-5/12}\,a^{-1/2}\,r^{3/4}\,V_f^{3/4} \tag{7.9}$$

where all other terms have their conventional meanings. The numerical constant represents a collection of constants. Note that both r *and* a *must have the units of cm when this constant is 1.59, and that solution flow is cited in units of* $cm^3\,s^{-1}$. *We will save ourselves much heartache by saying that equation (7.9) may be simplified to 'give' equation (7.6).*

Table 7.1 summarizes the various equations used to describe the convective systems discussed in this present chapter, relating the limiting currents and parameters such as flow rate.

SAQ 7.15

Amounts of Co^{3+} are being monitored at a wall-jet electrode. When a sample of concentration 3.23 μg cm^{-3} is squirted over the electrode, the limiting current is 152 μA. Keeping the flow rate and all other parameters constant, what is the concentration of a sample of Co^{3+} when the limiting current is 214 μA? Assume complete faradaic efficiency in both cases.

As always, a superior method would have been to construct a calibration curve, and read off the values of $c_{analyte}$ since we are more likely to discern non-faradaic currents in this way. It is also advisable to start any series of analyses by varying the flow rate over as wide a range as possible in order to observe those flow 'regimes' which are reproducible and can therefore be employed, and those which should be avoided.

DQ 7.13

What are the special advantages of employing a wall-jet electrode?

Answer

Like the flow and channel electrodes described above, the wall-jet electrode is not a batch-mode system, so it can be employed as the basis for

Table 7.1 Relationships between the limiting current and various convective parameters for a number of electrode types[a]

System	Equation
Rotated disc electrode[b,c]	$I_{\text{lim}} = 0.620nFAD^{2/3}\upsilon^{-1/6}\omega^{1/2}c_{\text{analyte}}$
Flow cell with a tubular electrode[d]	$I_{\text{lim}} = 5.43nFD^{2/3}x^{2/3}V_{\text{f}}^{1/3}c_{\text{analyte}}$
Flat channel electrode[e,f,g,h]	$I_{\text{lim}} = 1.165nFD^{2/3}\left(\frac{V_{\text{f}}}{h^2/d}\right)^{1/3}wX^{2/3}c_{\text{analyte}}$
Wall-jet electrode[i,j]	$I_{\text{lim}} = 1.59knFD^{2/3}\upsilon^{-5/12}a^{-1/2}r^{3/4}V_{\text{f}}^{3/4}c_{\text{analyte}}$

[a] c_{analyte}, concentration of analyte; n, number of electrons in electrode reaction; V_{f}, flow rate; F, Faraday constant; D, diffusion coefficient; A, area of electrode.
[b] υ, kinematic viscosity of the solvent.
[c] ω, angular frequency (rad s^{-1})
[d] x, length of the annular electrode, in the same direction as solution flow.
[e] w, width of the electrode, perpendicular to the direction of solution flow.
[f] X, length of the channel electrode, in the same direction as solution flow.
[g] h, internal height of the channel.
[h] d, internal width of the channel.
[i] k, a combination of constants, including υ.
[j] a, diameter of the jet.

continuous monitoring, provided that the electrode surface remains free of contamination. In addition, it is generally easier to clean a wall-jet electrode than to clean an electrode encased within a flow cell.

7.4 The Rotated Ring-Disc Electrode

Related to the rotated disc electrode (RDE) is the rotated ring-disc electrode (RRDE). Such an electrode is illustrated in Figure 7.9 and is seen to be, in effect, a modified RDE, insofar as the central disc is surrounded with a concentric ring electrode. The gap between the ring and the disc is filled with an insulator such as Teflon or epoxy resin. The face of the RRDE is polished flat in order to prevent viscous drag, which is itself likely to cause the induction of eddy currents.

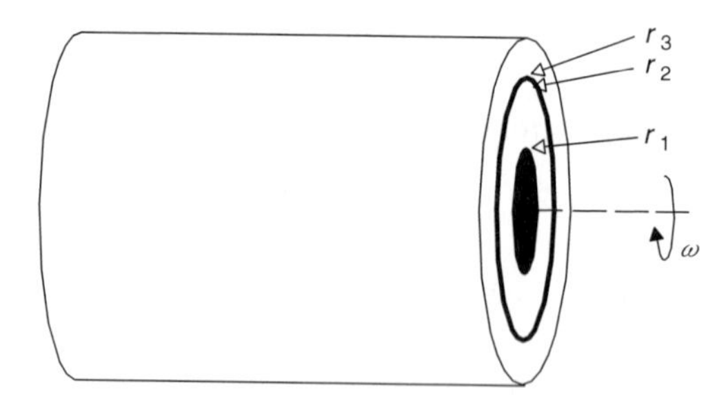

Figure 7.9 Schematic representation of a rotated ring-disc electrode, defining the radii r_1 (the radius of the disc), and r_2 and r_3 (the inner and outer radii of the ring, respectively).

DQ 7.14

What is the experimental advantage of operating with a ring as well as a disc?

Answer

The RRDE is one of the very best methods of determining rate constants (k) *for fast reactions in solution.*

This type of electrode is a particularly powerful analytical tool since by performing steady-state *measurements alone, it can measure faster rate constants than any other method. For a second-order reaction,*† *the RRDE can reliably and reproducibly determine rate constants as fast as* 10^9 *(mol* $dm^{-3})^{-1}$ s^{-1}, *while the maximum first-order rate constant measurable with the RRDE is about* 10^5 s^{-1}. *A further advantage of the RRDE is the way that* steady-state *currents are measured (see below), whereas other methods of determining such high values of* k *require the measurement of transients.*

DQ 7.15

How does the RRDE measure rates?

Answer

We will answer this question by first thinking about a typical undergraduate experiment designed to follow reaction kinetics. Imagine that we want to hydrolyse an ester with sodium hydroxide solution. At the start of the experiment, the extent of reaction, ξ, *is zero, so the concentration of NaOH is equal to its initial value.*

We start the reaction, and follow its course, e.g. by periodically measuring the concentration of reagents, and then plot the function ξ *(as* 'y'*) against time* t *(as* 'x'*). With the deesterification reaction here, the obvious way to monitor* ξ *is by periodically removing an aliquot of solution and then titrating it with acid to quantify the amount of NaOH remaining. The end of the reaction is indicated by the time at which [NaOH] remains constant. In order to obtain the rate constant* k, *we then plot a suitable function of* $[NaOH]_t$ *(as* 'y'*) against time (as* 'x'*) and measure the slope, from which the rate constant is obtainable.*

At heart, this titration method can be summarized as 'following the concentration of a reagent as a function of time'. Analyses with the RRDE are identical in concept. Analyte is formed at the central disc electrode, and because the disc rotates, such analyte is swept outward, past the ring electrode, and hence returned to the bulk solution. However, because the

† We will indicate the order of the chemical reaction with a subscript, so k_2 is the rate constant of a second-order reaction.

ring is concentric around the disc, analyte is swept over *the ring, the potential of which is pre-set in order to amperometrically determine how much of the analyte remains. We say that the bromine is an* ***intermediate****, because it is first formed and then reconsumed with the reactions occurring at either electrode.*

So we have two currents with the RRDE, where each relates to a concentration: the disc current I_D *relates to how much analyte is* formed *(via Faraday's laws), while the ring current* I_R *relates to how much analyte is reconverted back into starting material (again via Faraday's laws). If we take the example of the bromine–bromide couple, if* I_D *relates to the formation of bromine from bromide, then* I_R *relates to the 'back reaction' to reform bromide ion.*

In RRDE analyses, we refer to the fraction of the originally generated analyte which is retrieved as the **collection efficiency** N, *and define it according to the following equation:*

$$N = -\frac{I_R}{I_D} \qquad (7.10)$$

where the minus sign in the equation simply reminds us that if an oxidation reaction occurs at the disc (e.g. $2\,Br^- \rightarrow Br_2 + 2\,e^-$*) then by necessity the ring current must be reductive (e.g.* $Br_2 + 2\,e^- \rightarrow 2\,Br^-$*).*

The analogy between the RRDE and the deesterification reaction above goes further. The time t *required for analyte to travel from the disc (where it is formed) to the ring (where it is 're-reduced') is a function of both the distance between the disc and the ring, and the rate at which the RRDE is spun. Rapid rotation means that* t *is short, while slow rotation means that* t *is longer. We see that the time-scale of the observation is readily altered by variation of the rotation speed* ω*.*

DQ 7.16

How are kinetic effects followed with an RRDE?

Answer

In the previous chapter, we looked in detail at the kinetics of the reaction between electro-generated bromine and an electro-inert alcohol in solution. We recall that the product of this reaction (1,2-dibromo-3-hydroxypropane) was also electro-inert, i.e. we could monitor the course of the reaction by following the fractional loss of bromine.

When using an RRDE, no reaction occurs if there is no allyl alcohol in solution when bromine is formed at the disc and 're-reduced' to bromide at the ring. In such an instance, we would say that the collection efficiency was N_0 *(note the additional subscript). In practice, the value of* N_0 *is independent of rotation speed, so a graph of* $-I_R$ *(as 'y') against*

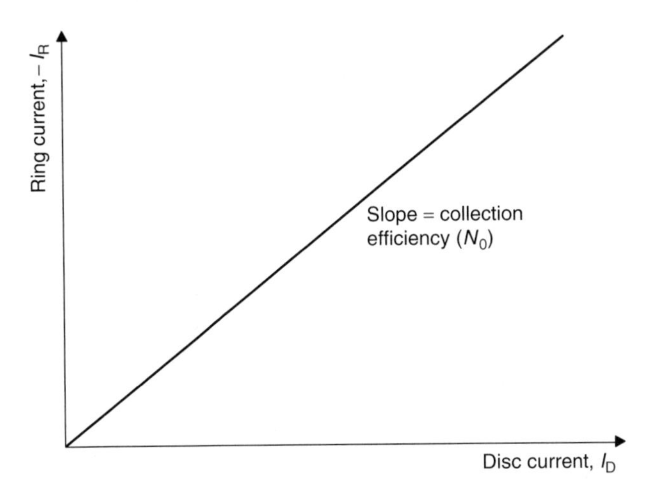

Figure 7.10 Plot of ring current, I_R, against disc current, I_D, to show how the collection efficiency, N_0, is determined.

I_D *(as 'x') is linear, with a gradient of* N_0. *In fact, the value of* N_0 *is usually determined by drawing such a graph (see Figure 7.10 above). A typical value would be about 0.2, with this value depending on the relative dimensions of the ring and disc and the distance between them. Alternatively,* N_0 *can be computed from theoretical equations or obtained from standard tables [1].*

SAQ 7.16

Calculate the collection efficiency N_0 if the current at the ring of an RRDE is 23.2 mA and the current at the disc is 90 mA.

DQ 7.17

Why is N_0 less than unity?

Answer

The streamlines over the surface of the RRDE are very similar to those at the face of an RDE (see Figure 7.2). Therefore, mixing occurs as solution is drawn across the surface of the electrode: solution is drawn in from *the bulk at the same time as solution containing electrogenerated material is lost* to *the bulk. In effect, we demonstrate the old saying, 'there's many a slip twixt cup and lip'.*

The extent of this mixing depends on the geometry of the ring relative to the disc. N_0 *can be maximized by constructing an RRDE with a large central disc and a narrow gap between the disc and ring electrodes.*

DQ 7.18

How is an RRDE analysis performed in practice?

Answer

(Note that the discussion here is much simplified, with a fuller treatment being given in reference [1].) The collection efficiency is N_0 *if no reaction occurs between the electrogenerated intermediate (such as bromine) and the material in solution.*

During a reaction, we call the measured collection efficiency N_κ *rather than* N_0 *(also note, by convention, that the subscript is now a Greek 'kappa' rather than a 'k'). The collection efficiency* N_κ *has the same value as* N_0 *if the RRDE is spun extremely fast because there is not enough time for the electrogenerated bromine to react during its lifetime, i.e. between the times it is formed and its subsequent re-reduction. In contrast, there is plenty of time for the electrogenerated bromine to react with other species in solution, e.g. allyl alcohol, when the RRDE is spun at a relatively slow speed. Because a slow rotation allows consumption of bromine, the ring current* I_R *will be much decreased, thereby causing the collection efficiency* N_κ *to decrease from its maximum value of* N_0.

In practice, the analyst measures the collection efficiency N_κ *as a function of time* t. *The 'time' we refer to here is the time required for the electrogenerated intermediate to be swept hydrodynamically from the disc and past the ring. Clearly,* t *is a function of the rotation speed, so in practice we determine* N_κ *as a function of* ω.

For a second-order reaction, the rate constant of reaction, k_2, *is obtained from the following somewhat gruesome-looking equation:*

$$\frac{I_R}{I_D} = N_\kappa = \frac{0.339\,(r_2/r_1)^2\,(D/\upsilon)^{1/3}\,(1 - F)\,\omega}{k_2\ c} \qquad (7.11)$$

where the two radii (r) *terms relate to the gap between the disc and ring –* r_2 *is the inside (that is, the smaller) radius of the ring electrode, while* r_1 *is the radius of the disc (as defined in Figure 7.9). The 'F term' in equation (7.11) is a function of the electrode geometry, which we can look up from tables [1] if we know the value of* r_1, r_2 *and* r_3. *Note that* N_κ *depends on* ω *rather than the* $\omega^{1/2}$ *term required for all of the other rotated electrode systems. All other terms have their usual meanings.*

*A word here about concentrations. The 'c term' in equation (7.11) relates to the electro-*inert *additive in solution, which, in the above discussion, is allyl alcohol. Therefore, we could have written this as* $c_{allyl\ alcohol}$. *Equation (7.11) does not contain a term for the initial concentration of the electroanalyte, since* I_D *and* I_R *are* both *functions of* $c_{analyte}$, *and so the respective concentration terms will cancel out.*

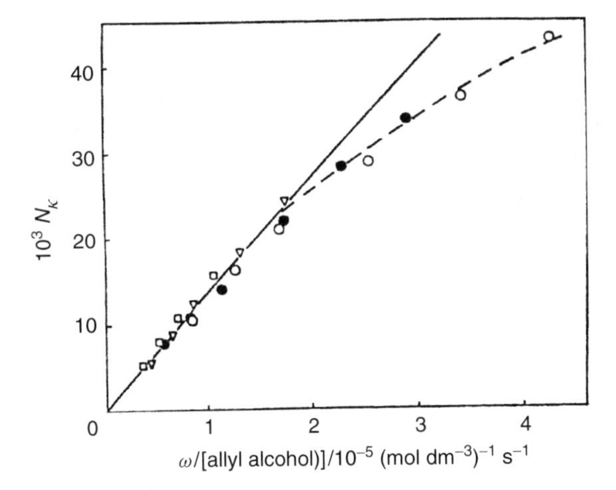

Figure 7.11 Plot of the kinetic collection efficiency N_K as a function of ω/[allyl alcohol], in which the concentration of bromide was always 0.2 mol dm^{-3}, and [allyl alcohol] was as follows: ○, 0.74×10^{-3}; ●, 1.10×10^{-3}; Δ, 1.44×10^{-3}; □, 1.80×10^{-3} mol dm^{-3}. From Albery, W. J., Hitchman, M. L. and Ulstrup, J., *Trans. Faraday Soc.*, **65**, 1101–1112 (1969). Reproduced by permission of The Royal Society of Chemistry.

We see from equation (7.11) that, for a simple second-order reaction in which one reagent is electrogenerated at the disc, a plot of N_K *against* (ω/c) *should be linear with a gradient proportional to* $1/k_2$. *Figure 7.11 above shows such a plot for the reaction between allyl alcohol and electrogenerated bromine. (In this figure, the deviation at higher concentrations relates to kinetic complications resulting from the tribromide ion,* Br_3^-.*)*

SAQ 7.17

From the gradient of the linear portion of Figure 7.11, calculate the rate constant k_2. The diffusion coefficient of bromine is 2×10^{-6} cm^2 s^{-1}, the kinematic viscosity is 0.239 cm^3 s^{-1}, $r_1 = 0.200$ cm and $r_2 = 0.226$ cm (take $F = 0.012$).

While we have discussed the RRDE in terms of an electron-transfer reaction followed by a chemical reaction (an EC reaction), it is such a versatile tool to the electroanalyst that extension to other mechanisms is straightforward and routine, although a further discussion is beyond the scope of this present text

7.5 Rate Constants of Electron Transfer

Every example we have looked at so far in this book has been simplified insofar as the rate of electron transfer was very fast – so fast, in fact, that we could assume that the electrode reaction was 'instantaneous'. Such an approach makes the life of an analyst much easier. Occasionally though, analysts need to appreciate that electrons must transfer between an electrode and an atom, ion or molecule of analyte, at a slower, finite speed. In short, such a transfer requires time.

The rate constant of electron transfer is symbolized as k_{et}, and is usually cited with the non-SI units of cm s^{-1}.

DQ 7.19

Why has the rate constant of electron transfer such unusual units?

Answer

We should note that the units of cm s^{-1} are similar to the typical velocity units of 'length per unit time'. Since the electron travels from an electrode to a molecule or ion of analyte (or back), it must traverse the distance between them during a certain interval of time. It is legitimate to take k_{et} as representing a function of the electronic velocity, as charge moves across the electrode | solution interface.

DQ 7.20

So why is k_{et} sometimes slow?

Answer

There are several possible reasons why the overall rate of electron transfer is slow. Occasionally, it is because the electrode is a poor conductor, such as **semiconductors** *like silicon or poor metals such as tungsten (see SAQ 2.6). Fabricating an electrode from metals such as platinum, gold, or from metallic conductors such as graphite or glassy carbon, will circumvent that possibility.*

Then again, sometimes the net rate at which electrons are taken from (or given to) the electrode is slow because the flux of analyte at the face of the electrode is low. We can overcome this problem by employing convective mass transport at an RDE, that is, by stirring at a sufficiently fast rate, ω.

However, in certain cases, the rate of electron uptake by a particular species just happens to be slow. *For example, electron transfer between the methyl viologen radical cation ($MV^{+\bullet}$) and hydrogen peroxide has a rate constant of 2.0 (mol $dm^{-3})^{-2}$ s^{-1}, while the reaction between $MV^{+\bullet}$ and just about any other chemical oxidant known is so fast as to be diffusion-controlled. The reason for this is simply not known at the present time.*

Therefore, if we as analysts need to quantify an analyte, but the electrode reaction is slow, we need to be aware of just how slow the reaction is, and know how to deal with such an electroanalysis if this type of situation arises.

7.5.1 The Tafel Approach to Electrode Kinetics

In simple chemical kinetics, the rate of a reaction is a simple function of temperature: increasing the temperature T causes an exponential increase in the rate constant k, as described within the Arrhenius and Eyring equations.

In a similar manner, the rate of electron transfer increases as the energy given to the electron increases. While k_{et} is temperature-dependent, its dependence on the potential of the electrode is far greater. At a fixed temperature, we find the following:

$$I = a + b \exp \eta \tag{7.12}$$

which is known as the **Tafel equation**. We recall that η is the overpotential, i.e. the difference between the potential at equilibrium and that applied by the driving power source (almost always a *potentiostat*), while a and b are empirical fitting constants.

> **Care** The Tafel equation is different to all the other equations we have discussed so far in this chapter, because in this case I is not a limiting current.

According to equation (7.12), a plot of $\log_{10} I$ (as 'y') against overpotential η (as 'x') should be linear. In practice, such a plot is not particularly linear, except over a relatively narrow range of potentials.

DQ 7.21

What is the cause of the non-linear portions of a Tafel plot?

Answer

The current is zero at equilibrium. Indeed, I $= 0$ *is one definition of equilibrium (see Chapter 2). As the potential is shifted away from* $V_{equilibrium}$*, so the electrode is polarized (cf. Section 6.1). We recall that the deviation of the potential from its equilibrium value is termed the overpotential* η *(as defined by equation (6.1)). The portion of the Tafel graph at extreme overpotentials represents insufficient flux at the electrode: in effect, the potential is so extreme that extra charge could flow if sufficient flux were available but, because of solvent viscosity, rate of solution stirring, etc., the flux is simply not large enough for the behaviour to follow the Tafel equation.*

At the opposite extreme, at very small overpotentials, the current is too small to follow the Tafel equation because both forward and reverse reactions (oxidation and reduction) occur, so we observe a **net current**. *While* $I_{forward}$ *might be sufficiently large to fall on a line extrapolated from larger values of* η, I_{back} *is also significantly large that we only see the remaining current. In summary, we say that:*

$$I_{net} = I_{forward} + I_{back} \tag{7.13}$$

where we note that $I_{forward}$ *and* I_{back} *must have different signs because one is oxidative while the other is reductive. We also see that* I_{net} *must always be smaller in magnitude than either* $I_{forward}$ *or* I_{back} *alone.*

DQ 7.22

How does I_{net} relate to electroanalysis?

Answer

Let us look at the consequences of equation (7.13).

- *(i) At equilibrium (i.e.* $\eta = 0$*),* $I_{net} = 0$ *because* $I_{oxidative} = -I_{reductive}$*. This is what we mean by our simplest definition of 'equilibrium'.*
- *(ii) The net electrode reaction is oxidation at large positive overpotentials (*$\eta > 0$*). In other words,* $I_{reductive}$ *is so minuscule that* $I_{net} \approx I_{oxidative}$*. This explains why we need to determine our analyses at large overpotentials, because* I_{net} *at large overpotentials is simply another way of saying* $I_{oxidative}$ = *the limiting current. In summary, any back reaction can be ignored if,* and only if, *we employ a sufficiently large overpotential.*
- *(iii) At the other extreme, the net electrode reaction is reduction at large negative overpotentials. When* $\eta < 0$*, the value of* $I_{oxidative}$ *is so minuscule that* $I_{net} \approx I_{reductive}$*. In other words, the current observed is wholly reductive.*

The above extremes are illustrated schematically in Figure 7.12. Only near the centre of the figure (at $\eta \approx 0$) are the two constituent currents comparable in magnitude.

7.5.2 The Butler–Volmer Approach

While the Tafel equation is useful, it is only semi-quantitative, and we therefore clearly need an approach which is more *quantitative*.

The **Butler–Volmer** approach is not empirical, so it therefore helps to explain the observed Tafel behaviour. Here, we start by saying that the observed current always has two components, i.e. oxidative and reductive, with both components depending strongly on the applied potential, V. The oxidative (anodic) current is

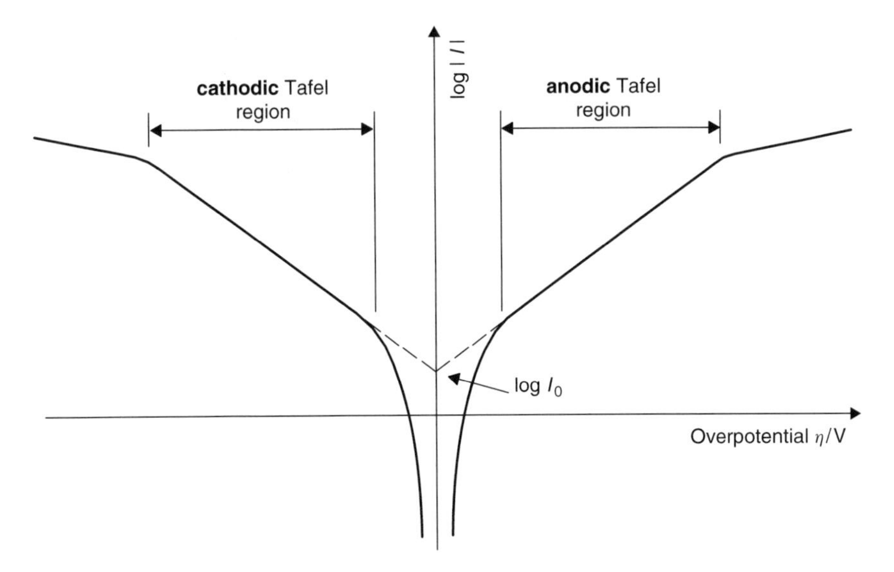

Figure 7.12 Schematic Tafel plot of log | I | (as 'y') against overpotential η (as 'x'). The linear regions yield the **Tafel** slopes, from which the transfer coefficients α can be determined. The intersection between the two Tafel regions occurs on the y-axis at log | I_0 |.

given by the following:

$$I_{\text{oxidative}} \propto \exp\left(\frac{\alpha n F \eta}{RT}\right) \tag{7.14}$$

while the reductive (cathodic) component is given by:

$$I_{\text{reductive}} \propto \exp\left(\frac{-\alpha n F \eta}{RT}\right) \tag{7.15}$$

where most terms have their usual meanings; α is called the **transfer coefficient** (or occasionally 'symmetry factor'), and relates to the symmetry of the transition state[†] generated between the electrode and the analyte. We will merely mention here that α commonly has a value of about 0.5, and will not discuss this parameter further in this text.

SAQ 7.18

Taking the constant of proportionality in equation (7.14) to be the area A (so anodic current *density* = exp ($\alpha nF\eta/RT$)), calculate the anodic current expected at an electrode of area 1.00 cm^2, when the applied overpotential was 0.120 V (take $\alpha = 0.5$, $T = 298$ K and $n = 1$).

[†] It is common to see the transition state called a 'complex'.

If we combine equations (7.14) and (7.15), we obtain the following:

$$I_{\text{net}} = I_0 \left[\exp\left(\frac{\alpha n F \eta}{RT}\right) - \exp\left(\frac{-\alpha n F \eta}{RT}\right) \right] \quad (7.16)$$

which is known as the **Butler–Volmer equation**. Equation (7.16) is simply a quantitative way of rewriting equation (7.13). The minus sign in the centre of the Butler–Volmer equation reminds us that the reductive and positive currents have different signs.

It is now time to define some terms. The **exchange current** (I_0) is best thought of as the rate constant of electron transfer at zero overpotential. This current is commonly expressed as a form of current density, I_0/A (cf. equation (1.1)), in which case it is called the **exchange current density**, i_0. (Incidentally, this also explains why the Butler–Volmer equation does not include an area term. This follows since both I_{net} and I_0 are functions of area, thus causing the two area terms to cancel out.)

SAQ 7.19

Use the Butler–Volmer equation to determine the net current at an overpotential η of 0.120 V (take $\alpha = 0.5$, $T = 298$ K, $n = 1$ and $I_0 = 0.034$ A).

In Chapter 2, Figure 2.1 showed a plot of I_{net} against overpotential,[†] itself as a function of i_0. We see that as i_0 increases, so the rate at which current changes with η also increases. If we were to take Figure 2.1 and plot the data in a slightly different way, we would obtain Figure 7.12. This new figure of log | I | (as 'y') against η (as 'x') reminds us of the Tafel equation, so we can call this a 'Tafel plot'. The two modulus signs, '|', either side of the current density reminds us to consider only the *magnitude* of the currents and to ignore their signs. This use of a modulus here is important since anodic currents are negative (cf. Section 1.2), so without this modulus sign, it would be impossible to take logarithms.

The Tafel plot shown in Figure 7.12 has three regions of interest, namely at (i) extreme (large), (ii) small, and (iii) intermediate values of the overpotential η. We will now discuss each of these regions in turn.

*(i) At **extreme** (large) overpotentials.* At very large η values, the plot in Figure 7.12 is linear and essentially almost horizontal at either wing, implying that I is independent of η. Here, the rate of mass transport is too small to bring sufficient material to the electrode | solution interface, thus meaning that the same amount of analyte is reduced (left-hand side) or oxidized (right-hand side) whatever the value of η. We say that these horizontal wings of Tafel plots represent the regions of *mass-transport control.*

[†] Figure 2.1 is actually drawn with an x-axis of $(E - E^{\ominus})$, but we recognize this as being the same as overpotential.

DQ 7.23

Why then is the Butler–Volmer plot not *exactly* horizontal at either end?

Answer

We have met the concept of IR *drop several times before. Even if the flux of electroanalyte material is maintained at a constant level, the potential of the electrode becomes progressively more cathodic (on the left-hand side of the plot) or more anodic (on the right). The analyte solution has a fixed and finite resistance, so from Ohm's law (see Section 2.1) we see that* $I_{solution}$ *increases as* η *increases. From Ohm's law,* $V = IR$, *and therefore the gradient* $= 1/R_{solution}$.

Incidentally, the occurrence of *IR* drop also explains why the slight gradients at each of the wings of the graph are equal but opposite, since the resistance of the solution is the same whether η is anodic or cathodic.

DQ 7.24

Surely the non-zero gradient implies that the limiting current is not independent of the potential?

Answer

Yes – the limiting current I_{lim} *does depend (albeit very slightly) on the potential. For this reason, it is the usual practice to determine* I_{lim} *at a fixed potential as well as at a fixed rotation speed (because* I_{lim} *is a function of* ω *from the Levich equation).*

(ii) At ***small*** *overpotentials.*

DQ 7.25

Why do the two currents dip so steeply at the centre of the graph rather than following a linear trend?

Answer

We recall that both of the currents, i.e. I_{anodic} *and* $I_{cathodic}$ *(see equations (7.14) and (7.15)), follow a logarithmic relationship with* η. *Next, we recall from the relationship in equation (7.13) that* $I_{net} = I_{forward} + I_{back}$, *so all the values of the current are* net *currents, thus explaining the observed decrease at the centre of the graph. The two constituent currents are comparable in magnitude, but of opposite signs, and so one current partially cancels out the other, and* I_{net} *is small.*

(iii) At ***intermediate*** *overpotentials.* By 'intermediate' here, we mean values of η in the ranges +(100 to 300) mV and −(100 to 300 mV) or so. In this potential

range, we find that log | I_0 | is directly proportional to the overpotential. For this reason, we call this the **Tafel region** of the plot, since the relationship log | I_0 |$\propto \eta$ reminds us of the Tafel relationship given in equation (7.12).

DQ 7.26

How do we obtain the exchange current, I_0, from a Butler–Volmer graph?

Answer

The central 'Tafel' portions of Figure 7.12 haven't been discussed in much detail yet. We sometimes call these linear Tafel regions the potentials 'under kinetic control', since the magnitude of I_{net} *is a function of the rate constant of electron transfer (itself a function of the overpotential* η*).*

We saw above that mass transport is insufficient at extreme potentials. Conversely, at small overpotentials the two currents, $I_{reductive}$ *and* $I_{oxidative}$ *are comparable in magnitude near the middle of the graph, so* $I_{reductive} > I_{oxidative}$ *on the left-hand side, while* $I_{oxidative} > I_{reductive}$ *on the right-hand side. However, in the Tafel regions at intermediate overpotentials (of* $\eta \approx \pm$ *100 mV or more), we see that although the net current comprises two components, only* one *component of that current is manifested. These two Tafel slopes are crucial, for they describe the rates of electron transfer, i.e. for reduction on the left-hand side and for oxidation on the right-hand side.*

SAQ 7.20

Show from equation (7.16) that the intercept of the Tafel plot in Figure 7.12 is a logarithmic function of | I_0 |.

Because I_0 is the rate of charge flow when the overpotential is zero, it is often referred to as an **idle rate**, much in the same way as we control the speed of a car engine with the idle screw, where the latter controls the speed at which it 'ticks over' while the car is stationary.

Worked Example 7.4. Use the following data to calculate I_0, and hence i_0. These data refer to a voltammetry solution containing $[Fe^{II}(CN)_6]^{4-} = [Fe^{III}(CN)_6]^{3-} = 20$ m mol dm^{-3} and [KCl] $= 0.1$ mol dm^{-3} at $T = 298$ K at an electrode of area $A = 0.60$ cm^2. For simplicity, only those data corresponding to kinetic control are included, i.e. I obtained at intermediate overpotentials (both anodic and cathodic η).

η/V	−150	−120	−90	−60	60	90	120	150
I/mA	18.6	10.3	5.77	3.21	−3.21	−5.77	−10.3	−18.6

Strategy. Because of the form of the Butler–Volmer equation, the data need to be plotted as log $| I |$ (as 'y') against η (as 'x'). The linear portions are then extrapolated toward $\eta = 0$. The ordinate where they cross is taken as log $| I_0 |$

Results. The data are plotted in Figure 7.13. The two linear portions, when extrapolated to zero overpotential, intercept on the y-axis at log $| I_0 | = -0.15$, so $I_0 = 10^{-0.15}$ mA = 0.708 mA.

From the way we defined the exchange current density i_0 (cf. equation (1.1)), we therefore obtain i_0 as:

$$\frac{I_0}{A} = \frac{0.708 \text{ mA}}{0.65 \text{ cm}^2} = 1.09 \text{ mA cm}^{-2}$$

DQ 7.27

How do we obtain the rate constant of electron transfer, k_{et}, from I_0?

Answer

The exchange current density, i_0, *and the rate constant of electron transfer,* k_{et}, *are related according to the following equation:*

$$I_0 = nFAk_{et} \qquad (7.17)$$

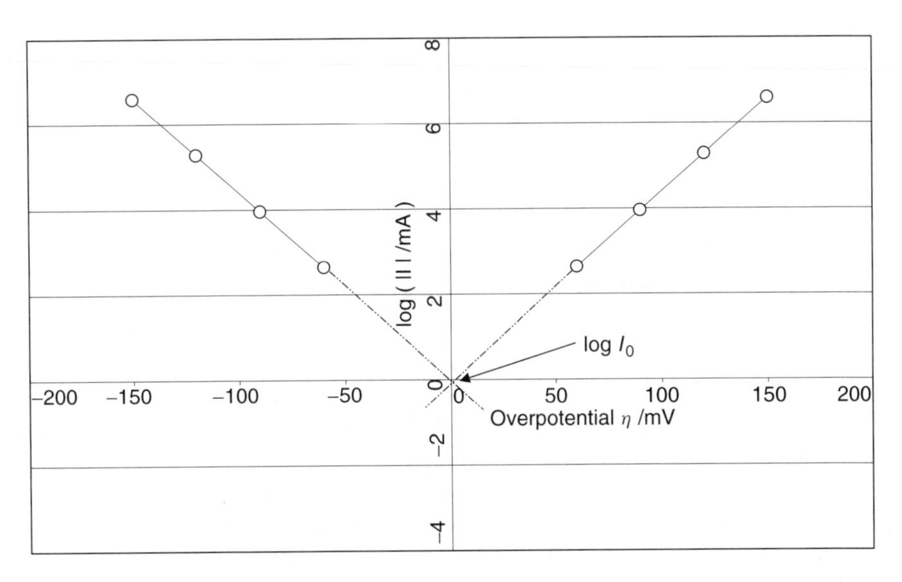

Figure 7.13 Tafel plot constructed with the data from worked example 7.4. These data refer to a solution containing $[Fe^{II}(CN)_6]^{4-}] = [Fe^{III}(CN)_6]^{3-}] = 20$ mmol dm^{-3} and [KCl] = 0.1 mol dm^{-3}. Only those data (both anodic and cathodic) corresponding to kinetic control are included. The logarithmic current where the two lines intercept is log $| I_0 |$.

where, because k_{et} *is a function of potential, we must appreciate that the value here is the 'implicit' value. From the way that the exchange current is defined,* k_{et} *here relates to the rate at zero overpotential.*

Worked Example 7.5. The one-electron reduction of the heptyl viologen dication (HV^{2+}) (to form the radical cation $HV^{+\bullet}$) occurs at a platinum electrode with an area of 1.50 cm^2. Calculate the value of k_{et}, given that the intercept on a Tafel plot of log $| I |$ (as 'y') against η (as 'x') is -2.2.

The intercept on the Tafel plot is -2.2, so $I_0 = 10^{-2.2}$ A $= 6.3 \times 10^{-3}$ A.

Next, we rearrange equation (7.17) by saying that $k_{et} = I_0/nFA$.
By inserting the appropriate values:

$$k_{et} = \frac{6.3 \times 10^{-3}\ \text{A}}{1 \times 96485\ \text{C mol} \times 1.50\ \text{cm}^2}$$

which gives:

$$k_{et} = 4.35 \times 10^{-8}\ \text{cm s}^{-1}$$

Note that this rate constant is quite slow.

SAQ 7.21

Calculate the rate of electron transfer, k_{et}, from the value of I_0 given earlier in Worked Example 7.4.

7.5.3 Koutecky–Levich Plots and Measurement of the Rates of Electron Transfer

The Levich equation (equation (7.1)) implies that faster rotation speeds allow for larger disc currents (see above), itself implying that it would be better to employ data obtained at higher rotation speeds ω when constructing Tafel plots such as those described in the previous sections, because the larger currents decrease the attendant errors associated with measurement of i_0 and α.

We will now combine the Levich and Butler–Volmer approaches. The Levich relationship (equation (7.1)) is written in terms of the limiting current I_{lim}, where 'limiting' here means 'proportional to $c_{analyte}$' – in other words, the electrode reaction is so fast that the magnitude of the current is controlled only by the flux of analyte to the electrode | solution interface, i.e. I_{lim} is **mass-transport controlled**.

In order to achieve this limiting current, we ensure that the overpotential is significantly anodic or cathodic, e.g. 500 mV beyond the standard electrode potential, $E^{\ominus}$. The Butler–Volmer equation (equation (7.16)) clearly shows that we can also measure currents at smaller overpotentials. Although the Levich equation suggests $I_{\text{lim}} \propto \omega^{1/2}$, by extension we can also say that non-limiting currents are also proportional to $\omega^{1/2}$.

DQ 7.28

Can we interpret the data in figures such as those in Figure 7.13, where neither the flux nor the rate of electron transfer have reached their maximum values, i.e. is the magnitude of the current mass-transport or charge-transfer controlled?

Answer

If ω is not particularly fast, then there is no mass-transport control (we have not reached a horizontal plateau when we draw a Levich plot). At the same time, however, if η is not extreme, then neither do we have kinetic control; stated another way, I *is no longer proportional to the bulk concentration of analyte. We have too many variables, so we're incapable of discerning whether mass or charge transport dictate the magnitude of* I.

In order to resolve this conundrum, we draw a **Koutecky–Levich** plot of 1/(current) (as '*y*') against $1/\omega$ (as '*x*'). (Plots using these axes are also sometimes called 'inverse-Levich plots'.)

DQ 7.29

What is the benefit of such a Koutecky–Levich plot?

Answer

The value of x = 0 *in a Koutecky–Levich plot implies that the rotation speed is infinite. We should appreciate that any current obtained at $\omega = \infty$ would automatically represent a mass-transport-limited quantity. So let us now look at what the non-limiting currents along the* y*-axis* really mean *when* x *is 0.*

It can be shown by rearranging the Levich equation and inserting via the Butler–Volmer equation that the Koutecky–Levich equation takes the following form:

$$\frac{1}{I_{not\ limiting}} = \frac{1}{nFAk_{et}\,c_{analyte}} + \frac{1.61\,\upsilon^{1/6}}{nFAc_{analyte}\,D^{2/3}}\,\frac{1}{\omega^{1/2}} \qquad (7.18)$$

where all the terms have their usual meanings. At an infinite rotation speed, the term on the far right tends to zero. At $\omega = \infty$, the reciprocal of current becomes a simple function of k_{et} *via the frequency-independent*

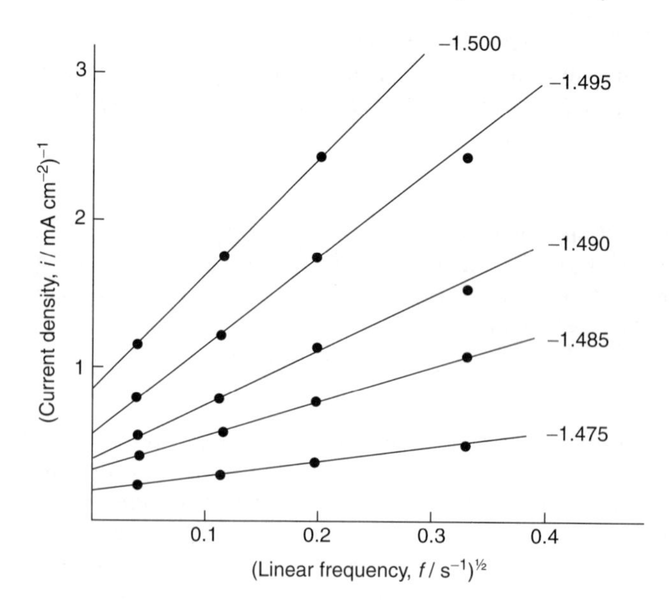

Figure 7.14 Koutecky–Levich plots of 1/(current) against $(1/\omega)^{1/2}$ as a function of applied potential (as indicated against each trace (V vs. SCE)). The data refer to the electro-oxidative dissolution of zinc immersed in aqueous NaOH (1.0 mol dm^{-3}). Reprinted from Armstrong, R. D. and Bulman, G. M., 'The anodic dissolution of zinc in alkaline solutions', *J. Electroanal. Chem.*, **25**, 121–130, copyright (1970), with permission of Elsevier Science.

term on the right-hand side. So, from equation (7.18), we see that plots of 1/(current) (as 'y') against 1/ω (as 'x') yield a series of intercepts on the y*-axis, one per overpotential, of 1/(*$nFAk_{et}\, c_{analyte}$*), with the values of* k_{et} *being potential-dependent. Figure 7.14 shows such a plot as a function of the applied potential.*

By constructing a Koutecky–Levich plot and obtaining the value of the current from the intercept, we have simplified the problems from two unknowns to one: the only limitation to the magnitude of the current on the y*-axis (strictly speaking, it is the* reciprocal *of the value on the* y*-axis) is the rate of electron transfer.*

The values of current as a function of rotation speed are then used to construct a Tafel plot such as that shown in Worked Example 7.4 (Figure 7.13), thereby allowing I_0 *and* α *to be calculated.*

7.6 Causes and Treatment of Errors

The principal cause of error in convective systems is non-laminar flow of solution over the face of an electrode. Turbulence and the attendant eddy currents can completely ruin an analysis. For this reason, it is wise to play safe by stirring with rotation rates (RDEs) and solution velocities (flow cells and channel electrodes) which are known to be laminar. In order to find out which concentrations, stirring rates and flow rates give a laminar flow regime, a Levich plot of I_{lim} against $\omega^{1/2}$ or a suitable function of V_{f} is drawn and those conditions which yield a linear graph must then be strictly adhered to. It is always a good idea to plot similar graphs of log (I_{lim}) against log (flow rate) for flow cells and channel electrodes.

A second cause of irreproducibility is not maintaining constant flow or rotation rate during analysis. Since convection is such an efficient form of mass transport, small variations in ω or V_{f} can cause I_{lim} to vary greatly. The variation in flux caused by a poor-quality pump can be negated if the solution is delivered from a well-observed constant-head tank. Vigilance is recommended in the absence of computer feedback systems,.

A related problem is occasional fluctuations in the flux of analyte at the electrode caused by the RDE 'wobbling' about its central axis – a common problem with home-made RDEs. In practice, a slight wobble will merely limit the range of rotation speeds, so the RDE cannot be allowed to spin fast, with the maximum rotation speed being a complicated function of the 'extent of wobble'.

Finally, we should reiterate that adsorption (particularly of organic material) on platinum is always to be expected.

Summary

Convection-based systems fall into two fundamental classes, namely those using a moving electrode in a fixed bulk solution (such as the rotated disc electrode (RDE)) and fixed electrodes with a moving solution (such as flow cells and channel electrodes, and the wall-jet electrode). These convective systems can only be usefully employed if the movement of the analyte solution is *reproducible* over the face of the electrode. In practice, we define 'reproducible' by ensuring that the flow is laminar. Turbulent flow leads to irreproducible conditions such as the production of eddy currents and vortices and should be avoided whenever possible.

Provided that the flow is laminar, and the counter electrode is larger than the working electrode, convective systems yield very reproducible currents. The limiting current at a rotated disc electrode (RDE) is directly proportional to the concentration of analyte, according to the Levich equation (equation (7.1)), where the latter also describes the proportionality between the limiting current and the square root of the angular frequency at which the RDE rotates.

The analyst conveniently employs a constant rotation speed for the RDE, and should note that $I_{\text{lim}} \propto c_{\text{analyte}}$. In this respect, analysts may prefer to determine limiting currents and then work from a previously constructed calibration graph.

Convective measurements using stationary electrodes again show that $I_{\text{lim}} \propto c_{\text{analyte}}$, provided that the solution flow is laminar. In this case, the limiting current follows a complicated relationship with the flow rate V_f, with the exact nature of this depending on what type of electrode is employed (flow, channel or wall-jet). It is again likely that the analyst will choose to employ calibration graphs of I_{lim} as a function of $c_{analyte}$.

One of the factors that an analyst must appreciate is that experiments involving the use of flow and channel cells can prove to be time consuming if the electrode requires frequent cleaning. Conversely, the RDE is not an ideal analytical instrument for work where continuous measurements need to be made.

The rotated ring-disc electrode (RRDE) has been shown to be an ideal tool for measuring the rate constants of very fast homogeneous reactions. In this method, we start with one reagent in the solution while the other is electrogenerated at the disc electrode, with the proportion of the latter that remains after reaction being monitored at the ring electrode.

Occasionally, the analyst is required to determine the rate of electron transfer, k_{et}, and can then use the Butler–Volmer equation (equation (7.16)) to determine I_0, from which k_{et} is readily calculated by using equation (7.17). The preferred method of obtaining the exchange currents in such cases is under conditions of 'infinite rotation speed' i.e. via a Koutecky–Levich plot.

Reference

1. Albery W. J. and Hitchman M. L., *Ring-Disc Electrodes*, Oxford University Press, Oxford, 1971.

Chapter 8

Additional Methods

Learning Objectives

- To understand the difference between *in situ* and *ex situ* methods.
- To appreciate that *in situ* spectroelectrochemistry (that is, simultaneous electrochemistry and spectroscopy) is a powerful method for carrying out an analysis.
- To recall that UV–visible spectroscopy monitors the energy required to excite (outer-shell) electrons within an analyte.
- To understand why, strictly speaking, all materials change their UV–visible spectrum in accompaniment with an electrode reaction, but appreciate that the majority of such changes are not discernible by the human eye.
- To learn to determine the faradaic efficiency of an electrode reaction by concurrent use of the Beer–Lambert law and straightforward coulometry.
- To learn that *in situ* spectroelectrochemistry requires optically transparent electrodes, be aware of the usual materials for making them, and how and why an *in situ* cell is constructed.
- To appreciate that *in situ* electron paramagnetic resonance (EPR) spectroelectrochemistry only monitors paramagnetic (radical) species.
- To learn how to use *in situ* EPR spectroelectrochemistry to tell if a radical is stable.
- To understand what impedance is, learn its nomenclature, and understand the methodology involved in obtaining impedance data.
- To learn what a Nyquist plot is, and what such a plot looks like for simple electrical components, plus appreciate that the Nyquist plot for an actual electrochemical cell can be mimicked by constructing an 'equivalent circuit' comprising arrangements of various components.
- To learn how to interpret an impedance plot for an electrochemical cell and be able to recognize the electrical components implied.

- To learn that the most common errors encountered with *in situ* UV–visible spectroelectrochemistry are those resulting from total internal reflection (TIR), which causes 'ringing', and by working at wavelengths beyond the band edge.
- To learn that the most common problems experienced with *in situ* EPR spectroelectrochemistry work result from employing a solvent that itself absorbs microwave radiation, and from using polymers which contain radical species.

8.1 Spectroelectrochemistry

8.1.1 Introduction: What is Spectroelectrochemistry?

DQ 8.1

What is 'spectroelectrochemistry'?

Answer

We can deduce the meaning of the word **spectroelectrochemistry** *by dissecting it piece by piece. Spectroelectrochemistry follows an electrochemical process by the use of electromagnetic radiation (hence 'spectro-'). In principle, any form of spectroscopy can be used to follow the progress of an electrode reaction, but in practice we tend to concentrate on two, namely UV–visible ('UV–vis') spectroscopy and a form of microwave spectroscopy known as electron paramagnetic resonance (EPR), as described below.*

If the spectroscopic investigation is performed at the same time as the electrochemical changes occur, then we say that this is an ***in situ*** measurement, where '*in situ*' is the Latin for 'same place'. The opposite of *in situ* is ***ex situ***, which mean literally 'from the place', i.e. *ex situ* implies that the measurement occurs after the analyte has been moved.

An example of an *ex situ* spectroscopic measurement would be the electrochemical generation of red bromine from colourless bromide ($2Br^- \rightarrow Br_2$), taking an aliquot of the sample from the electrolysis cell, and then taking a spectrum.

DQ 8.2

What dictates the choice of spectroscopic method?

Answer

All branches of spectroscopy follow the way that photons of light interact with matter, e.g. by absorption. The main difference between the distinct types of spectroscopy that we are interested in here, i.e. EPR or UV–vis, is that each monitors photons of different energy. The question of which spectroscopic method to use is akin to asking, 'at what energy does a process occur?'.

As an example, the energy needed to excite a valence electron is equivalent to the energy of a photon of visible light. This explains why UV–vis spectroscopy is sometimes called 'electronic spectroscopy', since in effect it is the electrons that impart the colour which is observed. Conversely, the energy needed to excite an unpaired electron between quantized spin states in a radical (that is, a species having one or more unpaired electrons) is equivalent to values lying within the microwave region of the electromagnetic spectrum.

We will now look at UV–vis and EPR (microwave) spectroscopy in the context of redox reactions at the electrode in an analytical investigation.

8.1.2 Electrochemical UV–Visible Spectroscopy

Everyone's eyes are different, so some can see colours that others cannot. By convention, we need to *define* ranges of radiation, by using the following:

Ultraviolet (UV) 200–350 nm;

Visible (vis) 350–700 nm;

Near-infrared (NIR) 700–1000 nm.

In electrochemical UV–vis spectroscopy, we monitor the colour of a material at the same time as we monitor the current. When we say that we monitor 'the colour', we mean here that we note both the wavelength at which the maximum of the absorption band(s) occurs, λ_{max}, together with the absorbance at each of these wavelengths. We define the optical **absorbance**, *Abs*, according to the following equation:

$$Abs = \log_{10}\left(\frac{T_{\text{with no sample}}}{T_{\text{with sample}}}\right) \tag{8.1}$$

where T is the transmittance of the light beam following its passage through the cell. Note that old-fashioned texts also refer to the absorbance as the **optical absorbance** or the **optical density**. Any changes in absorbance during *in situ* spectroelectrochemistry relate to changes in the amount of electroactive material, as converted by the flow of current.

The magnitude of the current tells us how many electrons transfer in unit time, while the absorption spectrum tells us the nature of the material generated. Discrepancies between the current ($\mathrm{d}Q/\mathrm{d}t$) and the rate of change of absorbance ($\mathrm{d}Abs/\mathrm{d}t$) can also tell us a lot about various side reactions that might take place.

DQ 8.3

Why is a spectroelectrochemical experiment worthwhile?

Answer

Electrode reactions are not particularly useful for telling us what *a 'something' is, but are excellent at telling us* how much *of that 'something' is present (if a potentiometric measurement) or has been formed or changed (if an amperometric measurement). Conversely, UV–vis spectroscopy is one of the better ways of identifying an analyte. We can therefore see that combining UV–vis spectroscopy with electrochemistry generates an extremely powerful way of answering both of the questions,* 'What *is it?'* *and* 'How much *of it is present?'*

We have already seen (Section 4.2) that a redox indicator *changes its colour in accompaniment with a redox reaction. This is a general phenomenon –* all *electroactive materials change their UV–vis spectrum (i.e. change colour) following an electron-transfer reaction, because the two different redox states possess differing numbers of electrons. A material having or imparting a colour is termed a* **chromophore** *(the word is Greek and, when split up, says 'colour imparting').*

If we can identify the chromophore, then we can say with some certainty what the product of the electron transfer actually is.

DQ 8.4

So why is there a colour change during a redox reaction?

Answer

The colour of a chromophore depends on the way its ***valence-shell electrons*** *interact with light, i.e. its colour depends on the way it absorbs photons. Photons are absorbed during the promotion of an electron between wave-mechanically allowed (i.e.* ***quantized****) energy levels. The magnitude of the energy required to achieve this,* E, *is given by the* ***Planck equation****, as follows:*

$$E = h\nu = \frac{hc}{\lambda} \tag{8.2}$$

where ν *is the frequency,* λ *is the wavelength of light absorbed, and* h *is the Planck constant. The value of* E *relates to the colour since* λ *is the wavelength maximum (usually denoted as* λ_{max}*) of the absorption band in a spectrum of the chromophore.*† *To obtain the energy in* $kJ\ mol^{-1}$*, we simply multiply the value of* E *by the Avogadro constant,* L.

† The colour actually seen is in fact the **complementary** colour since light is *absorbed*, so, for example, a blue colour is seen if the material absorbs red light.

SAQ 8.1

What is the energy of a photon of wavelength 600 nm? Express the answer in J (joules) per photon and kJ mol^{-1}.

The occurrence of an electron-transfer reaction ensures that an electroactive species has a different number of electrons before and after reaction, so different redox states must of necessity display different spectroscopic transitions, and hence will require different energies E for electron promotion. Accordingly, the colours of electroactive species before and after electron transfer will differ. It follows that *all* materials will change their spectra following a redox change.

SAQ 8.2

Look up the colours of the following redox states in a suitable reference source (the only difference between each is the number of electrons, with all species being aquo ions):

(a) vanadium(II) and vanadium(III);
(b) iron(II) and iron(III);
(c) methyl viologen dication and methyl viologen radical cation (e.g. see Table 4.1).

The colour change occurring on a redox change may be too subtle to be perceived by the human eye, for example, if the absorption maxima for the two respective redox states both reside either in the UV or in the NIR regions of the spectrum. In fact, it is quite rare for such changes to be *useful*, e.g. for the purpose of indicating the redox states under standard laboratory illumination.

Such changes in colour are only worth considering if one of the colours is markedly different from the other, for example, if the absorption band for the oxidized form of the couple is in the UV region, while the band for the reduced form is in the visible region. For convenience, we say that the material in such a case is **electrochromic**.

SAQ 8.3

Take apart the word 'electrochromic' to indicate its meaning.

DQ 8.5

Why is an electrochromic colour change of this sort useful to the electroanalyst?

Answer

If a material absorbs UV–visible light, then we can monitor its concentration by using the familiar ***Beer–Lambert*** *relationship, as* *follows:*

$$\text{Abs} = \varepsilon c_{analyte} l \qquad (8.3)$$

where the absorbance is determined at fixed wavelength λ, ε is the ***extinction coefficient*** *(cited at the same value of λ), and l is the optical pathlength. We should be aware that ε is also called the* ***molar decadic absorptivity****, particularly when the IUPAC units of concentration ($mol\ m^{-3}$) are employed.*

If we know the magnitude of the extinction coefficient at λ_{max}, we can quantify the amount of analyte simply by determining the optical absorbance and inserting values into equation (8.3).

Worked Example 8.1. A reductive current is passed through an aqueous solution of methyl viologen dication to effect the reaction $MV^{2+} + e^- \rightarrow MV^{+\bullet}$. The optical absorbance at 600 nm (λ_{max} for the radical cation $MV^{+\bullet}$) increased by 1.200 AU.† The pathlength l is 1 cm and the extinction coefficient ε for the radical in water is 13 300 $cm^{-1}\ mol^{-1}\ dm^3$. How much radical was formed?

We first rearrange equation (8.3) to make concentration the subject, as follows:

$$c_{\text{analyte}} = \frac{Abs}{\varepsilon l}$$

and then insert the appropriate values:

$$[MV^{+\bullet}] = \frac{1.200}{13\,300\ \text{cm}^{-1}\ \text{mol}^{-1}\ \text{dm}^3 \times 1\ \text{cm}} = 9.02 \times 10^{-5}\ \text{mol dm}^{-3}$$

We could have performed this experiment in reverse in order to determine ε.

When performing spectroelectrochemistry, we now have two laws which relate to the concentration of analyte, namely (i) The Beer–Lambert law, which relates optical absorbance and the concentration of a chromophore, i.e. $Abs \propto c_{\text{analyte}}$, and (ii) Faraday's law, which relates the amount of material formed (n) to the charge passed in forming it (Q) (see Section 5.1). However, if the amount of material (n) formed electrochemically is dissolved in a constant volume of solvent, then its *concentration* is proportional to Q.

By combining these two equations, we can say that if a material is electrochromic and the electrolysis is performed within a constant volume of solution, then the (faradaic) charge is proportional to the optical absorbance. This relationship of $Abs \propto Q$ is illustrated in Figure 8.1, where the absorbance, Abs, of the electrochromic colour (as 'y') is seen to increase linearly as the charge increases (as 'x'). The linearity of the graph indicates that both Q and Abs relate to the *same*

† AU is a dimensionless term known as the **absorbance unit**. The term AU was originally introduced because some analysts thought that all parameters should have units!

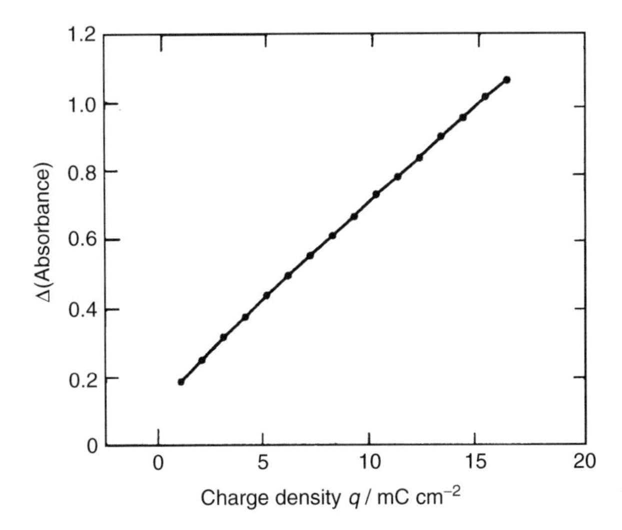

Figure 8.1 Beer's law-type plot of change in optical absorbance against charge density q for the cell WO_3 | polymer electrolyte | Prussian Blue. Reprinted from Inaba, H., Iwaka, M., Nakase, K., Yasukawa, H., Seo, I. and Oyama, N., 'Electrochromic display device of tungsten trioxide and Prussian Blue films using polymer gel electrolyte of methacrylate', *Electrochim. Acta*, **40**, 227–232 (1995), Copyright 1995, with permission from Elsevier Science.

concentration. (For convenience, Figure 8.1 shows a plot of absorbance change as a function of charge density).

SAQ 8.4

Calculate the extinction coefficient ε of the tungsten trioxide 'bronze', H_xWO_3, formed if a charge of 10^{-3} F is passed through H_0WO_3 to increase the optical absorbance by 0.52 AU, taking $l = 0.01$ cm and $n = 1$. (Hint – take the concentration of bronze from the charge passed, i.e. the concentration of bronze within the WO_3 'host' is x.)

DQ 8.6

Why would an *electroanalyst* want to use the Beer–Lambert law?

Answer

*The Beer–Lambert law is an ideal means of determining the faradaic efficiency (see Section 5.1.1). First, the analyst employs Faraday's laws by using the charge (Q) passed in order to calculate how much material was formed. Then, either at the same time (*in situ*), or straight afterwards* *(*ex situ*), the analyst spectroscopically monitors the amount of material formed via the Beer–Lambert law, or from a calibration graph such as Figure 8.1.*

Provided that the optical bands do not overlap, the optical measurement tells the analyst how much product was formed, while Faraday's laws show how much charge was passed and hence how much product should have been formed. The faradaic efficiency is then the ratio of these two amounts.

Worked Example 8.2. Following on from Worked Example 8.1, calculate the faradaic efficiency if the charge passed to generate the $MV^{+\bullet}$ radical cation was 10 C.

Strategy. First, we determine what were the number of moles of electrons passed (i.e. how many farads (F) of charge). Then, knowing how much $MV^{+\bullet}$ *was* formed and how much $MV^{+\bullet}$ *could* have been formed, we take the ratio of the two numbers.

1 F of charge comprises 96 485 C, so 10 C represents 1.036×10^{-4} F. The electrode reaction to form the radical cation involves one electron, so 1.036×10^{-4} F of charge generates 1.036×10^{-4} mol of $MV^{+\bullet}$. Therefore, we can write:

$$\text{faradaic efficiency} = \frac{\text{amount of MV}^{+\bullet}\text{ formed (from Worked Example 8.1)}}{\text{amount of MV}^{+\bullet}\text{ that could have been formed}}$$

$$= \frac{9.02 \times 10^{-5}}{1.036 \times 10^{-4}} = 87\%.$$

SAQ 8.5

50 C of charge is passed reductively through 1 l of a solution of anthracene (An^0) to generate the radical anion ($An^{-\bullet}$). Calculate the faradaic efficiency if the optical absorbance of the same solution increased (at the λ_{max} of 523 nm) by 0.970 AU. Take $\varepsilon = 10\,400$ cm^{-1} mol^{-1} dm^3 and $l = 0.2$ cm.

DQ 8.7

How do we perform an *in situ* spectroelectrochemical experiment?

Answer

Experimentally, ex situ *spectroelectrochemistry is very easy to carry out. We generate a chromophore at an electrode, and then transfer it to a standard UV–vis spectroscopy cell and measure the absorbance. (We may need a modified spectroscopy cell if the product is air-sensitive or otherwise too reactive.)*

The main problem we encounter when performing an in situ *UV–vis spectroelectrochemical experiment is an obvious one, i.e. the light must be able to pass* through *the electroanalysis cell, so everything – cell walls, solution and electrodes – must be highly transparent. To construct a cell from quartz or silica is easy, and to work with transparent solutions is generally problem-free. However, transparent electrodes are a greater challenge.*

DQ 8.8

How can we see *through* an electrode?

Answer

In general, two (similar) approaches are used to construct an ***optically transparent electrode*** *(OTE).*

Wire 'mini-grid' OTEs. *A 'mini-grid' is constructed with an array (or 'mesh') of microscopically thin wires criss-crossing the face of a sheet of glass, silica or quartz. The wires are themselves too thin to see, but as soon as product is formed, it diffuses away from the wire. Since diffusion is entropy-driven (i.e. random), electrogenerated material does not diffuse in straight lines, but moves in all directions at once. In practice, as soon as material is formed, it is seen between the wires, and hence can be detected by the light beam of a spectrometer.*

Thin-film OTEs. *In this method, a thin layer of conductor is deposited on to a transparent slide of glass or silica. If the layer is thin enough, it will be relatively transparent, yet will still maintain an appreciable extent of electronic conductivity.*

In order to perform spectroscopy with such an OTE we need to use a cell such as that shown in Figure 8.2, where the conducting face of the OTE is positioned facing in toward the solution bulk, and with the light beam passing through it at right angles. The working and counter electrodes are carefully positioned in such a way that the optical beam is not interrupted, thus ensuring that only changes at the thin-film OTE are registered by the optical detector of the spectrometer (probably a photocell).

DQ 8.9

Why is the counter electrode in Figure 8.2 depicted as a 'two-pronged fork'?

Answer

We must appreciate that a thin film of conductor (on a glass substrate for use as a working electrode) will not have the same high electronic

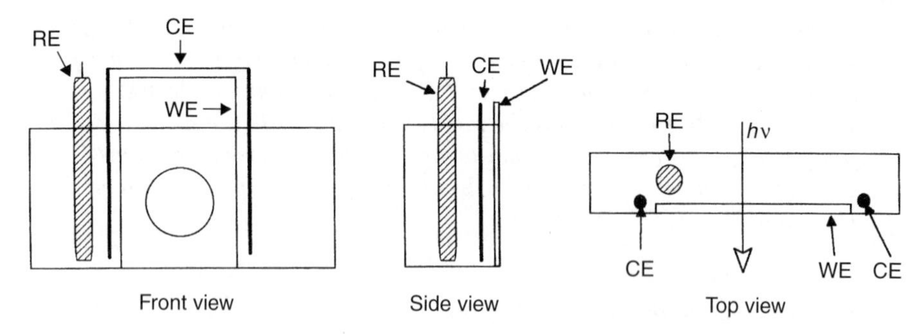

Figure 8.2 Schematic representation of a cell used for *in situ* spectroelectrochemistry. Notice how the counter electrode (CE) has two prongs, one either side of the optically transparent working electrode (WE). Neither the reference electrode (RE), nor the CE can be allowed to interrupt the path of the light beam (as indicated by the circle on the front view).

conductivity as does a normal 'slab' of conductor. For this reason, it is possible for a gradient of potential *to develop across the thin film. This effect occurs because of our 'old enemy', the* IR *drop – this time due to the resistance of the semiconductor layer.*

The most immediate effect of such a gradient is that more material will be electromodified at one side of the OTE and less at the other.

In order to circumvent this problem, it is a good idea to use a 'dual' counter electrode with 'prongs' on either side of the OTE, as shown in Figure 8.2. While the gradient of potential is not completely eradicated with such a dual CE it is nevertheless diminished to a significant extent.

DQ 8.10

What are the best materials to use to construct such OTEs?

Answer

The best choice of metal for making a minigrid is gold, because it is so malleable.

The first thin-film OTEs were made with a thin film of gold deposited on glass or silica. (This transparent substrate should be optically flat.) If thin enough, the gold reflects very little light[†] *so the beam can be transmitted (curiously, gold in thin-film form will only transmit* green *light).*

Nowadays, most electroanalysts have moved away from gold and employ a thin film of a semiconductor (and by 'thin', we mean about

[†] Metals reflect by a process known as **specular reflection**, with this effect being caused by the 'sea' of delocalized electrons on the surface of the metal. If thin enough, this 'sea' becomes more localized, thereby diminishing the extent to which specular reflection can occur.

0.3 μm). The two semiconductors of choice are tin dioxide doped with fluoride (often symbolized as SnO_2:F) and indium oxide doped with tin dioxide, otherwise known as ***indium–tin oxide*** *(ITO). Both can be purchased relatively cheaply under a variety of trade names.*

Figure 8.3 shows a set of spectra of Prussian Blue generated at an ITO electrode. Prussian Blue is the mixed valence compound, $K^+(Fe^{3+}[Fe^{II}(CN)_6])$. The colour formed, as shown by the increasing absorbance in the figure, represents the generation of coloured Prussian Blue from the colourless Everitt's salt, $K_2Fe^{2+}[Fe^{II}(CN)_6]$.

8.1.3 Electrochemical EPR Spectroscopy

As we have seen, the microwave region of the spectrum has just the right energy to excite the electronic spins in species possessing one or more unpaired electrons, i.e. radicals. Paired electrons are stabilized relative to unpaired electrons, so the

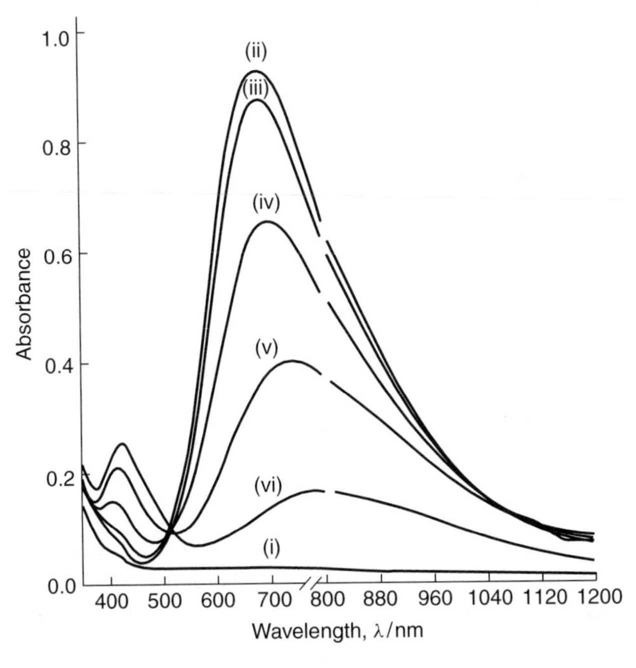

Figure 8.3 Illustration of *in situ* spectroelectrochemistry, showing a set of UV–vis ('electronic') spectra of solid-state Prussian Blue (iron(II,III) hexacyanoferrate(II)) adhered to an ITO-coated optically transparent electrode. The spectra are shown as a function of applied potential: (i) −0.2; (ii) +0.5; (iii) +0.8; (iv) +0.85; (v) +0.9; (vi) +1.2 V (all vs. SCE). From Mortimer, R. J. and Rosseinsky, D. R., *J. Chem. Soc., Dalton Trans.*, 2059–2061 (1984). Reproduced by permission of The Royal Society of Chemistry.

microwave region cannot readily probe the excitations between quantized spin states of paired electrons.

SAQ 8.6

What is the energy of a single microwave photon? Take $\lambda = 1$ cm.

Most of us have encountered nuclear magnetic resonance (NMR) spectroscopy many times in the past, for example when analysing the products of preparative organic chemistry. In NMR spectroscopy, the nucleus of an atom is excited following absorption of a photon (in the radiofrequency region of the electromagnetic spectrum).

NMR spectroscopy is particularly powerful because of the way that atomic nuclei **couple**, that is, their spins interact through bonds (or sometime through space). The way that these spin couplings occur, in terms of the number and types of adjacent atoms, gives rise to a complicated **splitting pattern** in an otherwise simple absorption spectrum. We call this 'forest' of interdependent peaks the **fine structure**. Careful analysis of the intensities and multiplicity of the coupling within the fine structure yields a wealth of data about the molecular geometry of the absorbing species.

Electron paramagnetic resonance (EPR), also known as **electron spin resonance** (ESR), has rightly been called 'the NMR of unpaired electrons'. A sample of a radical (generally in solution) is irradiated with a beam of microwave radiation and the proportion of the radiation that is absorbed is then determined.

To complicate matters, the spectrum cannot be determined by placing a sample cell in the path of a beam of microwave radiation. Like NMR, the EPR sample must first be placed within a strong magnetic field (conventionally symbolized as **B**).[†]

DQ 8.11

Why is a magnetic field needed in EPR spectroscopy?

Answer

An unpaired electron has spin, s*. The magnitude of this spin can have two values, i.e. +1/2 or −1/2. In the absence of a magnetic field, there is no energy difference between these two values of* s *– we say that they are* ***degenerate****. However, the degeneracy vanishes (we say it is 'lifted') when a strong magnetic field is applied, thereby allowing the two values of the spin to be seen. The difference in energy between the two spins is proportional to* **B** *(see Figure 8.4), so a large value of* **B** *increases the precision*

[†] Note that the magnetic field (strength) is a vector quantity, i.e. it has both magnitude and direction, and is therefore shown here in bold, italic script.

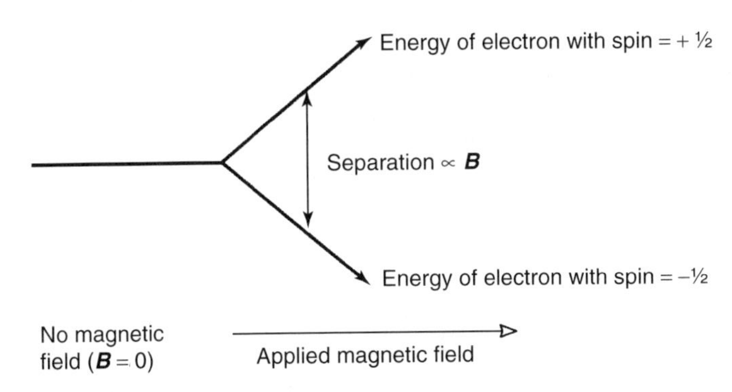

Figure 8.4 Energy levels for the two spin states of an unpaired electron in the absence of any net magnetic field (left-hand side) and in an applied magnetic field $\boldsymbol{B}$ (right-hand side). The frequency of the microwave radiation is assumed to be appropriate and constant.

of measurement. It is the interaction between the s $= +1/2$ *and* s $= -1/2$ *electrons that an analyst monitors during EPR spectroscopy, which also explains why the other name for EPR is electron* spin *resonance (ESR), as mentioned above.*

DQ 8.12

So how is the EPR experiment performed in practice?

Answer

Figure 8.5 shows a schematic representation of a cell used for obtaining in situ *spectroelectrochemical EPR spectra. This cell has to be constructed from silica or quartz since normal glass or plastic will itself have a paramagnetic signal. The counter electrode (CE) needs to be*

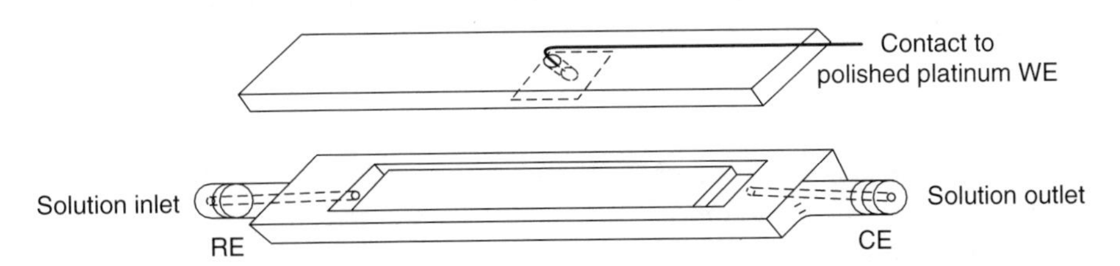

Figure 8.5 Schematic representation of a typical EPR cell – in this case, the 'Compton–Waller' flow cell – used for *in situ* spectroelectrochemistry. The working electrode is placed outside of the cavity of the EPR spectrometer, with the counter electrode being normally an SCE or 'AgCl,Ag'. The working electrode is a flat polished plate of platinum, positioned parallel to the direction of the electric field. Reproduced from Compton, R. G. and Waller, A. M., *Comprehensive Chemical Kinetics*, Vol. 29, p. 173, Copyright (1989), with permission from Elsevier Science.

placed beyond the area of greatest spectroscopic sensitivity in order to prevent the products of electrode reaction at the CE disturbing the spectrum of the products at the working electrode (WE).

This long, thin cell is lowered into the centre of the EPR spectrometer, with the WE at the focus of the magnetic field. Great care and skill are needed during this operation because otherwise the signal will be distorted or, worse still, the EPR trace of the electrons within the electrodes *may be picked up.*

With the cell in place, the usual procedure for obtaining in situ *spectroelectrochemical data with an EPR spectrometer is to irradiate the radical sample with microwaves at a constant, fixed wavelength and vary the strength of the magnetic field* **B**. *In UV–vis spectroscopy, we characterize an absorption peak in terms of where it occurs (cited as λ_{max}) and its absorbance (in absorption units (AU)). In EPR spectroscopy, we identify where the peak occurs in terms of the magnetic field (usually in terms of the old-fashioned unit of gauss (G)). Generally, though, we do not talk about a peak's absorbance – rather, because of the need to obtain clear distinct EPR peaks, the usual practice is to record an EPR spectrum as a* first derivative.

Figure 8.6 shows an in situ *EPR spectrum of the dibenzo[a,e]cyclooctene radical in organic solution, with the radical having been generated inside the cavity of an EPR spectrometer. The structure of the radical can be determined from detailed analysis of the fine structure. (Because of the energies involved in EPR analyses, the separation between peaks – leading to coupling constants – is so small that we tend to refer to such a situation as* ***hyperfine*** *coupling.)*

DQ 8.13

What can the EPR spectrum tell us about the molecular structure of a radical?

Answer

First, we should not lose sight of the obvious – if the product of an electrode reaction can be detected by an EPR experiment, then we have generated a radical.

The second thing that EPR can tell us is concentration. As with all forms of spectroscopy, the intensity of an EPR absorption peak is proportional to the concentration of the analyte giving rise to this absorption. (Care – because we record an EPR spectrum as a first derivative, *the concentration of analyte is in fact proportional to the* integral *of the EPR peak height.) However, EPR is so powerful a technique that to determine a concentration alone would be wasteful.*

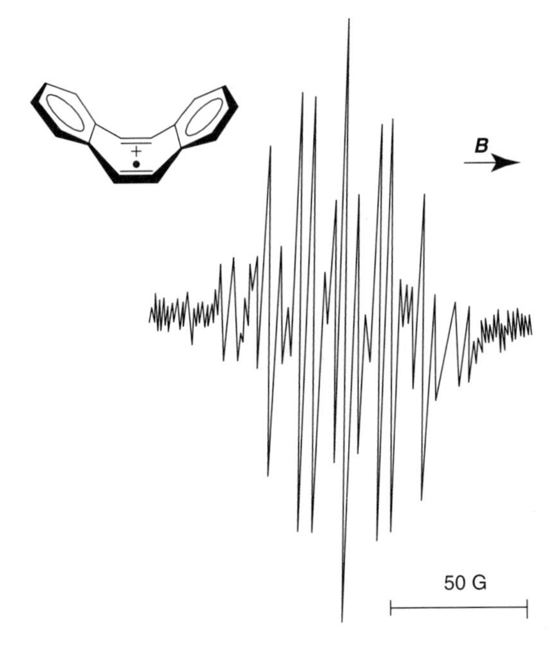

Figure 8.6 EPR spectrum of the dibenzo[a,e]cyclooctene radical cation showing hyperfine structure. The radical was generated by UV irradiation of the neutral parent molecule. From Gerson, F., Felder, P., Schmidlin, R. and Wong, H. N. C., *J. Chem. Soc., Chem. Commun.*, 1659–1660 (1994). Reproduced by permission of The Royal Society of Chemistry.

Thirdly, from the EPR hyperfine structure, it is possible to compute the structure *of the radical (both the* atomic *and the* electronic *structures). For example, we can tell that the paramagnetic cation giving rise to the EPR signal in Figure 8.6 is non-planar, which is quite unusual for such an aromatic radical cation, a fact we know from additional computations based on the hyperfine splitting pattern.*

However, for the electroanalyst, the most important tool in the EPR spectroscopist's arsenal is the ability to answer the question, 'Is this radical stable?'.

DQ 8.14

How can the stability of a radical be gauged from EPR spectroscopy?

Answer

We looked at in situ *flow cells in the previous chapter. It is easy to modify the EPR cell shown in Figure 8.5 to allow it to function as a flow cell. When used in this way, the counter electrode (CE) must be placed* downstream *in order to prevent the product of electron transfer at the CE being swept past the detector within the EPR chamber, while the reference electrode (RE) is best placed upstream.*

Next, we pass analyte through the EPR cell at a constant flow rate, V_f*, and determine the intensity of the EPR signal,* S. *(Remember that the EPR spectrum is in fact a first derivative, so the concentration of the radical that is generated is proportional to the* integral *of the peak.) We then vary the flow rate* V_f *and monitor the corresponding peak intensities.*

The magnitude of the peak height S *and the flow rate* V_f *are related by the following equation:*

$$V_f = k\left(\frac{S}{I_{lim}}\right)^{-2/3} \tag{8.4}$$

where k *is a constant of proportionality, which is itself a function of* n, D, *the electrode geometry and the bulk concentration of analyte,* $c_{analyte}$. *Accordingly, a plot of ln* (S/I_{lim}) *(as 'y') against ln* (V_f) *(as 'x') should be linear with a gradient of* $-2/3$, but only if the radical is stable. *In other words, if a radical obeys the flow behaviour defined by equation*

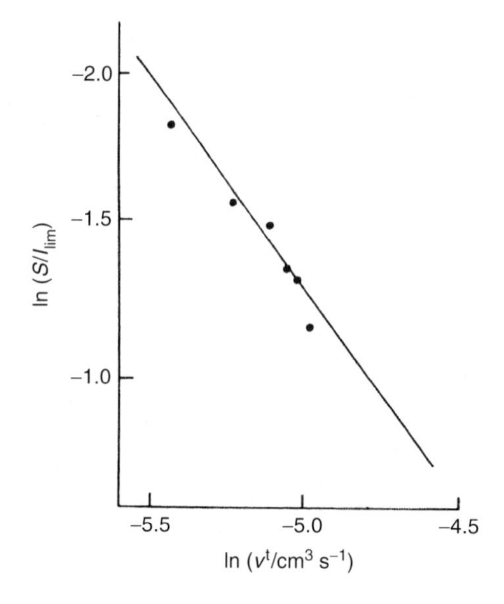

Figure 8.7 EPR testing for the stability of a radical, showing a plot of $\ln(S/I_{\text{lim}})$ against $\ln(V_f)$ for the viologen radical cation $CPQ^{+\bullet}$ (where 'CPQ' = 1, 1′-bis(*p*-cyanophenyl)-4, 4′-bipyridilium bis(tetrafluoroborate) in propylene carbonate solution. This radical was formed by one-electron reduction of CPQ^{2+} at a platinum electrode; V_f is the solution flow rate and S is the magnitude of the EPR peak. The stability of the CPQ radical cation is shown by the plot being linear, with a gradient of $-2/3$. From Compton, R. G., Monk, P. M. S., Rosseinsky, D. R. and Waller, A. M., *J. Chem. Soc., Faraday Trans.*, **86**, 2583–2586 (1990). Reproduced by permission of The Royal Society of Chemistry.

(8.4), then it is stable. The radical is unstable if the gradients of graphs based on equation (8.4) are much less that the theoretically predicted value of $-2/3$.

SAQ 8.7

From equation (8.4), show why the above logarithmic analysis is suggested.

Figure 8.7 shows such a flow plot for the radical cation of the viologen species, 1,1′-bis(*p*-cyanophenyl)-4,4′-bipyridilium ('CPQ'), as the bis(tetrafluoroborate) salt in propylene carbonate solution, with the radical having been formed at a polished platinum electrode. The plot is seen to be linear, implying that, once formed, the CPQ radical cation is chemically stable.

SAQ 8.8

Why is it emphasized here that the platinum be 'polished'?

8.2 Electroanalytical Measurements Involving Impedance

8.2.1 *What* is *Impedance?*

Increasingly often, **impedance** analysis is now being used as an electrochemical technique. Such analyses are different to all of the dynamic electroanalyses we have looked at so for in this text, because in all of the previous techniques, the potential was either constant or was ramped† at a constant rate of ν ($=\mathrm{d}E/\mathrm{d}t$).

An alternative approach, which we will introduce here, is to apply a small *perturbing* potential across a cell or sample. An additional advantage of this approach over potentiometric analyses is that current is generated, since the potential is different from the equilibrium value – there is an overpotential η. An additional advantage over amperometric analyses is that the potential is only 'perturbing', and so any concentration changes within the cell or sample are minimized.

DQ 8.15

Why does the concentration not change if a non-equilibrium potential is applied?

Answer

First, the potential is tiny – by a 'small' potential referred to above, we meant about 5 mV or so. However, more importantly, the perturbing

† Such potential–time profiles (e.g. in Figures 6.2(a) and 6.5) still represent direct-current (DC) situations.

voltage changes in a cyclic sinusoidal *manner, so we can say that the time-averaged overpotential is zero. In effect, we set up a type of steady-state situation. Accordingly, we have some of the advantages of a dynamic technique, although the actual changes do not occur on the macroscopic level.*

Because the voltage changes in such a cyclic manner, we say that the induced current *alternates*, and hence the term **alternating current** (AC).

While introducing this new way of obtaining electroanalytical data, we will need to rely on the analogies between an electrochemical cell (or sample) and an electrical circuit made up of resistors and capacitors assembled in order to mimic the current–voltage behaviour of the cell. All the time, though, we need to bear in mind that the ideas and attendant mathematics are for interpretation only, although they are fundamentally very simple.

Unfortunately, impedance has amassed quite a few different names. For example, it is common now to see the acronym 'EIS', meaning electrochemical impedance spectroscopy. Strictly speaking, impedance is not a form of spectroscopy because the sample is not absorbing photons of any kind. A further linguistic problem is that 'AC impedance' is tautologous since the word 'impedance' itself implies an 'AC' measurement. In this present discussion, we will simply say 'impedance'.

DQ 8.16

So, what *is* impedance?

Answer

Applying a constant voltage V *across a resistance* R *induces a constant current* I *according to* Ohm's law *(see equation (2.3)). In a similar way, application of a sinusoidally varying potential (see Figure 8.8) across an electrochemical cell induces an alternating current (AC). The AC analogue of Ohm's law is given by the following:*

$$\overline{V} = \overline{I}Z \tag{8.5}$$

where Z *is the* **impedance**, *and the over-bars merely imply that these quantities are time-dependent. At its simplest, impedance is a resistance that varies in a cyclical manner, and therefore,* Z *has the units of ohms* (Ω), *just like any other type of resistance determined with a direct current (DC).*

SAQ 8.9

Calculate the impedance of a cell if the current at an instant in time is 12 mA and the potential at the same instant is 430 mV.

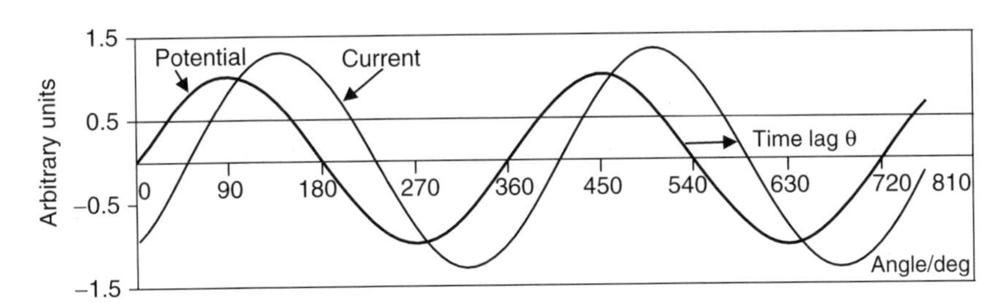

Figure 8.8 In impedance analysis, a sinusoidally varying potential $\overline{V}$ is applied across a sample, and the time-dependent current I is measured as a function of the frequency ω. The current induced in response to the varying potential will be out of phase, by a time lag θ, and of different magnitude.

We will give the symbol Z^* to the overall (or 'complete') impedance. In fact, Z^* comprises two components, which we will term **real** and **imaginary** (Z' and Z'', respectively). Z^*, Z' and Z'' are related as follows:

$$Z^* = Z' - jZ'' \tag{8.6}$$

where $j = \sqrt{-1}$, and hence the term 'complex' impedance.[†]

A mathematician would say that a plot of Z'' (as 'y') against Z' (as 'x') forms an **Argand** diagram (or 'Argand plane'). As electroanalysts, we will call such a set of axes a **Nyquist** plot or simply an **impedance plot** (see Figure 8.9).

DQ 8.17

What actually happens during the impedance experiment?

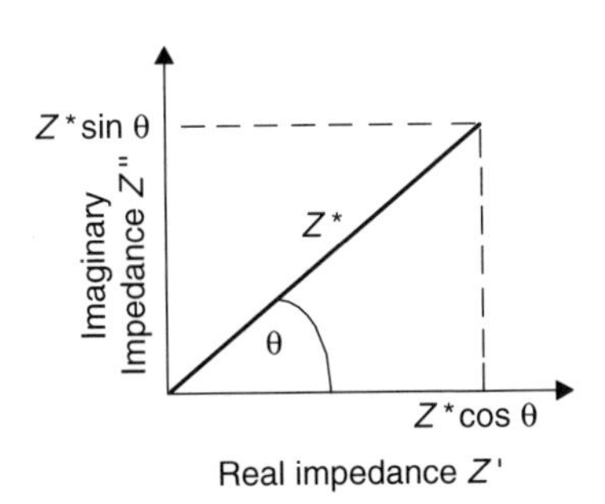

Figure 8.9 A Nyquist plot of the imaginary impedance Z'' against the real impedance Z', showing how Z^* and θ are defined.

[†] Note that the term 'complex AC impedance', although often encountered, is *doubly* tautologous.

Answer

The frequency-dependent resistance Z* *of a cell or sample is measured over as wide a frequency range as possible (typically* $10^6 - 10^{-2}$ *Hz). The apparatus measuring the impedance is known as a* ***frequency analyser****. Note that the latter may also be termed a* ***phase-gain-*** *or* ***signal-response analyser****. (The exact mode of measurement, in terms of the internal electronics, defines which name is applicable – but we will not bother with such details here.)*

The analyser applies a tiny voltage $\overline{V}$ *across the sample (perhaps superimposed on a pre-set or* ***offset*** *voltage). The magnitude of* $\overline{V}$ *varies with time since it is sinusoidally modulated. The analyser measures the respective time-dependent currents* $\overline{I}$*, and hence calculates* Z* *and the* ***time lag*** θ *experienced between the current and voltage.*

The frequency analyser then alters the frequency at which the voltage oscillates and Z* *is determined once more, with this procedure being performed for as many as 50 different frequencies. (It is the usual practice to start at the upper frequencies and progress down to the lower frequencies.)*

From these values of Z* *and* θ*, the components* Z′ *and* Z″ *are obtained, and so a Nyquist plot can be generated.*

DQ 8.18

So, why are we looking at impedance at all?

Answer

At the heart of impedance analysis is the concept of an ***equivalent circuit****. We assume that any cell (and its constituent phases, planes and layers) can be approximated to an array of electrical components. This array is termed 'the equivalent circuit', with a knowledge of its make-up being an extremely powerful simulation technique. Basically, we mentally dissect the cell or sample into resistors and capacitors, and then arrange them in such a way that the impedance behaviour in the Nyquist plot is reproduced exactly (see Section 10.2 below on electrochemical simulation).*

Any real *electrical component has three inherent properties, namely a resistance* R*, a capacitance* C *and an inductance*[†] L *(although we will not look at inductance any further since the frequency ranges employed here do not require it).*

To construct such a circuit, we further assume that a series of 'pure' components exist, each with only one attribute, so a pure resistor has

---

[†] Inductance can be thought of as an electrical analogue to inertia, so current flow is opposed at short times. Strictly speaking, a potential V is induced in the circuit immediately after the circuit is switched on, with V of the opposite polarity to that being applied. The current caused by the application of V tries to stop the *net* flow of current.

no capacitance or inductance, while a 'pure' capacitor has nothing but capacitance.

We will now look at the AC behaviour of these pure electrical components as a function of the frequency ω. (Remember from Chapter 7 that a *linear* frequency of f cycles per second corresponds to an *angular* frequency of $2\pi\omega$ radians per second.)

> **Care** Most AC instruments cite the frequency in f (in cycles per second) The impedance equations used here are cited in terms of ω (in radians per second)

The **resistance** (R) may be defined as an impediment to the flow of electronic charge. Consider a 'pure' resistor (that is, one having no capacitance whatsoever): its resistance when determined with a continuous current[†] is R, and its impedance is frequency-independent. We can say that:

$$Z^*(R) = Z' = R \tag{8.7}$$

Note that R is not described by ω at all, so a resistor is represented on a Nyquist plot by a single point on the x-axis (see Figure 8.10(a)).

Worked Example 8.3. What is the impedance of a pure resistor having a resistance R of $1.2 \times 10^5\ \Omega$.

From equation (8.7), the resistance and impedance of a *pure* resistor are the same, so $Z^* = Z' = 1.2 \times 10^5\ \Omega$.

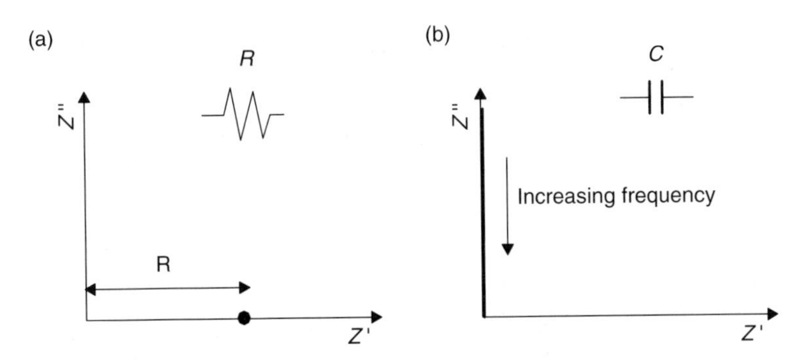

Figure 8.10 Sample Nyquist plots of 'pure' electrical components, shown for (a) a resistor R and (b) a capacitor C.

[†] We refer to a continuous current as a 'direct current' (DC), as opposed to an alternating current (AC).

SAQ 8.10

Platinum wire is used to make an electrode, with the metal behaving as a pure resistor. If the resistance of platinum is 10^{-3} Ω per cm (in length) and the electrode is 7 mm long, what is (a) its resistance and (b) its impedance?

The **capacitance** (C) can be defined as the ability to retain or store charge. The impedance of a capacitor $Z^*_{(c)}$ is given by the following equation:

$$Z^*_{(c)} = Z'' = \frac{1}{j\omega C} \tag{8.8}$$

From our definition of a 'pure' capacitor (i.e. one having no resistive component), we can say that the real impedance Z' is zero. We see straightaway from equation (8.8) that the impedance is a function of frequency. The impedance of a capacitor is infinite when a DC voltage is applied (just put $\omega = 0$ into equation (8.8)), while the imaginary impedance Z'' decreases as the frequency ω is increased.

Worked Example 8.4. What is the impedance of a pure capacitor with a capacitance of 10^{-12} F at a frequency of 5×10^5 Hz?

We first convert from linear to angular frequency by saying $\omega = 2\pi f$ (equation (7.4)), which gives:

$$\omega = 2 \times 3.142 \times 5 \times 10^5 \text{ s}^{-1} = 3.14 \times 10^6 \text{ rad s}^{-1}$$

Next, we insert the appropriate values into equation (8.8) as follows:

$$Z^*_{(c)} = \frac{1}{j \times 3.142 \times 10^6 \times 10^{-12}} \ \Omega$$
$$= 3.18 \times 10^5 j\Omega$$

Notice that if we are strict with ourselves, we should really include the 'j' term, although we should note that most workers don't do this. If we had written Z'', then the 'j' term would not have been needed since an imaginary impedance presupposes the inclusion of this term.

SAQ 8.11

The electrode | solution double-layer sometimes behaves as a capacitor. By assuming that it behaves as a pure capacitor with a capacitance of 10^{-7} F, what is its impedance at a frequency of 10^2 Hz?

On a Nyquist plot of Z'' against Z', the impedance of a capacitor is represented by a vertical line along the y-axis, with its value descending this axis as the frequency increases, going from $Z' = 0$ when $\omega = \infty$ to $Z'' = \infty$ at $\omega = 0$. This behaviour is illustrated in Figure 8.10(b). It is common to see this feature of a pure capacitor described as a **capacitive 'spike'**, because of the vertical nature of this line.

As a rule, when thinking in terms of genuine electrical components, rather than 'pure' ones, we can still say that the real impedances Z' behave largely like resistors while the imaginary impedances Z'' behave largely like capacitors.

We shall now consider how these components of an 'equivalent circuit' behave in combination. There are two types of circuit we need to think about here, i.e. with the components in series and with the components in parallel.

The total resistance of several resistors placed **in series** (Figure 8.11(a)) is simply the sum of the individual resistances. Similarly, the total impedance of a

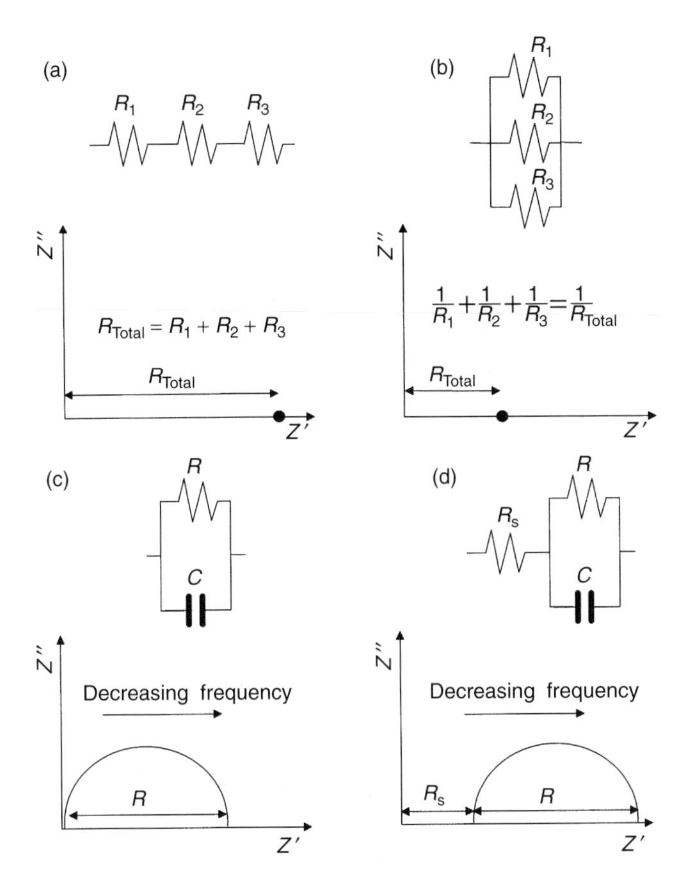

Figure 8.11 Nyquist plots for circuits comprising more than one electrical component: (a) pure resistive impedances in series; (b) pure resistive impedances in parallel; (c) an RC element; (d) an RC element in series with a resistance.

circuit comprising impedances in series is, although conceptually more complicated, the same type of calculation, as follows:

$$Z_{\text{total}} = Z_1 + Z_2 + Z_3 + \cdots + Z_n = \sum Z_i \tag{8.9}$$

Worked Example 8.5. Three pure resistors are joined together in series. What is the total impedance if $R_1 = 10$, $R_2 = 210$ and $R_3 = 3.1\ \Omega$?

We first note that all the resistances are pure, so each impedance is frequency-independent. Secondly, we note that the impedances Z are the same as the resistances R: $R_1 = Z_1$, $R_2 = Z_2$ and $R_3 = Z_3$.

By applying equation (8.9) we then simply add up the values of the three impedances:

$$Z_{\text{total}} = Z_1 + Z_2 + Z_3 = (10 + 210 + 3.1)\ \Omega = 223.1\ \Omega$$

We notice that impedances, when combined in series, generate a *bigger* impedance than any of the constituents. We should also note that these values of Z may be a function of frequency if Z represents components other than pure resistors.

SAQ 8.12

The impedances of a cell's leads, contacts and electrolyte solution can be approximated to a situation of impedances in series. Typical values will be $Z_{\text{leads}} = 1$, $Z_{\text{contacts}} = 5$ and $Z_{\text{electrolyte solution}} = 40\ \Omega$ (with all Z values being cited at fixed frequency). Calculate the total impedance of these three components when combined together.

The total impedance of a circuit comprising impedances placed **in parallel** is obtained by taking reciprocals, as follows:

$$\frac{1}{Z_{\text{total}}} = \frac{1}{Z_1} + \frac{1}{Z_2} + \frac{1}{Z_3} + \cdots = \sum \frac{1}{Z_i} \tag{8.10}$$

Worked Example 8.6. The three impedances, $Z_1 = 12.5$, $Z_2 = 100$ and $Z_3 = 0.5\ \Omega$ (at fixed frequency), are placed in parallel. What is their combined impedance, i.e. Z_{total}?

By using equation (8.10), and inserting the appropriate values:

$$\frac{1}{Z_{\text{total}}} = \frac{1}{12\ \Omega} + \frac{1}{100\ \Omega} + \frac{1}{0.5\ \Omega} = (0.08 + 0.01 + 2)\ \Omega^{-1} = 2.09\ \Omega^{-1}$$

which gives:

$$Z_{\text{total}} = \frac{1}{2.09\ \Omega^{-1}} = 0.48\ \Omega$$

Again, we note that the values of Z are a function of frequency. We also note how these impedances, when combined in parallel, generate a *smaller* impedance than any of the constituent components.

SAQ 8.13

An electrode bears a layer of indium–tin oxide (ITO) having an impedance of 25 Ω, on which is a layer of adsorbed chromophore having an impedance of 1.0 Ω (all values of Z being cited at fixed frequency). In addition, between the chromophore layer and the bulk electrolyte is the Helmholtz double-layer (see Section 5.1.2), which has an impedance of 120 Ω. By assuming that these three layers act as impedances in parallel, calculate the total impedance, Z_{total}.

We must now consider more complicated arrangements of electrical components. One of the most common arrangements we will encounter is that of a resistor and a capacitor placed in parallel. This combination is so common that we give it a special name, calling it an **RC element**. This arrangement is found often in electrochemistry, e.g. as a useful first approximation to the electric double-layer or other thin films (cf. SAQ 8.13). Such an arrangement of components yields a Nyquist plot with a semicircular arc of diameter R (see Figure 8.11(c)). Higher frequencies are represented on the left-hand side of the arc, with lower frequencies on the right. In fact, this is a *general* observation – frequency always decreases on going from left to right on a Nyquist plot.

Remember The left-hand side of a Nyquist plot represents *higher* frequencies
The right-hand side of a Nyquist plot represents *lower* frequencies

The apex of the semicircle obeys the **Maxwell relationship**, as follows:

$$1 = \omega_{max} RC \tag{8.11}$$

The time corresponding to one cycle of frequency, ω_{max}, is occasionally termed the **time constant** of the arc (or of the RC element causing the arc).

SAQ 8.14

What is the time constant of the RC element constructed when $R = 1$ kΩ and $C = 50$ nF? (Note – remember to use angular frequency.)

A more realistic picture of the double-layer has an RC element (that is, a capacitor and resistor in parallel) itself in series with a second resistor R_s (see Figure 8.11(d)). This circuit yields a similar Nyquist plot to that of an RC element alone (Figure 8.11(c)), but the semicircular arc in this case is offset from the origin by an impedance of R_s.

We will now look at a real electrochemical cell, and see how an equivalent circuit yields information about the electron-transfer processes occurring within it.

8.2.2 Impedances of Real Cells: Quantification of Diffusion Phenomena and the Warburg Impedance

To summarize the impedance discussion so far: an electrochemical cell is constructed, and its impedance Z^* determined as a function of frequency. From these impedance values, the real and imaginary impedances, Z' and Z'', respectively, are computed and hence a Nyquist plot is drawn.

It is now time to look at the Nyquist impedance plot of a real cell. Figure 8.12(a) shows such an impedance plot for the all-solid-state cell, ITO/WO_3/PEO–H_3PO_4/ITO(H),† at 8°C. The two ITO layers are needed as transparent electronic conductors (cf. Section 8.1.2).

We recall that frequency drops as we go from the left-hand to the right-hand side of the plot. In other words, the impedances nearest to the origin were determined at high frequencies.

This plot may be taken to consist of five regions, as follows:

(i) A high-frequency intercept arising in the main from ohmic resistances of the ITO electrodes, but which also comprises the resistance of the leads and contacts.
(ii) A semicircular arc representing the polymer electrolyte (PEO–H_3PO_4) layer.
(iii) Another (less well defined) semicircular arc corresponding to charging of the electrolyte–electrode interface.
(iv) A linear region inclined to an angle of c. 45° which indicates that diffusional processes of some kind are involved, e.g. of H^+ moving through the WO_3. This feature is termed a **Warburg impedance** (Z_w).
(v) A steep spike suggesting **blocking** electrode behaviour, as below.

The above features are assigned on the basis of a detailed examination of the experimental data, and by reference to the equivalent circuit for such a call (Figure 8.12(b)). We will now discuss each of these regions in turn.

Region (i). First, we note that there is a slight offset near the origin of the Nyquist plot, as caused by the resistances of the ITO layers, leads and contacts. We will call this an **ohmic resistance** because we do not know any further details about the features in the gap. Note that it is merely a 'gap' because the frequencies

† PEO–H_3PO_4 is a viscous eutectic electrolyte of relatively high ionic conductivity; ITO(H) is indium-tin oxide containing protonic charge.

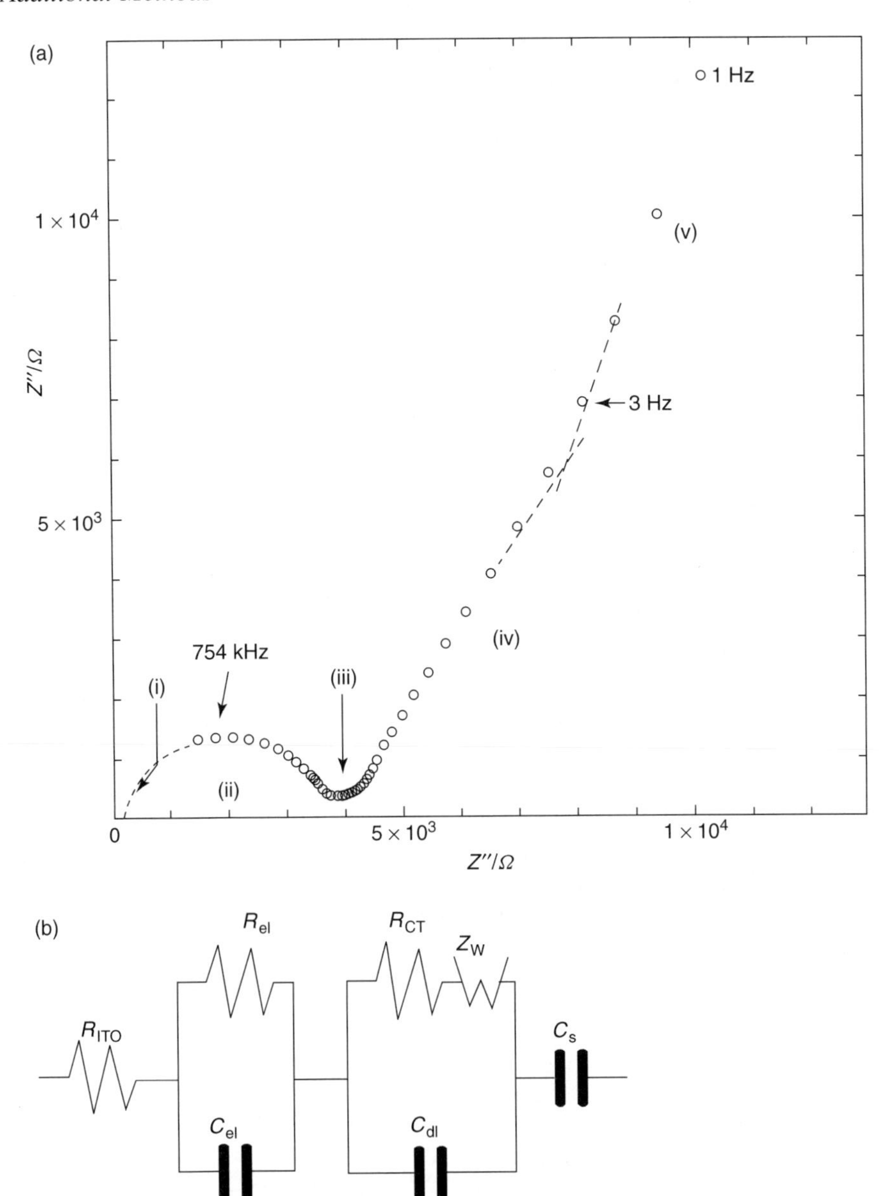

Figure 8.12 (a) Nyquist plot obtained for the all-solid-state cell, ITO/WO_3/PEO–H_3PO_4/ITO(H) at 8°C, with the electrolyte being unplasticized. The WO_3 layer was 0.3 μm in thickness (as gauged during vacuum evaporation with a thin-film monitor), while the electrolyte thickness was 0.24 mm (achieved by using 0.3 mm spacers of inert plastic placed between the two ITO electrodes). (b) Schematic representation of the equivalent circuit for this cell.

obtainable with the frequency analyser are too low, with the maximum of such an apparatus being usually about 10^6 Hz. Access to higher frequencies would have allowed this gap to be filled.

We will denote this resistance in the equivalent circuit as R_{ITO}.

Region (ii). The next feature to be seen in the Nyquist plot is the first semicircular arc, which falls at frequencies in the range 1 MHz $> \omega >$ 50 kHz, and represents the polarization of the electrolyte layer.

DQ 8.19

How do we know that this region represents the electrolyte layer?

Answer

We can show that the resistance of the electrolyte corresponds to the arc in region (ii) as follows. First, we realize that an arc is generated by an RC element, and then that the diameter of the arc has the value of the resistance R *for the RC element which generates it.*

Next, we show that this value of $R_{diameter}$ *is the same as the resistance of the electrolyte layer* R_{el}*, calculated by considering the electrolyte conductivity* σ *and the cell constant* κ *(both of which are known), by using the following relationship:*

$$R_{el} = \frac{\kappa}{\sigma} \tag{8.12}$$

Worked Example 8.7. Show that $R_{diameter}$ from the RC element and R_{el} are the same. The electrolyte conductivity[†] is 1.4×10^{-6} S cm^{-1} at 8°C, and the electrolyte layer has a thickness l of 0.03 cm and an area A of 4.2 cm^2. The diameter of the arc in region (ii) of Figure 8.12(a) is c. 4000 Ω.

In order to calculate the resistance due to the electrolyte, we recall that the cell constant κ is defined as l/A, and so its value is:

$$0.024/4.2 \text{ cm}^{-1} = 5.71 \times 10^{-3} \text{ cm}^{-1}$$

Then, from equation (8.12):

$$R_{el} = \frac{5.71 \times 10^{-3} \text{ cm}^{-1}}{1.4 \times 10^{-6} \text{ S cm}^{-1}} = 4081 \ \Omega$$

It can therefore be seen that the agreement between the calculated and experimental values of R_{el} is actually quite good, with the probable cause of the calculated value being too large resulting from uneven electrolyte thickness.

[†] Here, S is the SI unit of siemen, defined as reciprocal ohm, Ω^{-1}.

We can say then that the feature yielding the semicircular arc in region (ii) is *likely* to be the electrolyte layer, with the latter having both resistance and capacitance, R_{el} and C_{el}, respectively.

DQ 8.20

Can the frequency of the arc maximum give us any kinetic information?

Answer

Yes – the duration of one AC cycle is called the ***cycle life****, τ. The time required for a process to occur during the AC experiment is represented by the frequency at ω_{max}, as follows:*

$$\tau = \frac{1}{f} = \frac{1}{2\pi\omega_{max}} \tag{8.13}$$

For feature (ii), the cycle life τ relates to the time required for the proton to completely move across the electrolyte layer, from one side to the other.

SAQ 8.15

What is the duration of one cycle if in Figure 8.12(a) $\omega_{max} = 754$ Hz?

DQ 8.21

What does the time calculated in SAQ 8.15 represent physically?

Answer

From SAQ 8.15, it can be seen that the mobile H^+ ions move very fast through the electrolyte layer. Remember that the AC voltage is sinusoidal, so the polarity of each electrode is changing sign at a frequency of $2 \times f$, i.e. twice per cycle from the way that a cycle is defined (see Figure 8.8).

During the part of the cycle when the ITO is negative, the H^+ ions align themselves with the ITO–electrolyte interface. Similarly, during the other half of the cycle when the WO_3 is negatively polarized, H^+ moves through the electrolyte in the opposite direction to realign with the WO_3–electrolyte interface.

In fact, many semicircular features in Nyquist plots represent layers that have both resistive and capacitive components, i.e. they behave as RC elements.

Region (iii). After the semicircular arc representing the charging of the electrolyte layer, the Nyquist plot in Figure 8.12(a) shows a second, much smaller, semicircular arc in the frequency range >50–80 kHz. This second arc is sufficiently small that we merely see here a levelling of the trace between features (ii) and (iii). Accordingly, the apex of the arc is not easily to discern with any accuracy, but falls at a value of about $\omega_{max} \approx 100$ Hz. This feature represents the movement of electrons across the WO_3 | electrolyte interface.

DQ 8.22

Why should the movement of electrons across an interface generate a semicircular arc?

Answer

The double-layer consists of ions juxtaposed with the electrode, so it resembles a plate capacitor (e.g. see Figure (5.3)) – we will describe this in terms of its ***double-layer capacitance****,* C_{dl} *(cf. Chapter 5).*

The movement of the electron is activated, i.e. it requires energy. Stated another way, movement is restricted – there is a resistance. We call this the ***resistance to charge transfer****,* R_{CT}*. The latter parameter and* C_{dl} *behave as though being in parallel, and hence the appearance of the semicircular arc in the Nyquist plot.*

DQ 8.23

How is the resistance to charge transfer related to the kinetic parameters discussed in the previous chapter?

Answer

The resistance to charge transfer is related to the exchange current density, i_0*, by the following relationship:*

$$R_{CT} = \left(\frac{nF}{RT}\right)\frac{1}{i_0} \tag{8.14}$$

We recall from Section 7.5.2 and equation (7.17) that the exchange current relates to the rate constant of electron transfer, k_{et}*. The redox process described is the flow of electrons into and out of the layer of solid* WO_3 *via the electrode double-layer.*

SAQ 8.16

The magnitude of R_{CT} is about 200 Ω (from the diameter of the arc in Figure 8.12(a)). Calculate the value of i_0 for the electron-transfer reaction occurring in the cell. (Remember to convert the temperature of 8°C into its thermodynamic temperature; take $n = 1$.)

Region (iv). This region of the Nyquist plot displays a linear portion at an angle of about 45°, occurring at frequencies in the range $50 > \omega > 15$ Hz. While the rate-limiting process was the movement of *electrons* in region (iii), in this region of the plot, the rate-limiting feature is the movement of *ions* into and out of the solid layer of WO_3.

We recall again that the frequency f decreases when passing from left to right on a Nyquist plot. As f decreases, so the cycle life τ increases. Longer cycle lives clearly allow the ions a progressively longer time to diffuse into and through the solid layer of WO_3.

Electrons are small, and move comparatively fast through WO_3, that is, when compared with ionic motion. Therefore, a concentration gradient of H^+ soon forms, curving quite steeply from the WO_3–electrolyte interface and decreasing through the layer of solid WO_3 in much the same way as we saw previously in Section 6.2.1 (Figures 6.3(b) and 6.4).

The value of $[H^+]$ at the WO_3–ITO interface will be effectively zero, but because the proton enters from the electrolyte layer, $[H^+]$ can be quite high at the surface of the WO_3 in contact with the electrolyte.

DQ 8.24

How does a concentration gradient give rise to the feature observed in region (iv) of the Nyquist plot?

Answer

Although there is a smooth *gradient of* $[H^+]$ *inside the film of* WO_3*, we will continue with our discussion as though the* WO_3 *is composed of an infinite sandwich of many millions layers, each with a slightly different concentration of* H^+.

Each of these layers behaves just like an RC element (that is, a capacitor and resistor in parallel) within the equivalent circuit (see Figure 8.13). The respective values of R_i *and* C_i *will be unique to each RC element since each layer has a distinct value of* $[H^+]$*. In order to simplify the equivalent circuit, this infinite sum of RC elements is given the symbol* Z_W *or* W̶ *and is termed a* **Warburg impedance**, *or just 'a Warburg'. The Warburg in Figure 8.12 extends from about 50 down to 15 Hz.*

As a good generalization, the linear portions of a Nyquist plot (the line inclined to an angle of about 45°) always implies diffusion, usually of ions.

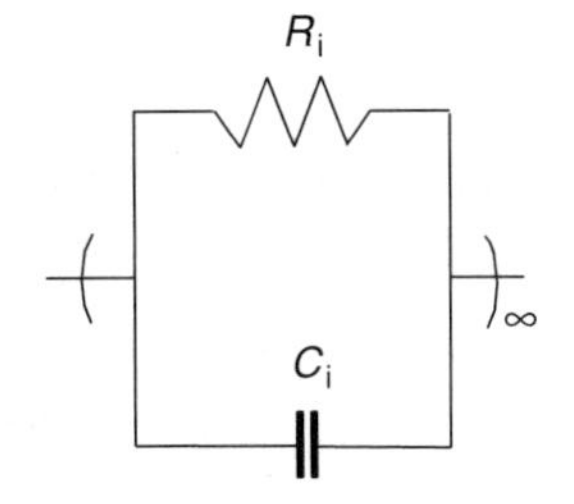

Figure 8.13 Schematic representation of a Warburg impedance Z_W as an infinite sum of RC elements, which is commonly employed in an equivalent circuit as a simple model for a concentration gradient.

DQ 8.25

If the Warburg relates to diffusion, can the diffusion coefficient D be calculated from it?

Answer

Yes, in principle, D *can indeed be determined from analysis of the time-dependence of* Z^**, by using only those frequencies represented by the Warburg, as follows. First, we draw a graph of* Z' *and* Z'' *(both on the 'y' axis) against* $\omega^{-1/2}$ *(as 'x'). The Warburg is seen in such a plot by the frequencies at which the two lines, for* Z' *and* Z''*, are parallel (see Figure 8.14).*

Next, we use the following relationship, which we sometimes call the ***Huggins*** *equation [1]:*

$$Z^* = \left(\frac{d\,\mathrm{emf}}{d[H^+]}\right) \times \frac{V_m \omega^{-1/2}}{nFAD^{1/2}} \tag{8.15}$$

where V_m *is the molar volume of the solid (which is usually available from textbooks – we use the value for the bulk solid), and* $(d\,\mathrm{emf}/d[H^+])$ *is the gradient of a coulometric titration curve of cell* emf *(as 'y') against* $[H^+]$ *in the* WO_3*. The value of* $[H^+]$ *is obtained from the faradaic charge passed. All other terms in equation (8.15) have their usual meanings.*

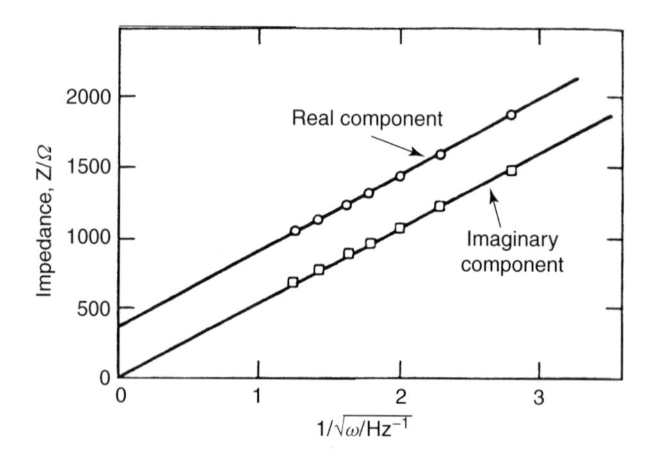

Figure 8.14 'Huggins analysis' of a Warburg element in a Nyquist plot such as that shown in Figure 8.12(a), for the diffusion of Li^+ ions through solid-state WO_3. The traces for Z' and Z'' against $\omega^{-1/2}$ will not be parallel for features other than that of the Warburg. From Ho, C., Raistrick, I. D. and Huggins, R. A., 'Application of AC techniques to the study of lithium diffusion in tungsten trioxide thin films', *J. Electrochem. Soc.*, **127**, 343–350 (1980). Reproduced by permission of The Electrochemical Society, Inc.

Obtaining a coulometric titration curve is often far from straightforward, since the requirement for faradaic *charge insertion makes for slow and painstaking work.*

SAQ 8.17

The gradient of a graph of Z' and Z'' against $\omega^{-1/2}$ is 85.4 Ω (rad s^{-1})$^{-1/2}$ where the Z' and Z'' portions are parallel. By taking the molar volume of WO_3 to be 42.2 cm^3 mol^{-1} and the gradient of a coulometric titration curve of (d*emf* /d[H^+]) to be 5.2×10^{-3} V, calculate the diffusion coefficient D for the mobile protons as they move through the WO_3 (the area A of the cell was 4.0 cm^2, and $n = 1$).

Region (v). Finally, we will look at region (v) in the Nyquist plot of the ITO/WO_3/PEO–H_3PO_4/ITO(H) cell, which displays a near-vertical **spike** similar to that of a capacitor alone (e.g. see Figure 8.10(b)). As the applied frequency decreases, so the spike gets steeper, i.e. Z'' gets bigger. Such a spike represents frequencies which are lower than about 10 Hz.

DQ 8.26

The nearest impedance feature to a 'spike' we have seen so far was shown in Figure 8.10(b), which represented the Nyquist plot for a capacitor. Why is the feature displayed in region (v) called a 'spike'?

Answer

This portion of the AC plot was obtained when applying quite small frequencies corresponding to cycle lives, τ, in the range 0.1–10 s or so. The feature in region (v) tells us that the WO_3 layer (or at least its surface) has become saturated with protonic charge, so any further charge for insertion into the WO_3 must enter through a layer which already contains a high concentration of H^+.

However, although the WO_3 surface is 'filled', ions still move from the electrolyte 'reservoir' toward the WO_3–electrolyte interface. Such a situation results in the accumulation of charge at this interface. In effect, we have a structure which is physically very similar to that of a typical plate capacitor (see Figure 5.3). For this reason, the equivalent circuit (see Figure 8.12(b)) also contains a capacitor C_s *(where the subscript denotes 'surface').*

The electrical components within the impedance plot are listed in Table 8.1. In summary, we see that a Nyquist plot of imaginary against real impedances can be dissected piece by piece, with each component representing a physical part of the cell or a kinetic phenomenon. We see that impedance analysis is a powerful and versatile tool which is capable of discerning the individual processes

Table 8.1 Interpretation of the features displayed in Figure 8.12(a), the Nyquist plot for the all-solid-state cell, ITO/WO_3/PEO–H_3PO_4/ITO(H)

Region	Feature	Physical representation	Symbolism in equivalent circuit[a]
(i)	Offset on real axis	Ohmic resistance due to ITO, leads and contacts	Resistance, R_{ITO}
(ii)	Semicircular arc	Movement of ions through the electrolyte layer	RC element made up of R_{el} and C_{el}
(iii)	Semicircular arc	Movement of electrons across the electrical double-layer	RC element made up of R_{CT} and C_{dl}
(iv)	Straight line at an angle of 45° (a 'Warburg' impedance)	Diffusion of protons through the WO_3 layer	Infinite array of RC elements[b]
(v)	Near-vertical 'spike'	Build-up of charge at the electrode–electrolyte interface	Capacitor C_s

[a] See Figure 8.12(b).
[b] Abbreviated as W̶.

(as a function of the time-scale), and furthermore it also provides a method of quantifying these.

8.3 Causes and Treatment of Errors

8.3.1 Discontinuities in an OTE Conductor

The most common cause of trouble when working with optically transparent electrodes (OTEs) is scratches in the thin film of conductor. Since the electrode is coated with a thin layer of semiconductor, a scratch can in effect cut right through the conductive layer, thus causing an insulatory channel. The two parts of the semiconductor on either side of the scratch are therefore prevented from 'communicating' with each other, and so the portion of the electrode beyond the scratch is rendered useless.

Prevention is better since cure is impossible, and therefore care during experimental use is recommended.

8.3.2 Total Internal Reflection in UV–Visible Spectroscopy

A second problem that occurs with OTEs is the need to avoid **total internal reflection** (TIR) within the layer of glass substrate. TIR is best avoided by ensuring that the incident light beam of the spectrometer (which is horizontal) strikes the glass at a perpendicular angle, which is achieved by positioning the OTE exactly vertically within the cell housing.

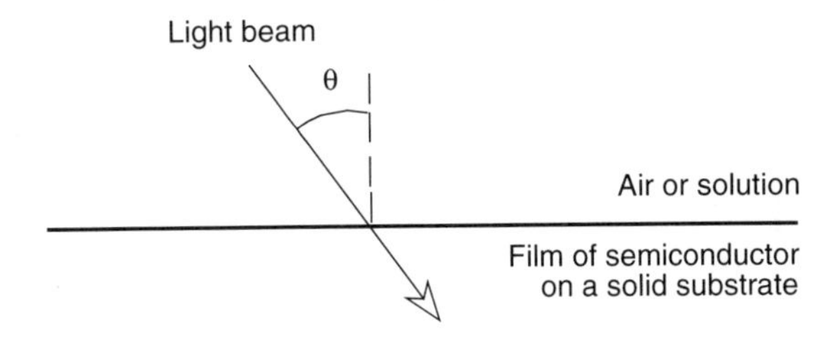

Figure 8.15 The angle of incidence between the incoming light beam and the glass of an OTE is defined as θ.

DQ 8.27

Why is total internal reflection a problem?

Answer

Total internal reflection (TIR) occurs if the angle of incidence θ between the light beam and the glass substrate is greater than 0° from perpendicular. Figure 8.15 shows the way that θ is defined. TIR in the glass causes a phenomenon called ***ringing****, which produces a nasty 'ripple' superimposed on top of the spectrum of the analyte. Quantitative work can be severely impeded, if not made impossible, by this ringing. Figure 8.16 shows a spectrum of thin-film tungsten trioxide on ITO, itself on glass. This spectrum clearly shows the ringing effect.*

8.3.3 Large Absorbances and the Optical Band Edge

During *in situ* UV–vis spectroelectrochemical work, it is easier to obtain spectra by using a single-beam instrument. At the start of the experiment, the analyst sets the absorbance to zero with the *in situ* cell placed in the path of the beam, so the cell then acts as a spectroscopic **blank** or ('**reference**'). Any changes in absorption will relate to the changes in the amounts of each of the redox states within the cell, rather than from the cell itself.

Glass appears transparent because none of the components in the glass absorb at wavelengths corresponding to the sensitivity of the eye. However, in fact normal 'soda' glass has an intense optical absorption below about 250 nm owing to the presence of oxide ions, which have a large value of ε. Similarly, indium–tin oxide has a massive absorbance at wavelengths less than about 320 nm due to the indium(III) ion. Most glass also contains traces of iron too.

The result of having indium ions on the glass is therefore to prohibit all light from transmitting through the optical electrode at wavelengths less than 320 nm. Stated another way, the OTE is wholly opaque below 320 nm. We say that the OTE has a **band edge** of 320 nm, meaning that it can be used above $\lambda = 320$ nm but not at lower wavelengths.

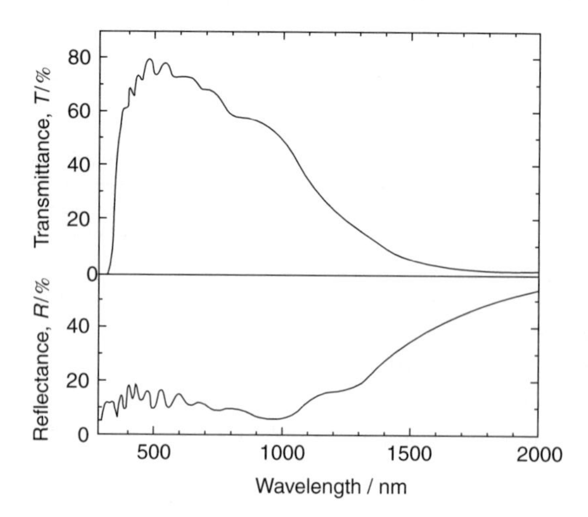

Figure 8.16 Illustration of the problem of total internal reflection (TIR) in UV–vis spectroscopy, causing 'ringing', shown by the spectrum of the cell, ITO | WO_3 | solid-state electrolyte | ITO. Note how this figure is shown as transmittance T against wavelength λ to emphasize the oscillating nature of TIR; transmittance and absorbance are related as $Abs \propto -\log T$. For comparative purposes, the corresponding reflectance (reflectivity) (R) trace is also shown. Reprinted from Kamimori, T., Nagai, J. and Mizuhasi, M., 'Electrochromic devices for transmissive and reflective light control', *Solar Energy Mater. Solar Cells*, **16**, 27–38 (1987), Copyright 1987, with permission from Elsevier Science.

DQ 8.28

Why does working beyond such a 'band edge' cause errors?

Answer

*Absorbance may be defined loosely as the amount of light that passes through a sample or, more specifically, we can define the absorbance (*Abs*) according to equation (8.1). If* T $\approx$ *0 before the sample is generated, then the electrochemical generation of chromophore will only cause* T *to be closer to zero – this is equivalant to passing a hand before your eyes if there's a metal screen in the way and expecting to see less light!*

From equation (8.1), if both the numerator and denominator terms are very similar, then small fluctuations in either can have a profound influence on Abs, *even allowing this to be zero if the error in reading* T *is appreciable. Such errors are likely to be huge. In effect, below the band edge, absorbances are worthless.*

As a generalization, we can also say that the accuracy of an absorbance measurement (e.g. for use within the Beer–Lambert law (equation (8.2))) decreases as the absorbance of the blank increases.

SAQ 8.18

The same amount of chromophore was electrogenerated in two experiments, as monitored within an *in situ* spectroelectrochemistry cell. Calculate the optical absorbance changes, ΔAbs, in the following two cases: (a) using a cell with a large initial absorbance (with $T \approx 0$), such that $T_{\text{with no sample}} = 0.04\%$, and $T_{\text{with sample}} = 0.039\%$; (b) using a cell with a small initial absorbance, such that $T_{\text{with no sample}} = 100\%$, and $T_{\text{with sample}} = 50\%$.

8.3.4 Stray EPR Absorptions

Solvent effects. As any modern cook knows, a microwave oven is a superb means of cooking: any foodstuff containing water, when irradiated with microwaves of sufficient frequency, will absorb energy and so the food gets hot and therefore 'cooks'. Unfortunately, microwaves are at the very heart of EPR spectroelectrochemistry experiments, so care is therefore required to remove all traces of water.

The best solvents for *in situ* spectroelectrochemical EPR analysis are aprotic ones, such as acetonitrile, dimethylformamide (DMF) and propylene carbonate (PC).

Polymers containing radicals. A related problem arises from the way in which an *in situ* EPR cell is constructed. Many monomers polymerize via a radical mechanism and, although 'set' and hard, the polymer may still contain traces of paramagnetic radical monomer species. For this reason, EPR cells tend to be constructed from silica and inorganic adhesives and cements.

DQ 8.29

Why are radicals still present within a hardened 'set' polymer?

Answer

It is rare for a polymerization reaction to proceed to completion *because the viscosity of a polymer increases during chain growth – this is why monomers are usually liquid, while the polymers produced are hard solids. The last step during polymer formation is a termination reaction known as 'radical annihilation', in which two radicals meet and then spin-pair to form a covalent bond.*

Complete annihilation is not possible during polymerization because the polymer becomes so rigid that radicals are immobilized, thereby preventing annihilation. It is these immobilized paramagnetic radicals that preclude the use of most polymers in EPR work.

Summary

Having defined *in situ* and *ex situ* methodology, we have seen that *in situ* spectroelectrochemistry (simultaneous electrochemistry and spectroscopy) is a powerful technique for studying electrode processes.

In situ UV–visible spectroscopy monitors the energy of electrons within the analyte. While, strictly speaking, all materials change their UV–visible spectrum in accompaniment with electrode reactions (they are said to be *electrochromic*), the majority of these changes are not discernible by the human eye, and therefore may not be useful to the analyst. Electrodes must be optically transparent for *in situ* work, with the most commonly used being a thin film of the semiconductor, indium–tin oxide, on glass.

In situ methods can help to identify an analyte, but also enable the calculation of electrode faradaic efficiencies by concurrent use of the Beer–Lambert law and coulometry.

The most common errors encountered *in situ* UV–vis spectroelectrochemistry result from total internal reflection (TIR) (causing 'ringing') and carrying out analyses at wavelengths beyond the band edge.

In situ EPR spectroelectrochemistry monitors paramagnetic species, usually radicals in solution. The chemical stability of such species can be readily determined by this technique. It was seen that the most common problems encountered with *in situ* electrochemical EPR work emanate from the use of absorbing solvents and polymers containing paramagnetic impurities.

Having introduced the concept of impedance, Z (and its nomenclature), the methodology involved in obtaining impedance data was outlined. It was seen that the most powerful way to interpret impedance data of an electroanalytical nature was to plot the imaginary impedance Z'' (as 'y') against the real impedance Z' (as 'x'), thus constructing a so-called Nyquist plot.

Having discussed the shapes of Nyquist plots for simple electrical components, and for common groupings of these components, it was seen that the Nyquist plot for an actual electrochemical cell can be mimicked by constructing an 'equivalent circuit' made up of suitable arrangements of such components.

Reference

1. Ho, C., Raistrick, I. D. and Huggins, R. A., *J. Electrochem. Soc.*, **127**, 343–350 (1980).

Chapter 9

Electrode Preparation

Learning Objectives

- To recall from previous chapters that the quality of an analysis decreases (if it is not invalidated altogether) when the electrode surface is dirty.
- To appreciate that the most common causes of contamination for solid metallic electrodes are coatings of oxide and adsorbed organic materials.
- To learn that abrasion, e.g. with diamond-dust rubbing compound or alumina, is the best way of removing these surface-bound contaminants.
- To learn why removal of adsorbed species by soaking the electrode in an organic solvent, flaming in air or by dipping in aqua regia will often cause additional problems with contamination.
- To appreciate that it is advisable to immerse the whole electrode in solution, rather than just its tip, because of surface tension effects, and to learn that encasing the metal of the electrode within a glass sleeve is the best way to avoid ambiguity in this respect.
- To learn how to make electrical contacts to electrodes, e.g. with conductive silver paint.
- To learn how to test if an optically transparent electrode made of indium–tin oxide (ITO) is dirty or over-reduced, and therefore unsuitable for electroanalyses, just by visual examination.
- To learn how to construct your own silver–silver chloride reference electrode, and learn how to test its quality with a simple test of its standard electrode potential $E^{\ominus}_{AgCl,Ag}$.
- To learn how to make a microelectrode, and how to clean it.

9.1 Preparation and Characterization of Solid Electrode Surfaces

9.1.1 Cleaning Electrode Surfaces

We start by assuming that the electroanalyst has a ready supply of freshly distilled mercury. The usual contaminants in impure mercury are other metals, so the mercury represents a dilute amalgam. Distillation removes these metals.

However, most of the electroanalyst's experiments require solid electrodes, which naturally fall within two categories, namely either inert electrodes such as platinum, gold and glassy carbon, or redox electrodes such as copper, lead or magnesium. We will also consider optically transparent electrodes in this discussion.

Platinum electrodes. Whatever the method chosen, it is important to be aware that *most* organic species will adsorb to platinum, particularly if the adsorbate is charged, unsaturated or contains oxygen or sulfur moieties. Dipping the platinum in a concentrated oxidizing acid such as sulfuric or nitric, or a quick dip in aqua regia, is certain to remove all adsorbed organic species. It is likely, however, that such a powerfully oxidizing environment will also oxidize the surface of the platinum to form a layer of the oxide, PtO_2 (this layer can be as thick as 0.4 μm if immersion is prolonged). Such chemical methods are also unlikely to remove some inorganic adsorbants.

SAQ 9.1

In the context of cleaning electrodes, why do we need to be so careful with aqua regia?[†]

In fact, platinum oxide itself is the most likely contamination on the surface of a platinum electrode. Such layers of PtO_2 are too thin to see under standard laboratory illumination, so these electrodes may still look bright and shiny.

In addition to the chemical removal of adsorbants by using acid, PtO_2 is also formed by 'flaming' the surface of a platinum electrode by a Bunsen burner – another traditional method of removing adsorbed organic materials. The cause of the oxide coating is thermal oxidation, as follows:

$$Pt + O_2 \longrightarrow PtO_2 \tag{9.1}$$

Although PtO_2 is rather inert chemically, its presence causes many problems. First, it represents part of a redox couple in the presence of Pt (albeit a rather irreversible couple), which itself causes errors, particularly during potentiometric

[†] Aqua regia is a mixture of 1 volume of concentrated nitric acid and 3 volumes of concentrated hydrochloric acid.

analyses. Secondly, the oxide is less conductive than Pt alone, thus causing possible irreversibility during voltammetry. Thirdly, the electrochemical reduction of a surface layer of PtO_2 can also cause spurious current peaks during voltammetry.

The layer of platinum oxide may be removed by several methods, as follows.

(i) By abrasion, for example with commercially available diamond-dust rubbing compound (a butter-like paste), which removes the surface of the electrode much as sandpaper does (the diamond dust acting as microscopic 'sand'). Alumina will perform the same process. Subsequent rinsing in methanol removes the hydrocarbon paste in which the abrasive dust is suspended, thus allowing for a clean, fresh surface of Pt metal.

Some workers repeat this procedure with a range of abrasives of different particle sizes, starting with the coarsest and finishing with the smallest.

SAQ 9.2

Why is the above modification to the standard abrasion method adopted?

(ii) By 'electrochemical cleaning'. Such cleaning is effected by immersing two Pt electrodes in dilute acid (sulfuric is the usual choice) and applying a potential between them, just sufficient to cause hydrogen gas to form at the (negative) cathode; oxygen may also form at the anode. The *atomic* hydrogen chemically reduces any PtO_2 at the surface of the polarized platinum. For best results, the polarity of the electrodes should be swapped a few times, so that the electrode is cathodized, anodized, and then cathodized again. However many cycles are performed, the last one must be 'cathodic' if an oxide-free surface is required.

As an additional advantage, any adsorbed materials – inorganic as well as organic – are 'blown away' from the surface at the same time as the oxide layer is removed by this electrochemical procedure.

(iii) Another common choice of cleaning method is to immerse the electrode in an organic solvent such as methanol. Although this method is adopted to remove layers of adsorbed material, it is not particularly efficacious, and can even introduce additional adsorbate. In addition, note that the majority of inorganic adsorbates are not soluble in organic solvents.

DQ 9.1

Are there occasions when we *do* want a layer of PtO_2 on the electrode?

Answer

Sometimes, during oxidative *voltammetry, we obtain a current peak caused by the oxidation of platinum to form* PtO_2*. In order to avoid the formation of such a peak, we can deliberately form a substantial* *layer of platinum oxide on the electrode prior to the voltammetric experiment, e.g. by the anodise–cathodise cycles above, but ending with anodising. By this means, the voltammetric peak owing to reactions such as* $Pt + 2H_2O \rightarrow PtO_2 + 2H_2$ *can be avoided.*

Whether or not we want a layer of platinum oxide, we need to know the state of the electrode surface.

Platinum is the most popular choice of inert electrode for electroanalysis. It will also be clear from the discussion above that electrodes made of platinum should be cleaned thoroughly and as often as possible. Unfortunately, many analytical laboratories employ automated equipment, so cleaning occurs sporadically at best. Worse still, in many industrial laboratories, where the cost and number of each sample are always a consideration, the frequency of electrode cleaning will be a *commercial* decision.

Gold electrodes. Gold is generally less reactive than platinum, so surface layers of oxide are less likely to be present. The major contaminant with gold electrodes is sulfur-containing species such as thiols, mercaptans and even such common inorganic sulfur species as hydrogen sulfide. The S–Au bond is immensely strong.

Gold is too soluble in aqua regia to risk dipping a contaminated electrode in it, so the best way to clean the gold surface is by abrasion, e.g. with diamond-dust rubbing compound or graded alumina, as described above.

SAQ 9.3

What is the reaction occurring as gold dissolves in aqua regia?

We can sometime flame the gold to remove an organic contaminant. Gold is not likely to form an oxide coating by such treatment, although it could melt!

Glassy carbon. We have not really mentioned glassy carbon (GC) before since so few analysts use it. GC is a conductive form of carbon made by pyrolysing carbon or graphite. For this reason, it is sometimes called ‘vitreous carbon’ or ‘pyrolytic carbon’. GC is very hard – as diamond is – thus making it difficult to machine and hence to make electrodes. This hardness also explains why glassy carbon is relatively expensive.

Should the analyst ‘inherit’ a piece of apparatus with GC electrodes, then these electrodes will require cleaning. Oxide formation is not a likely cause of contamination, although adsorption can be extensive. The best way to clean the GC surface is by abrasion with diamond-dust rubbing compound or graded alumina, as described above.

Redox electrodes. The majority of redox electrodes are transition metals, although some p-block elements such as tin and lead and s-block elements such as magnesium, can also function as redox electrodes. The usual contaminant with all of these metals is a layer of oxide, MO_x, which forms on the metal surface. Since '$MO_x|M$' represents a redox couple, the presence of an oxide layer will inevitably lead to errors. The best ways to remove these oxides to yield a clean, fresh metal surface, are by the abrasion and 'electrochemical cleaning' methods described above. (Note, however, that s-block elements tend to dissolve violently in acid, so magnesium or barium redox electrodes should only be cleaned by abrasion.)

When cleaning with acid, the usual choice is nitric (of relatively high concentration), since virtually all nitrate salts are water soluble, thereby ensuring that the products of contamination are removed readily by rinsing.

SAQ 9.4

Why is it advisable to wash a redox electrode with acid only infrequently?

Optically transparent electrodes. In situ spectroelectrochemistry was discussed in the previous chapter. The most common materials for constructing optically transparent electrodes for use in such analyses are thin films of semiconducting oxide deposited on to glass. Such materials are readily available commercially.

Indium oxide doped with tin oxide (indium–tin oxide (ITO)) or fluoride-doped tin oxide (often expressed as SnO_2:F) are the two favoured layers of choice for most electrochemists and analysts. The thickness of such oxide layers is typically 0.3–0.6 μm, so the ITO-coated side of the electrode displays a pleasant purple 'sheen'.

SAQ 9.5

Why do these thin-layer electrodes have a *purple* sheen?

If the ITO-coated side of the electrode does not have a purple sheen, then the electrode is either dirty or damaged. The best way to clean an ITO electrode is with a low-molecular-weight alcohol such as isopropyl alcohol (IPA), which does not adsorb to the oxide layers. Some modern laboratories have advanced facilities for 'burning off' the surface contaminants by using a low-pressure plasma. Few analysts would employ ITO-coated electrodes if they always required such stringent conditions prior to routine analyses.

A second reason for the absence of a purple sheen could be damage to the ITO layer. Thin-film indium–tin oxide is stable in most basic solutions unless the pH is extremely high, but in aqueous acidic solutions the tin oxide will leach out to leave a porous 'honeycomb' of indium oxide, which soon collapses and falls off the electrode. This dissolution process is termed **etching**.

Even if the remaining In_2O_3 persisted, is it relatively non-conductive and so worthless for accurate electroanalysis. However, such etching can be turned to our advantage; if we require a patterned electrode surface, rather than a single expanse, then immersion of an ITO electrode in acid will effect *selective* etching. Those portions of the conductive ITO that we wish to preserve are covered with a protective barrier such as wax, while the parts of the surface to be removed are left bare and are hence susceptible to dissolution. Dissolving in acid in this way is one means by which the patterning for a calculator or digital watch face is effected. The wax is then removed from the undamaged ITO.

A further problem is the tendency for the ITO itself to be reduced. If there is little or no analyte in solution, and the ITO is polarized cathodically in the presence of moisture, then the indium and tin oxides are themselves reduced to metal, according to the following:

$$In_2O_3 + 6e^- \longrightarrow 2In^0 + 3O^{2-} \qquad (9.2)$$

$$SnO_2 + 2e^- \longrightarrow 2Sn^0 + 2O^{2-} \qquad (9.3)$$

'Over-reduction' of this type is easily identified since the electrodes have a metallic sheen, particularly near the electrode edges. Such reduction is wholly irreversible, so the over-reduced ITO electrode has to be thrown away.

Fluoride-doped SnO_2 is more robust, and is less prone to electrochemical reduction of this sort.

9.1.2 Manufacturing Electrodes

Inert electrodes such as gold and platinum, and redox electrode such as copper and zinc, can be made simply and cheaply in the laboratory. In this process, a short cylinder of the metal is soldered to a length of conductive wire,[†] and then sealed within a glass tube (see Figure 9.1). Care is needed since none of the solder can be allowed to come into contact with the analyte solution. Traditionally, the glass is simple soda glass, but in truth it rarely matters what type of glass is used. Soda glass will stick to most metals to form a relatively strong bond.

DQ 9.2

Why should be solder be kept away from the analyte solution?

Answer

Most solder contains elements such as lead, tin, antimony, etc. Current passage through such metals could readily lead to their oxidation, with subsequent contamination of the analyte solution. Furthermore, such redox processes greatly complicate – if not invalidate – any coulometric analyses.

[†] It does not matter what material the wire is made of since it never touches the solution. Copper is the usual choice.

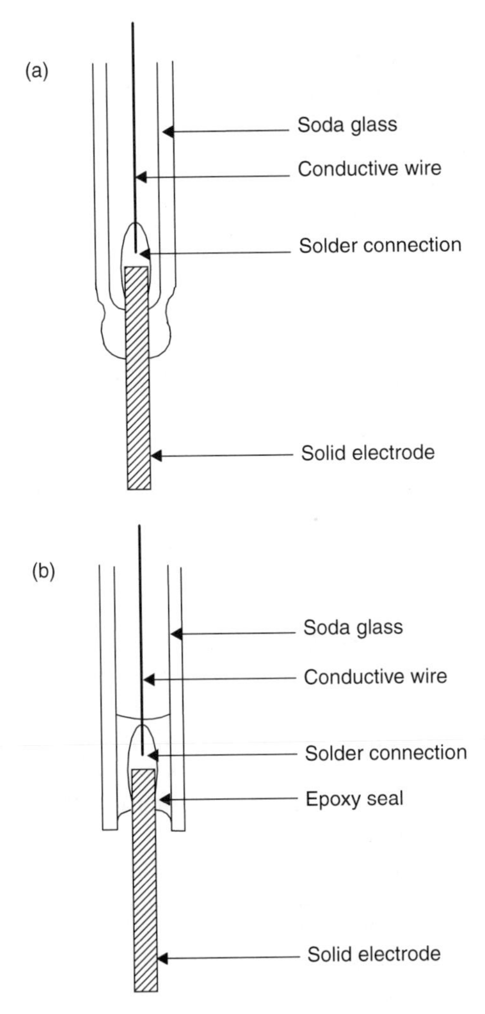

Figure 9.1 Cross-sections of a typical inert or redox electrode, embedded (a) within glass, and (b) within epoxy resin. Type (b) is preferred if the metal is liable to melt or react when very hot.

DQ 9.3

Why use a protective sleeve of glass rather than just immerse a length of, say, copper metal into the analyte solution?

Answer

In most circumstances, the emf *or current from a redox or inert electrode will be the same whether or not the metal is encased within a protective* *sleeve of glass. For the very smallest currents, though, small differences will be apparent. The problem arises from the meniscus between the solution and the air, which possesses a very high surface tension (γ) – particularly if the solution is aqueous.*

A high γ will itself cause a build-up of ions near the air | solution interface. Such ionic accumulation would not be too much of a problem, except that the extent and strength of the build-up is itself a function of the potential of the electrode, thus causing the nature of the air | solution interface to change when the electrode is polarized.

DQ 9.4

How is such a change seen in practice by the analyst?

Answer

***Capillary action** is the term used to describe the way in which a liquid will 'climb' a solid object immersed in that liquid. This is why a tea spoon, when immersed in cup of tea, perturbs the tea | air meniscus. Perturbation of this sort would not be important except that the extent to which the meniscus climbs the spoon (or the analyst's electrode) itself depends on the potential applied. In effect, then, as the potential of the electrode changes, so the* area *of the electrode in contact with the solution changes.*

The area of an electrode in contact with the solution is termed the ***active area** (with the 'non-active' area being that portion* above *the liquid). Accordingly, we can see that the active area of the electrode changes with potential.*

Stated another way, the current density i *(see equation (1.1)) alters with potential, even if the absolute current* I *does not. Clearly, the scope for errors is huge. In order to circumvent this problem, we seal the metal of the electrode within glass, and immerse the electrode such that the metal is wholly submerged. In this way, no metal | air interface can form.*

So, in summary, the best electroanalytical results are obtained with electrodes made by encasing the metal of the electrode within a protective sleeve, and ensuring that the conductive tip of the electrode is wholly submerged.

DQ 9.5

How do we make a contact with an optically transparent electrode or a low-melting-point metal such as lead?

Answer

We cannot solder a wire to a lead redox electrode because the lead would melt. Furthermore, we cannot solder a wire to an indium–tin oxide (ITO) electrode because solder does not adhere to the oxide surface of the ITO. Similarly, solder is useless for making a contact to glassy carbon.

If high temperatures or poor adhesion preclude the use of solder when making an electrical connection, we generally use conductive ***silver paint****. Such 'paint' is in fact a colloidal suspension of silver in a fast-setting polymer matrix. Solvent evaporates during drying, thus causing a large decrease in volume – itself ensuring that the silver particles are brought closer together. When dry, the silver particles in the paint generate a highly conductive contact, generally characterized by a maximum resistance of 1 Ω, and sometimes much less. (The ITO itself has a resistance considerably higher than 1 Ω, so the resistance of the silver paint | electrode contact can be ignored.)*

DQ 9.6

Does the silver itself interfere with the analytical measurements?

Answer

A redox couple is formed as soon as the silver within the paint starts to corrode. Clearly, this can be disastrous. In addition, the silver can enter into solution if the contact is immersed in a reactive solution, again greatly complicating analytical measurements. Thirdly, the silver can be oxidized (e.g. to Ag_2O *or* Ag^+*), thus causing spurious voltammetry peaks and removing all accuracy or chance of computing the faradaic efficiency.*

In order to circumvent these problems, it is usual to cover the silver-paint contact with a thin layer of varnish or epoxy adhesive (see Figure 9.2). This procedure has three advantages, as follows:

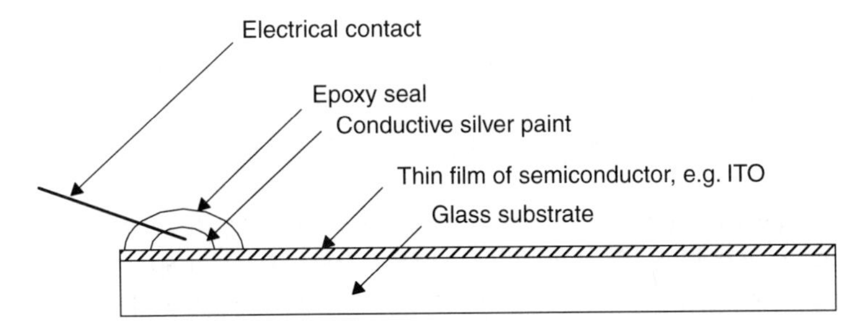

Figure 9.2 Schematic diagram showing how an electrical contact is fixed with silver paint to the conductive side of an optically transparent electrode. The outer layer of epoxy resin is necessary to impart strength, to insulate the silver paint from the analyte solution and to stop analyte solution seeping between the paint and the conductive layer.

(i) It provides an electrical insulation around the conductive silver paint.

(ii) It gives mechanical *strength to the contact, and thereby increases the lifetime of the electrode.*

(iii) (Following from (ii)) The silver | electrode contact itself is not strong, so solution can seep between the silver paint and the electrode, thus destroying all chance of reproducibility. A protective over-layer prevents such seepage.

9.1.3 Making Reference Electrodes

The majority of reference electrodes can be purchased readily and quite cheaply. The major exceptions are the standard hydrogen electrode (SHE) and the silver–silver chloride electrode.†

The SHE is experimentally cumbersome, and is hazardous to use owing to the involvement of elemental hydrogen gas, while the values of E_{H^+,H_2} can fluctuate quite badly during operation because of the cyclic nature of bubble formation. For these reasons, the SHE is avoided experimentally unless a secondary reference electrode requires calibration. We will not consider the SHE any further because it is so unlikely that an analyst would in fact wish to calibrate a new reference electrode.

The **silver–silver chloride electrode** (Ag | AgCl) is easily and cheaply made. Two silver electrodes are cleaned (see Section 9.1.1 above) and immersed in aqueous KCl solution‡ (a concentration of 0.1 mol dm^{-3} is convenient). Next, a potential of about 2 V is applied across them for c. 10 min, causing a thin outer film of silver chloride to develop on the positive electrode. Solid AgCl is formed by a two-step reaction, involving first the electro-formation of silver ion:

$$Ag_{(s)} \longrightarrow Ag^+_{(aq)} + e^- \qquad (9.4)$$

followed by rapid ion-association with chloride ion, with immediate precipitation to form AgCl:

$$Ag^+_{(aq)} + Cl^-_{(aq)} \longrightarrow AgCl_{(s)} \qquad (9.5)$$

This solid layer of AgCl is porous to chloride because of its relatively open and flocculent morphology. Accordingly, the two redox states of silver can 'communicate' and the Ag wire coated with AgCl thus acts as a good-quality reference

† We recall from earlier that this electrode is often termed the SSCE, an acronym we will not be using here because so many authors use it to refer to the sodium-chloride saturated calomel electrode.

‡ It is found, experimentally, that hydrochloric acid (as a source of chloride ion) will yield a film of poor physical strength, so AgCl films disintegrate rapidly.

electrode. The layer of AgCl has a pale beige appearance, and hence its common name of 'brown electrode'.†

SAQ 9.6

What is the Nernst equation for the AgCl | Ag electrode?

DQ 9.7

So, why do we make our own Ag | AgCl electrodes?

Answer

Silver chloride is the active component within black and white photography. When AgCl (in a suitable emulsion 'matrix') is exposed to light, it decomposes rapidly, as follows:

$$AgCl + h\nu \longrightarrow Ag^0 + Cl^- + e^- \qquad (9.6)$$

Atomic silver forms in illuminated regions of the film, and hence the dark colour; silver does not form in regions of the film shielded from light, which is why a photographic negative remains colourless in the dark.

Unfortunately, photochemical degradation of the silver also occurs in the analyst's laboratory, especially if fluorescent strip lighting is employed. (The extent of degradation is much less for an Ag | AgCl electrode than in a photographic film since the electrode is not in contact with the photographer's special mixture of emulsions and enhancers.) Such degradation causes the brown colour of the AgCl to change to purple, and eventually black, as elemental silver forms. For this reason, an Ag | AgCl electrode has a very short 'shelf life', and the layer of AgCl should be regenerated as often as practicable.

SAQ 9.7

Why is *strip* lighting more likely to cause degradation than normal light bulbs?

DQ 9.8

How do we assess the quality of Ag | AgCl reference electrodes made in this way?

---

† Originally, the use of 'brown' here was intended as a pun, since this form of Ag | AgCl was first made popular by Alfred Brown (see [1]).

Answer

By determining a value of $E^{\ominus}_{AgCl,Ag}$*, as follows. The redox reaction of an Ag | AgCl electrode is as follows:*

$$AgCl + e^- \longrightarrow Ag^0 + Cl^- \quad (9.7)$$

while the Nernst equation for this system is given in the answer to SAQ 9.6. AgCl and Ag are both pure phases, so the value of $E_{AgCl,Ag}$ *only depends on* $a(Cl^-)$*. To a good approximation, we can say* $a(Cl^-) = [Cl^-]$*, except at higher concentrations.*

Experimentally, the 'home-made' Ag | AgCl electrode and another reference electrode (e.g. an SCE) are immersed in solutions of varying $[Cl^-]$ *and the* emf *determined for each concentration, e.g. in the range 0.001–0.1 mol* dm^{-3}*. Hydrochloric acid is a suitable source of chloride ion for this experiment. Respective values of* $E_{AgCl,Ag}$ *are then obtained from each* emf *via the use of equation (3.3).*

Next, a graph of $E_{AgCl,Ag}$ *(as 'y') against* $\ln [Cl^-]$ *(as 'x') is drawn, which should be essentially linear, particularly at lower concentrations, thus allowing a reasonably accurate value of* $E^{\ominus}_{AgCl,Ag}$ *to be obtained. We then compare our value of the standard electrode potential with the literature value of 0.2223 V (see Appendix 3).*

SAQ 9.8

Starting with equation (3.8) and SAQ 9.6, show that the graph described in the above discussion should indeed have a y-axis intercept of $E^{\ominus}_{AgCl,Ag}$ and a slope of −59 mV.

9.2 Microelectrodes

Microelectrodes were mentioned previously in Chapter 5, where we saw how their small size increased the faradaic efficiency since the interfacial capacitance C_{dl} is decreased, itself minimizing the charging currents. Microelectrodes can be purchased relatively cheaply, and in a variety of types, e.g. hemispherical and flat circular rings or bands, with a wide range of diameters. Such electrodes were discussed previously in Section 5.3.

The best method of cleaning a flat microelectrode is to abrade the face with alumina or diamond-dust paste (as above) and rinse thoroughly with an organic solvent. A hemispherical microelectrode, however, requires electrochemical cleaning.

It is possible to make a microelectrode by using the following method. As fine a wire as possible is encapsulated within a matrix of non-conductive polymer, such as an epoxy resin, to form a tubular arrangement, with the wire running down its centre. When the epoxy has set, the end of the 'tube' is machined away and is then polished with successively fine rubbing compound (diamond-dust or alumina, as above) until smooth. A suitable electrical connection is then made to the top of the wire at the other end of the microelectrode 'tube'. Cleaning of such microelectrodes can be carried out by using the same procedures adopted for commercially available systems.

9.3 Screen-Printed Electrodes

Occasionally, we prepare a different type of electrode – we *screen-print* them. With modern screen-printing technology, a complicated arrangement of precisely arranged and well-defined electrodes can be prepared on the solid substrate of choice. In essence, these electrodes are simply 'dots' of conductive material, with suitable electrical contacts to ensure that each can be 'addressed' correctly. As an example, each electrode can be connected to a separate channel of a frequency analyser or potentiostat. In this way, a great many analyses can be performed simultaneously.

The 'ink', when screen-printing, is a slurry of conductive carbon particles (sometimes in a solution of conductive resin) that flows smoothly when wet and sets quickly without losing definition and shape. Alternatively, conductive silver paint (as above) will perform the same function.

The screen need not be of a specialized type, and could be a conventional fine-mesh gauze from a standard printing press. If the electrodes are small enough, we can also screen-print microelectrode systems.

Summary

Electroanalysts perform analyses by using electrodes, with the quality of the analysis being closely allied to the quality of the electrode surface. Whether the electrode was bought commercially or made by the analyst, it will require regular cleaning. Several cleaning strategies were recommended, together with a discussion of the various advantages and disadvantages of each of these.

Electrical contact between the electrode and connecting wires can be made with solder if the electrode is a refractory metal, while lower-melting-point metals such as lead, and reactive metals such as magnesium, should be joined to a connection lead with commercially available conductive 'silver paint'. Contact to ITO-coated electrodes will similarly require this conductive paint.

It was shown that both reference electrodes as well as metallic electrodes can be prepared 'in house' by the analyst. For such systems, the best results are obtained if the metal of the inert or redox electrode is first bonded within a protective glass sleeve, and with the metallic tip of the electrode fully immersed below the interface between the analyte solution and air.

It was also shown that silver–silver chloride electrodes are easily made by anodizing a silver wire (all but the tip encased in a glass sleeve) immersed in a chloride-containing solution. A few elementary tests of the quality of such a Ag | AgCl electrode were also described.

Microelectrodes were discussed briefly, including the methods of cleaning, and a procedure for making a simple microelectrode systems. Finally, electrodes prepared by screen-printing were mentioned.

Reference

1. Brown, A. S., *J. Am. Chem. Soc.*, **56,** 646–647 (1934).

Chapter 10
Data Processing

Learning Objectives

- To appreciate how the validity of a measurement is enhanced if it can be simulated mathematically with a reasonable model.
- To learn the basic ideology of how simulation is performed.
- To appreciate that a close fit between experimentally obtained data and a theoretically derived simulation is not proof that the model is correct.
- To learn the meanings of the standard terms of simulation.
- To learn about the market leaders in the field of electroanalytical simulation.

10.1 Simulation of Electrochemical Data

We must first appreciate that electrochemical simulations are probably more useful to the research chemist than the quality control analyst: an analyst involved in quality control is likely to repeat a large number of measurements on a known chemical system, each time asking 'how much', rather than needing to ponder complicated mechanistic questions.

DQ 10.1

What is 'simulation', in the context of electroanalysis?

Answer

To ***simulate*** *in this context means to compile a series of mathematical expressions, in terms of at least the current* I, *potential* E *and time* t, *which then allows an electrochemical trace to be mimicked.*

Worked Example 10.1. What data are required to simulate a *voltammogram*?

We first recall that a voltammogram is a graph of current as a function of applied potential, itself a function of time (according to Figure 6.2(a) above), so we require detailed knowledge for computing the cyclic voltammogram (CV), via the following equations:

(i) The Butler–Volmer equation (equation (7.16)), to describe current as a function of potential, so we need the number of electrons n, the exchange current I_0 and the transfer coefficients α. We will also require the standard electrode potential $E^{\ominus}$, since equation (7.16) is defined in terms of the overpotential η.
(ii) Fick's laws (see Section 2.2), to describe the flux of the analyte, i.e. the time-dependence of analyte as it diffuses toward the solution | electrode interface. We will need the diffusion coefficient D to do this.
(iii) The rate laws – if a homogeneous reaction is also part of the overall process(es) that are occurring, then, having decided whether the reaction(s) proceed by a CE, EC, ECE, etc. mechanism (see section 6.4.4), we need a fully balanced rate equation(s) and hence obtain the rate constant(s) k for the rate-limiting step of each reaction.
(iv) Finally, since D and k are both activated quantities, we also need to know the temperature.

So, in order to simulate a voltammogram, we require values of A, n, I_0, α, $E^{\ominus}$, D and T, and possibly a few values of the rate constant k.

DQ 10.2

So, why is it useful to simulate a voltammogram?

Answer

We can be satisfied that the variables we have (e.g. A, D, *etc., as above) are accurate if they allow us to model a system* precisely. *We say that the variables **fit** the data. Figure 10.1 shows a cyclic voltammogram (CV), with superimposed on this current–potential data simulated with the DigiSim® package (as described in more detail below). The fit between theory and experiment is seen to be good, so we say that the derived variables fit the data.*

Conversely, if we can only generate a fit of this sort at one single scan rate ν, but the fit is poor at other scan rates, then we must reluctantly conclude that our data are not yet right – it is merely a coincidence that we are able to model our system at this one value of ν.

So modelling is therefore a matter of verification: if we can simulate a trace with good agreement between theory and experiment, then we can confidently assert that the reaction is indeed following such and such a course, with such and such a series of variables.

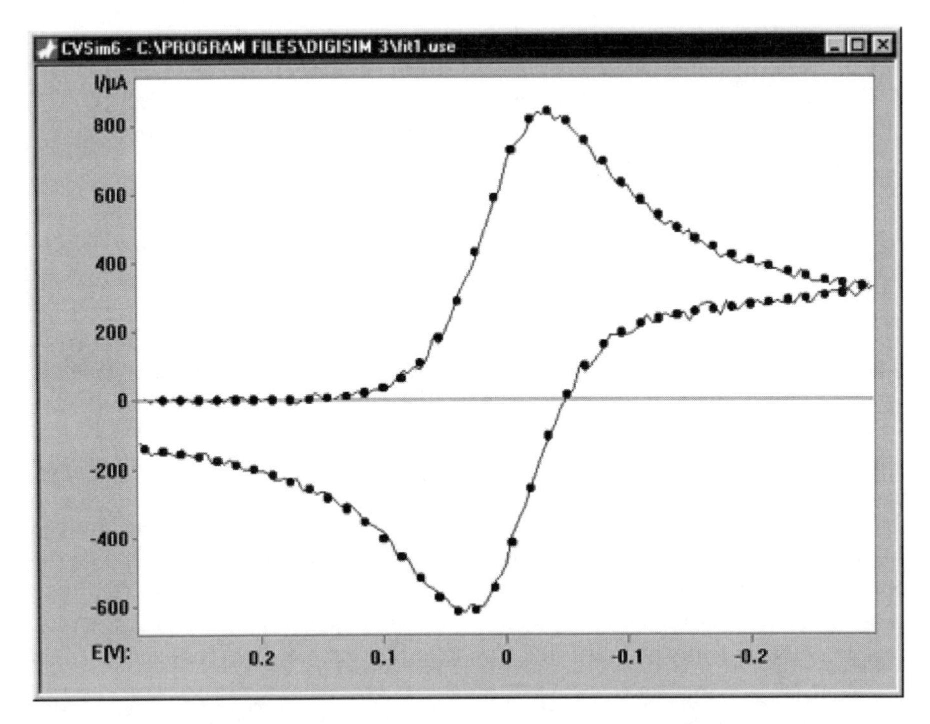

Figure 10.1 A cyclic voltammogram, normalized to make the standard electrode potential, $E^{\ominus} = 0$ V. The experimental trace is shown by as a continuous line, while data calculated by the DigiSim simulation package are represented as circles. Reprinted with permission from *Current Separations*, Vol. 15(2), 67–71, copyright Bioanalytical Systems, Inc. 1996.

DQ 10.3

How do we perform a simulation?

Answer

The first requirement for simulaton is a good quality computer with a suitable programme. Several such programmes are described in Section 10.1.1 below.

Next, we need to decide on what we think is occurring in terms of the system actually before us. Let's suppose that we have a CV which looks as though it describes a simple single reversible electron-transfer reaction. From the experimental trace of current against potential, it should be easy to obtain the standard electrode potential $E^{\ominus}$. *In addition, before we start, we measure the area of the electrode,* A, *and the thermodynamic temperature,* T. *Next, knowing* A, T *and* $E^{\ominus}$, *we estimate a value for the exchange current* I_0, *run a simulation, and note how similar (or not) are the experimental and calculated traces. No doubt, either the experimental or the calculated peak current will be too large, so we then alter* I_0 *accordingly, and recalculate. In essence, we are 'honing in' toward a set of variables which allow for a good fit. We give the name of* ***iteration*** *to this process of 'tweaking the variables'.*

SAQ 10.1

How do we obtain $E^{\ominus}$ from a CV?

A good simulation is as close to 'proof' as we can get to saying that a reaction proceeds according to a certain pathway. When the simulation and experimental traces are identical, we note the set of variables and reaction steps, and call this a **model**. Until proved otherwise, we then proceed by assuming that this model is correct.

However, we must always bear in mind that an electrochemical reaction can sometimes follow a completely different series of steps, and with different variables, and yet by coincidence *it can simulate our model quite closely*. The usual problem is that our model is too simple. A good example is to recall that an EC reaction always looks like an E reaction (see Section 6.4.4), except at slow scan

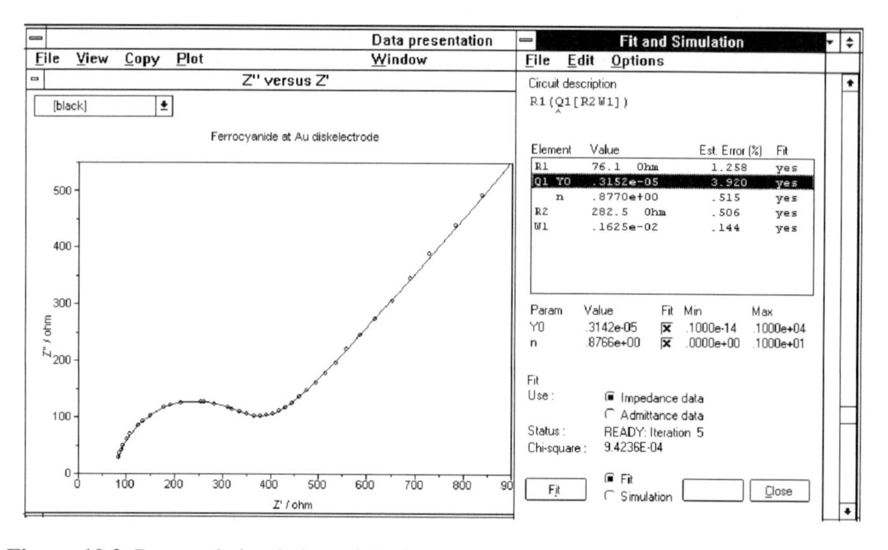

Figure 10.2 Data and simulation of the impedance behaviour of hexacyanoferrate(II) ion at a gold disc electrode. The experimentally obtained data are represented with circles while the simulation, performed with the Eco Chimie GPES software, is shown as a continuous line. Reprinted with permission of Dr K Dawes, Windsor Scientific, Slough, UK.

rates. We could argue this from Figure 6.20 (see above), since the ratio of the currents, $I_{p(back)}/I_{p(forward)}$, is unity above a certain scan rate ν. As an electroanalyst, we would further investigate by varying both the scan rate and scan limit as a matter of course when ascertaining the exact nature of an electrode process.

With practice, though, we can usually eliminate unlikely reaction schemes, and in most cases devise a model that can simulate a reaction over a wide range of reaction variables to the extent that a model is almost certain to be the correct one.

The technique of constructing an equivalent circuit for impedance analysis represents the exception to the general rule that a chosen model can be almost certain to be correct. It is all too easy to compile an equivalent circuit which fits the impedance data, *but is altogether wrong*. In fact, many practitioners would say that impedance studies are so susceptible to this fitting to a bogus model that another technique should always be applied as a form of 'validation'. It is much more unlikely for *two* techniques to fit a particular model, and the latter still be wrong!

Figure 10.2 shows data, both experimental and simulated, for hexacyanoferrate(II) ion at a gold disc electrode. The Warburg line at 45° indicates a diffusion process (see Section 8.2.2). The continuous line represents impedance data simulated by the Eco Chimie GPES software (as described below). Here, the fit between theory and experiment is seen to be very good.

10.1.1 Sample Programs

10.1.1.1 Simulation of a Potential-Step Experiment for a Reversible E Reaction [1]

We will consider the simple system;

$$A + e^{-} \longrightarrow B$$

with only A initially in the voltammetry solution.

This program requires a knowledge of the FORTRAN programming language.

```
      DIMENSION FANEW(150),FAOLD(150),FBNEW(150),FBOLD(150),
      RAT(100)
      DM=0.45
      L=100
C     INITIALIZE CONCENTRATIONS
      DO 10 J=1,150
      FAOLD(J)=1.
      FBOLD(J)=0.
10    CONTINUE
      DO 20 I=1,L
      JMAX=IFIX(4.2*SQRT(FLOAT(I)))+1
C     DIFFUSION SEGMENT
      DO 30 J=2,JMAX
```

```
      FANEW(J)=FAOLD(J)+DM*(FAOLD(J1)2.*FAOLD(J)+FAOLD(J+I))
      FBNEW(J)=FBOLD(J)+DM*(FBOLD(J1)2.*FBOLD(J)+FBOLD(J+1))
30    CONTINUE
C     FIRST BOX
      FANEW(1)=FAOLD(1)+DM*(3.*FAOLD(I)+FAOLD(2))
      FBNEW(I)=FBOLD(I)+DM*(2.*FAOLD(1)FBOLD(1)+FBOLD(2))
C     CALCULATE CURRENT
      Z=SQRT(FLOAT(L)*DM)*FANEW(1)*2.
C     CALCULATE TIME
      T=FLOAT(I)/FLOAT(L)
      CUR=1./(SQRT(3.1415*T))
C     CALCULATE RATIO OF SIMULATED TO COTTRELL CURRENT
      RAT(I)=Z/CUR
C     RESET CONCENTRATION ARRAYS
      DO 40 J=1,JMAX
      FAOLD(J)=FANEW(J)
      FBOLD(J)=FBNEW(J)
40    CONTINUE
20    CONTINUE

C     OUTPUT DATA
      DO 50 I=1,L
      WRITE(6,100) I,RAT(I)
50    CONTINUE
100   FORMAT(15,F 10.5)
      STOP
      END
```

10.1.1.2 Simulation of a Potential-Step Experiment for an ECE System [1]

The reaction scheme considered here is as follows:

$$A + e^- \longrightarrow B$$

$$B + C \longrightarrow D$$

$$D + e^- \longrightarrow E$$

(where C represents an electro-inert reagent in solution.)

The initial conditions are as follows:

(i) Only species A and C are present initially; for convenience, $[A]_{t=0} = [C]_{t=0}$.
(ii) Species B, D, and E are all absent at $t = 0$.
(iii) All electron-transfer reactions occur at a diffusion-controlled rate, i.e. current is controlled by mass transport rather than by kinetic control.

Again, this model requires a knowledge of FORTRAN.

```
      DIMENSION FAOLD(150),FANEW(150),FBOLD(150),F13NEW(150),
      FCOLD(150),KNEW(150),FDOLD(150),FDNEW(150),FEOLD(150),
      FENEW(150),NAPP(100)
C     SET CONSTANTS
      DM=0.45
      L=100
      RATE=0.02
C     INITIALIZE CONCENTRATIONS
      DO 10 J=1,150
      FAOLD(J)=1.
      FBOLD(J)=0.
      FCOLD(J) 1.
      FDOLD(J) 0.
      FEOLD(J) 0.
10    CONTINUE
      DO 20 I=1,L
      JMAX=IFIX(4.2*SQRT(FLOAT(I)))+1
C     DIFFUSION SEGMENT
      DO 30 J=2,MAX
      FANEW(J)=FAOLD(J)+DM*(FAOLD(J-1)-2.*FAOLD(J)+FAOLD(J+1))
      FBNEW(J)=FBOLD(J)+DM*(FBOLD(J-1)-2.*FBOLD(J)+FBOLD(J+1))
      FCNEW(J)=FCOLD(J)+DM*(FCOLD(J-1)-2.*FCOLD(J)+FCOLD(J+I))
      FDNEW(J)=FDOLD(J)+DM*(FDOLD(J-1)-2.*FDOLD(J)+FDOLD(J+I))
      FENEW(J)=FEOLD(J)+DM*(FEOLD(J-1)-2.*FEOLD(J)+FEOLD(J+1))
30    CONTINUE
C     FIRST BOX
      FANEW(1)=FAOLD(1)+DM*(3.*FAOLD(1)+FAOLD(2))
      FBNEW(1)=FBOLD(1)+DM*(2.*FAOLD(1)-FBOLD(1)+FBOLD(2))
      FCNEW(1)=FCOLD(1)+DM*(FCOLD(2)-FCOLD(1))
      FDNEW(1)=FDOLD(1)+DM*(-3.*FDOLD(1)+FDOLD(2))
      FENEW(1)=FEOLD(1)+DM*(2.*FDOLD(1)-FEOLD(I)+FEOLD(2))
C     KINETIC SEGMENT
      DO 40 J=l,JMAX
      REACT=RATE*FBNEW(J)*FCNEW(J)
      FBNEW(J)=RBNEW(J)REACT
      FCNEW(J)=FCNEW(J)REACT
      I'DNEW(J)=FDNEW(J)+REACT
40    CONTINUE
C     CALCULATE CURRENT
      ZA=2.*SQRT(FLOAT(L)*DM)*FANEW(1)
      ZD=2.*SQRT(FLOAT(L)*DM)*FDNEW(1)
      Z=ZA+ZD
      T=FLOAT(I)/FLOAT(L)
      ZCOTT=1./SQRT(3.1415*T)
C     APPARENT NO. OF ELECTRONS
      NAPP(I)=Z/ZCOTT
C     RESET CONCENTRATIONS
```

```
      DO 50 J=1,JMAX
      FAOLD(J)=FANEW(J)
      FBOLD(J)=FBNEW(J)
      FCOLD(J)=FCNEW(J)
      FDOLD(J)=FDNEW(J)
      FEOLD(J)=FENEW(J)
50    CONTINUE
C     OUTPUT DATA
      DO 60 I=1,L
      WRITE(6,100) I,NAPP
60    CONTINUE
100   FORMAT(15,F 10.5)
      STOP
      END
```

10.1.1.3 Simulation of a Cyclic Voltammogram for an EC Reaction [2]

The program listing below represents an interactive program written in TURBO PASCAL to simulate a cyclic voltammogram. The iterative procedure employs the method of fractional steps.†

If we prefer to simulate a CV for a simple E reaction rather than an EC reaction (see Section 6.4.4), then we should 'type in' a ludicrously small value, such as '10^{-30}', when the program asks for a value of the rate constant k to be inserted.

```
Program CV;
{A sample program for chemical kinetics}
{$N+} {enable numeric coprocessor}
Uses Crt;
Var
nt,ns,k,a,b,r,s,i:longint;
pot,ipot,spot,fpot,T,X:extended;
delx,delp,delt,E,scanr,area:extended;
Current,khet,kf,kr,kchem,rsc:extended;
C,Ct,Ctemp:array[1..3,1..200] of extended;
J:array[1..3] of extended;
outfile:text;

Procedure Setup;
begin
   write('What is the reduction potential in Volts?:');
   readln(E);write('What is the heterogeneous rate
      constant?;');
   readln(khet);writeln('Transfer coefficient set to 0.5')
   write('What is the k(in sec-1) of the following
      chemical reaction?:');
```

† Via a second-order Runge–Kutta method.

```
    readln(kchem);write('What is the initial
       potential?:');
    readln(ipot);
    write('What is the switch potential?:');
    readln(spot);
    writeln('final potential=initial potential');
    fpot:=ipot;write('What is the scan rate?:');
    readln(scanr);
    T:=2*abs(spotipot)/scanr;{Time of experiment}
    X:=6*sqrt(IE-5*T);{The diffusion layer assuming D IE-5}
    nt:=800;{A default number of time increments}
    area:=0.01;
    delt:=T/nt;
    if kchem>100 then begin
    delt:=0.3/kchem;{if kchem high, reset time increment}
    nt:=trunc(T/delt);
    end;
    kchem:=kchem*delt;{make it dimensionless}
    delx:=sqrt(IE-5*delt/0.45);
    ns:=trunc(X/delx);
    khet:=khet*delt/delx;{make it dimensionless}
 For a:=1 to ns+1 do
    begin
    C[1,a]:=1.0;
    C[2,a]:=0.00;
    C[3,a]:=0.00;
    Ctemp[1,a]:=1.00;
    Ctemp[2,a]:=0.00;
    Ctemp[3,a]:=0.00;
    end;
    for a:=1 to 3 do
    begin
    j [a]:=0.
    end;
 Delp:=2*(Spot-ipot)/nt;
 pot:=ipot+delp;
 assign(outfile,'data.pas');
 rewrite(outfile);
 end;

 Procedure Electrode;
 Begin
 kf:=khet*exp(-19.46*(pot-E));
 kr:=khet*exp(19.46*(pot-E));
 J[1]:=(kf*C[1,1]-kr*C[2,1])/(1+kf/0.9+kr/0.9);
 J[2]:=-J[1]*;J [3]:=0.00;
 Current:=J[1]*(delx/delt)*96484*1 E-6*area;
```

```
   {This is i/Area}
writeln(outfile,pot:9:5 '',current: 12);
end;

Procedure Diffusion:
Begin
For k:=1 to 3 do
begin
   Ctemp[k,1]:=C[k,1]+0.45*(C[k,2]-C[k,1])-J[k];
   For b:=2 to ns do
begin
Ctemp[k,b]:=C [k,b]+0.45*(C[k,b-1]-2.0*C[k,b]+C [k,b+1]);
end;
end;
end;

Procedure Chemreact;
Begin
If kchem>0.00 then begin
for i:=1 to ns do
ct[2,i]:=ctemp[2,i];
For k:=1 to ns do
   begin
      Ctemp[2,k]:=Ct[2,k]0.5*kchem*(Ct[2,k]+ctemp[2,k]);
   end;end;
   end;

   {***************The Main Program*****************}

begin
clrcr;
Writeln('CV Simulation of EC Mechanism');
writeln;

setup;
for a:=1 to nt do
   begin
      for s:=1 to 3 do
      begin
         for r:=1 to ns do
             begin
         rsc:=550/ns;
      C[s,r]:=Ctemp[s,r];
      end;
      end;
      electrode;
      diffusion;
      chemreact;
```

```
if a<nt/2 then pot:=pot+delp;
if a>=nt/2 then pot:=pot-delp;
end;close(outfile);writeln('simulation finished');
end.
```

10.2 Simulation Packages

The programs described above are limited in scope, and a superior approach is to employ a simulation package. There are three commercial programs at the forefront of the market, namely DigiSim, Condecon and GPES.

The **DigiSim**® simulation package was compiled by Manfred Rudolph and Stephen Feldberg, in collaboration with Bioanalytical Systems, Inc. (BAS).

The DigiSim program probably represents the current 'state of the art' which is achievable for simulating and analysing cyclic voltammograms. This package can perform cyclic voltammetry for a wide range of mechanisms at planar, spherical, cylindrical or rotated disc electrodes. It also computes concentration profiles.

Carlo Nervi considers DigiSim to be 'presently the best simulation package available', even though he has produced a package of his own, namely ESP, as described below.

The DigiSim software is presently available in Version 3 for Windows 95/98 [3], and is a general simulator for cyclic voltammetry. It combines the numerical stability of the fully implicit finite-difference method with the benefits of an exponentially growing space grid. The computational efficiency of the grid is maintained by employing an algorithm known as 'fast fully implicit difference' [4], which translates any user-defined mechanism into a linear system of tridiagonal matrix equations; the mechanism can comprise electron-transfer reactions, as well as first- and second-order chemical reactions. These matrix equations are solved in such a way that the computation time increases with the third power of the number of species, but *only linearly* with the number of grid points. This latter facet ensures that the speed of the simulation remains fast even when simulating mechanisms characterized by fast homogeneous kinetics, which itself causes the necessity for a high density of grid points near the electrode surface. Such a high density is needed to ensure accuracy.

The DigiSim program enables the user to simulate cyclic voltammetric responses for most of the common electrode geometries (planar, full and hemispherical, and full and hemicylindrical) and modes of diffusion (semi-infinite, finite and hydrodynamic diffusion), with or without inclusion of *IR* drop and double-layer charging.

Other features of the DigiSim package are as follows:

(i) The program enables the variation of the concentration profiles originating from the electrode processes to be visualized during the execution of the simulation.

(ii) The program is able to detect thermodynamically superfluous reactions. It calculates the equilibrium constants for such reactions and prevents the user entering thermodynamically incorrect values.
(iii) The program operates in terms of a non-linear regression strategy (based on the Gauss–Newton procedure) to fit simulated CVs for any user-defined mechanism to the experimental ones. The parameters that are to be fitted can be selected by the user.
(iv) The program takes advantage of the multi-tasking capabilities of the Win32 operating system, thus allowing a multi-document interface: more than one simulation document can be open while, at the same time, running simulations and/or data fittings in different documents.

Figure 10.3 shows the CV of a nickel(II) complex as a function of scan rate (continuous lines), together with data simulated by the DigiSim package (solid and open circles). The agreement between experiment and theory is seen to be very close.

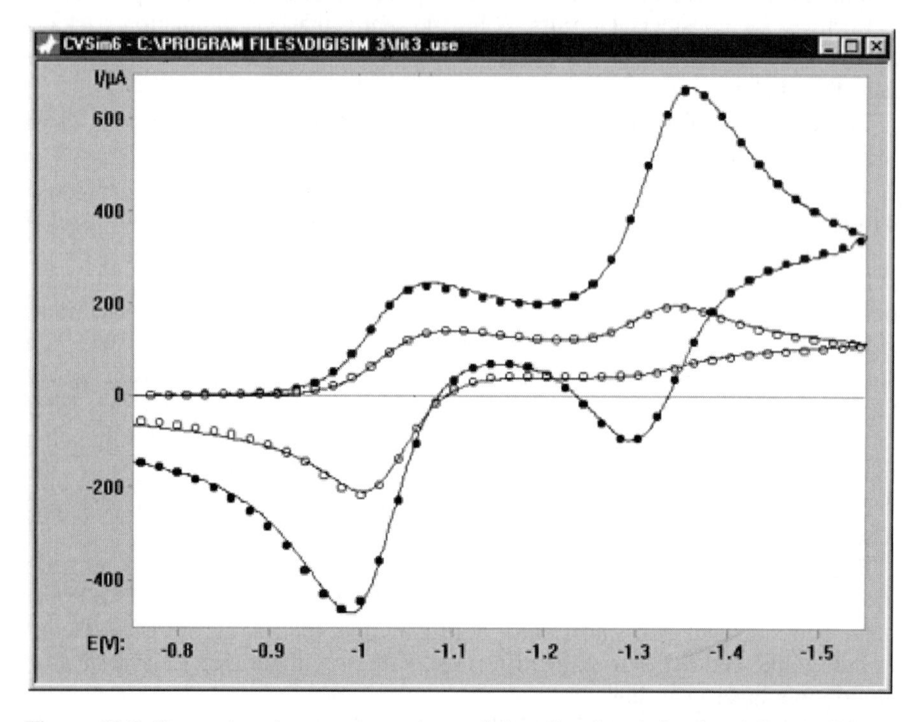

Figure 10.3 Comparison between experimental data (lines) and simulated data (solid and open circles), with the latter being obtained with the DigiSim package. These CV data were obtained for a nickel(II) complex in the presence of 2,2′-bipyridine: inner trace, $\nu = 40$ V s^{-1}; outer trace, $\nu = 333$ V s^{-1}. Reprinted with permission from *Current Separations*, Vol. 15(2), 67–71, copyright Bioanalytical Systems, Inc. 1996.

The **Condecon**® package is produced by E G & G Instruments (part of the Perkin Elmer and Princeton Applied Research Group), and was largely written by Norman Taylor of Leeds University, UK.

In any theoretical treatment of an electrochemical mechanism, the main requirement is to obtain a relationship between the current, potential and time, together with the homogeneous and heterogeneous parameters appropriate to the model, and then to devise a means of testing the experimental data and obtaining the magnitudes of each parameter involved. At a stationary electrode, this aim is largely frustrated (except in some simple cases, normally involving elementary mechanisms and extreme applied potentials).

A powerful alternative does exist, however. Instead of considering the current as a parameter in its own right, we can manipulate the experimentally determined current within certain integrals of time, each of which is a direct route to the concentration of product and reactant at the electrode | solution interface [5].

In addition, the time-dependence of these concentrations also contains (albeit in encoded form) the homogeneous parameters of the particular mechanism being considered. These latter techniques are termed **convolutions**. Convolution (and its reverse, i.e. **deconvolution**) are ideal for the electroanalyst because the theoretical calculation of current, and direct comparison with experimental data, is often not viable. This alternative of testing experimental currents via convolutions results in expressions for concentrations at the electrode which arise directly from the data rather than requiring iterations(s). The electrode concentrations thus estimated for a particular mechanism then allow for correlations to be drawn between potential and time, thereby assessing the *fit* between the data and the model.

Such a treatment, within its mechanistic restrictions, is general across the full range of experimental techniques. This convolution–deconvolution approach is the basis of the E G & G **Condecon** system.

Although the theoretical basis of the technique is different from other electrochemical-simulation packages, in operation it is just as powerful as DigiSim, allowing parameters to be computed for a variety of mechanisms, electrode geometries and experiment types. More information and representative references to Condecon, can be found at its website:

http://www.condecon.com

The **GPES**® package [6] (produced by Eco Chemie) provides the electroanalyst with virtually all electroanalytical techniques via an Eco Chimie 'Autolab' system. The GPES (General Purpose Electrochemical Software) transforms a conventional instrument into an electrochemical work station capable of simultaneously capturing data in real time but allowing subsequent manipulation to retrieve kinetic parameters, both of the electrode reaction, and also for homogeneous chemical processes.

An adjunct within the GPES software allows for the manipulation of impedance data, thus leading to the rapid assembly of an equivalent circuit for mimicking experimental data.

Further information is available at the Eco Chemie website:

http://www.ecochemie.nl

10.3 Web-Based Resources

In addition, there are now several web-based resources available to aid data simulation. The internet addresses (URLs) below are correct at the time of writing (Spring 2000), but it should be noted that such URLs do have a habit of changing often.

One of the best web-based directories for gauging the status of electrochemical simulation is:

http://chpc06.ch.unito.it/chemistry/electrochemistry_right.html

which is compiled and maintained by Carlo Nervi and Mauro Ravera of the Universit del Piemonte Orientale. The links given here seem to be updated fairly regularly, and the site is well worth a visit.

Electrochemical Simulation Package (ESP) is a *free* program which allows a PC to simulate virtually any mechanism by the following pulse techniques, i.e. cyclic voltammetry, square-wave voltammetry, chronoamperometry and sample DC polarography. The program can also be used in conjunction for 'fitting' experimental data at solid and DME electrodes. It is the only package to explicitly claim to be 'bug-free'.

This software was written by Carlo Nervi. ESP (Version 2.4) can be accessed at the following website:

http://www.netsci.org/Resources/Software/Modeling/Tools/esp.html

while the on-line 'help' manual may be accessed at:

http://chpc06.ch.unito.it/esp_manual.html

Polar (for Windows), written by Weiguang Huang, can be downloaded free from the following:

http://come.to/visualmath

although you will have to pay for the actual use of this package. The program digitally simulates voltammograms (or polarograms at a HMDE) from within a range of about 20 mechanisms at eight electrode geometries (planar, spherical, semi-spherical, cylindrical, semi-cylindrical, micro-disc, thin-film and rotating electrodes) and with five techniques (linear-sweep and cyclic voltammetry, and the following pulse techniques: normal pulse, differential pulse, and square-wave voltammetry).

Polar will also simulate the effects of charge current, resistance, noise, electrolyte, stripping time and stripping potential.

Transient is a C-program for solving systems of generally non-linear, parabolic partial differential equations in two variables (that is, space and time), in particular, reaction–diffusion equations within the generalized Crank–Nicolson Finite Difference Method.

Newton's method is used for non-linear equations. The program requires the user to compile program modules and then link them to the libraries provided.

The Transient program can be accessed at the following:

http://pangea.ca/~kolar/software/T.html

and a comprehensive on-line help file is also available.

ELSIM ('Electrochemical Simulations', Version 3.0) was written by L. K. Bieniasz. It claims to be a user-friendly means of simulating transients for electrochemical techniques. The program runs on Windows 95 (or greater) under MS-DOS. The user simply has to type in electrochemical reaction mechanism and then specify the parameter values. The built-in 'reaction compiler' then generates the text of the corresponding equations governing the mathematical treatment adopted, and then verifies the correctness of the reaction mechanisms. Additional options include homogeneous reaction–diffusion with electrochemical and homogeneous reactions, in the presence of bulk transport.

The ELSIM program may be accessed at the following address:

http://www.cyfkr.edu.pl/~nbbienia/elsim3ad.html

Summary

The validity of an electroanalytical measurement is enhanced if it can be simulated mathematically within a reasonable 'model', that is, one comprising all of the necessary elements, both kinetic and thermodynamic, needed to describe the system studied. Within the chosen model, the simulation is performed by first deciding which of the possible parameters are indeed variables. Then, a series of mathematical equations are formulated in terms of time, current and potential, thereby allowing the other implicit variables (rate constants of heterogeneous electron-transfer or homogeneous reactions in solution) to be obtained.

It is not possible to say that a model *is* correct; but a good 'fit' between theory and experimental data is taken as confirming the *likelihood* of the model being the correct one.

While several sample programs are available, nowadays a preferred and simpler process is to simulate the system via the use of more sophisticated packages such as DigiSim, Condecon or GPES.

References

1. Greef, R., Peat, R., Peter, L. M., Pletcher, D. and Robinson, J., *Instrumental Methods in Electrochemistry*, Ellis Horwood, Chichester, 1990, pp. 433–438.
2. Gosser, D. K. Jr., in *Modern Techniques in Electroanalysis*, Vanýsek, P. (Ed.), Wiley, New York, 1996, pp. 323–325.
3. Rudolph, M. and Feldberg, S. W., DigiSim® 3.0, Bioanalytical Systems Inc., West Lafayette, IN 47906, USA.
4. Rudolph, M., in *Physical Chemistry: Principles, Methods and Applications*, Rubinstein, I. (Ed.), Marcel Dekker, New York, p. 81.
5. Britz, D., *Anal. Chim. Acta*, **122**, 331–346 (1980).
6. Autolab®, Eco Chemie B. V. Kanaalweg 18/J, 3526 KL Utrecht, The Netherlands: [htpp://www.ecochemie.nl].

Appendices

1 Named Electroanalysis Equations Used in the Text

All symbols are defined in the *Acronyms, Abbreviations and Symbols* section on pages xv–xix, and in the relevant text.

Debye–Hückel (extended) $$\log_{10}\gamma_{\pm} = \frac{0.509\mid z^{+}\times z^{-}\mid\sqrt{I}}{1+\sqrt{I}} \qquad (3.15)$$

Heyrovsky–Ilkovic $$E = E_{1/2} + \frac{RT}{nF}\ln\left(\frac{I_{\mathrm{d}}-I}{I}\right) \qquad (6.6)$$

Huggins $$Z^{*} = \left(\frac{\mathrm{d}emf}{\mathrm{d}[\mathrm{H}^{+}]}\right)\times\frac{V_{\mathrm{m}}\omega^{-1/2}}{nFAD^{1/2}} \qquad (8.15)$$

Koutecky–Levich $$\frac{1}{I_{\text{not limiting}}} = \frac{1}{nFAk_{\mathrm{et}}c_{\mathrm{analyte}}} + \frac{1.61\upsilon^{1/6}}{nFAc_{\mathrm{analyte}}D^{2/3}}\frac{1}{\omega^{1/2}} \qquad (7.18)$$

Levich $$I_{\mathrm{lim}} = 0.620nFAD^{2/3}\upsilon^{-1/6}\omega^{1/2}c_{\mathrm{analyte}} \qquad (7.1)$$

Nernst $$E_{\mathrm{O,R}} = E^{\ominus}_{\mathrm{O,R}} + \frac{RT}{nF}\ln\left(\frac{a_{\mathrm{O}}}{a_{\mathrm{R}}}\right) \qquad (3.8)$$

Randles–Sevčik $$I_{\mathrm{p}} = 0.4463nFA\left(\frac{nF}{RT}\right)^{1/2}D^{1/2}\nu^{1/2}c_{\mathrm{analyte}} \qquad (6.13)$$

2 Writing a Cell Schematic

There are many ways of representing an electrochemical cell: the simplest, but most laborious, is simply to draw it. A more sensible alternative is to write a **cell schematic**. Such a schematic is simply a shorthand method of saying what is present, and which part of the cell is the positive, negative, electrolyte, etc.

A cell schematic is a convenient abbreviation for a cell, and can be taken as a 'cross-section' through the cell, showing all interfaces and phases. Since it is across phase boundaries and interfaces that the potential is dropped, the correct indication of their relative positions within the cell is vital. Accordingly, a series of simple rules may be used when constructing such a schematic, as follows:

1. The redox couple associated with the positive electrode is always written on the right-hand side.
2. The redox couple associated with the negative electrode is always written on the left-hand side.
3. A salt bridge is written as a double vertical line, i.e. $||$.
4. If one redox form is conductive and can function as an electrode, then it is written on one extremity of the schematic.
5. The phase boundary between this electrode and the solution containing the other redox species is represented by a single vertical line, i.e. $|$.
6. If both redox states of a couple are in the same solution (e.g. Fe^{2+} and Fe^{3+}), then they share the same phase. Such a couple is written conventionally with only a comma for separation, i.e. Fe^{3+}, Fe^{2+}.
7. Following from (6), it will be seen that there is now no electrode in solution to measure the energy at equilibration of the two redox species. An inert electrode is therefore placed in solution: almost universally, platinum is the material of choice.
8. A single vertical line, $|$, or better still, a dotted vertical line, $\vdots$, is used if the salt bridge is replaced by a simple porous membrane.

As an example, a simple cell containing copper and zinc redox couples is therefore written as follows:

$$Zn_{(s)} \mid Zn^{2+}_{(aq)} \mid\mid Cu^{2+}_{(aq)} \mid Cu_{(s)}$$

3 The Electrode Potential Series (against the SHE)

The electrode potential series is an arrangement of reduction systems in ascending order of their standard electrode potential, $E^{\ominus}$.

Couple	$E^{\ominus}$/V
$Sm^{2+} + 2e^- = Sm$	−3.12
$Li^+ + e^- = Li$	−3.05
$K^+ + e^- = K$	−2.93
$Rb^+ + e^- = Rb$	−2.93
$Cs^+ + e^- = Cs$	−2.92
$Ra^{2+} + 2e^- = Ra$	−2.92
$Ba^{2+} + 2e^- = Ba$	−2.91
$Sr^{2+} + 2e^- = Sr$	−2.89
$Ca^{2+} + 2e^- = Ca$	−2.87
$Na^+ + e^- = Na$	−2.71
$Ce^{3+} + 3e^- = Ce$	−2.48
$Mg^{2+} + 2e^- = Mg$	−2.36
$Be^{2+} + 2e^- = Be$	−1.85
$U^{3+} + 3e^- = U$	−1.79
$Al^{3+} + 3e^- = Al$	−1.66
$Ti^{2+} + 2e^- = Ti$	−1.63
$V^{2+} + 2e^- = V$	−1.19
$Mn^{2+} + 2e^- = Mn$	−1.18
$Cr^{2+} + 2e^- = Cr$	−0.91
$2H_2O + 2e^- = H_2 + 2OH^-$	−0.83
$Cd(OH)_2 + 2e^- = Cd + 2OH^-$	−0.81
$Zn^{2+} + 2e^- = Zn$	−0.76
$Cr^{3+} + 3e^- = Cr$	−0.74
$O_2 + e^- = O_2{}^-$	−0.56
$In^{3+} + e^- = In^{2+}$	−0.49
$S + 2e^- = S^{2-}$	−0.48
$In^{3+} + 2e^- = In^+$	−0.44
$Fe^{2+} + 2e^- = Fe$	−0.44
$Cr^{3+} + e^- = Cr^{2+}$	−0.41
$Cd^{2+} + 2e^- = Cd$	−0.40
$In^{2+} + e^- = In^+$	−0.40
$Ti^{3+} + e^- = Ti^{2+}$	−0.37
$PbSO_4 + 2e^- = Pb + SO_4{}^{2-}$	−0.36
$In^{3+} + 3e^- = In$	−0.34
$Co^{2+} + 2e^- = Co$	−0.28
$Ni^{2+} + 2e^- = Ni$	−0.23
$AgI + e^- = Ag + I^-$	−0.15

(*continued overleaf*)

(*table continued*)

Couple	$E^{\ominus}$/V
$Sn^{2+} + 2e^- = Sn$	−0.14
$In^+ + e^- = In$	−0.14
$Pb^{2+} + 2e^- = Pb$	−0.13
$Fe^{3+} + 3e^- = Fe$	−0.04
$Ti^{4+} + e^- = Ti^{3+}$	0.00
$2H^+ + 2e^- = H_2$ (*by definition*)	**0.000**
$AgBr + e^- = Ag + Br^-$	0.07
$Sn^{4+} + 2e^- = Sn^{2+}$	0.15
$Cu^{2+} + e^- = Cu^+$	0.16
$Bi^{3+} + 3e^- = Bi$	0.20
$AgCl + e^- = Ag + Cl^-$	0.2223
$Hg_2Cl_2 + 2e^- = 2Hg + 2Cl^-$	0.27
$Cu^{2+} + 2e^- = Cu$	0.34
$O_2 + 2H_2O + 4e^- = 4OH^-$	0.40
$NiOOH + H_2O + e^- = Ni(OH)_2 + OH^-$	0.49
$Cu^+ + e^- = Cu$	0.52
$I_3^- + 2e^- = 3I^-$	0.53
$I_2 + 2e^- = 2I^-$	0.54
$MnO_4^- + 3e^- = MnO_2$	0.58
$Hg_2SO_4 + 2e^- = 2Hg + SO_4^{2-}$	0.62
$Fe^{3+} + e^- = Fe^{2+}$	0.77
$AgF + e^- = Ag + F^-$	0.78
$Hg_2^{2+} + 2e^- = 2Hg$	0.79
$Ag^+ + e^- = Ag$	0.80
$2Hg^{2+} + 2e^- = Hg_2^{2+}$	0.92
$Pu^{4+} + e^- = Pu^{3+}$	0.97
$Br_2 + 2e^- = 2Br^-$	1.09
$Pr^{2+} + 2e^- = Pr$	1.20
$MnO_2 + 4H^+ + 2e^- = Mn^{2+} + 2H_2O$	1.23
$O_2 + 4H^+ + 4e^- = 2H_2O$	1.23
$Cl_2 + 2e^- = 2Cl^-$	1.36
$Au^{3+} + 3e^- = Au$	1.50
$Mn^{3+} + e^- = Mn^{2+}$	1.51
$MnO_4^- + 8H^+ + 5e^- = Mn^{2+} + 4H_2O$	1.51
$Ce^{4+} + e^- = Ce^{3+}$	1.61
$Pb^{4+} + 2e^- = Pb^{2+}$	1.67
$Au^+ + e^- = Au$	1.69
$Co^{3+} + e^- = Co^{2+}$	1.81
$Ag^{2+} + e^- = Ag^+$	1.98
$S_2O_8^{2-} + 2e^- = 2SO_4^{2-}$	2.05
$F_2 + 2e^- = 2F^-$	2.87

Notes for Table

1. The more positive the $E^{\ominus}$ value, then the more readily the half reaction occurs in the direction from left to right. Correspondingly, the more negative the value, then the more readily the reaction occurs in the direction from right to left.
2. F_2 is the strongest oxidizing agent and Sm^{2+} is the weakest. The oxidizing power increases from Sm^{2+} to F_2.
3. Sm is the strongest reducing agent and F^- is the weakest. The reducing power increases from F^- to Sm.

Responses to Self-Assessment Questions

Chapter 1

Response 1.1

By writing the oxidized form of the redox couple first within the subscript, we obtain the following:

(a) E_{Br_2,Br^-};
(b) $E_{Ag^+,Ag}$;
(c) $E_{Fe(cp)_2{}^{+\bullet},Fe(cp)_2}$.

Response 1.2

By using equation (1.1), we obtain 3.43×10^{-2} A cm^{-2}.

Response 1.3

By using equation (1.2), 595 mol m^{-3} = 0.595 mol dm^{-3}.

Response 1.4

By using equation (1.3), we obtain values of 700 mol m^{-3} and 7×10^{-4} mol cm^{-3}, respectively.

Response 1.5

By using equation (1.4), we obtain 215 A m^{-2} and 0.0215 A cm^{-2} (or 21.5 mA cm^{-2}), respectively.

Chapter 2

Response 2.1

By using equation (2.2), we obtain 0.233 C.

Response 2.2

By using equation (2.3) and summing the two resistances, we obtain 3.02×10^{-3} A.

Response 2.3

By using equation (2.4), we obtain 0.362 V.

Response 2.4

By using equation (2.3), we obtain $1.4 \times 10^8\ \Omega$.

Response 2.5

For (a) a reduction reaction has occurred because the oxidation number of the product is lower than that of the initial redox state (Br^0 is converted to Br^{-1}), while (b) and (c) are both oxidation reactions because the oxidation number increases during reaction.

Response 2.6

From a suitable textbook such as [1] (in Chapter 2), we find values of $\sigma(\text{Ag}) = 63 \times 10^6$, $\sigma(\text{Au}) = 45 \times 10^6$, $\sigma(\text{Hg}) = 1 \times 10^6$, $\sigma(\text{Pt}) = 9.3 \times 10^6$, and $\sigma(\text{graphite}) = 0.07 \times 10^6$ S m^{-1}. Therefore, the conductivity of graphite is a thousandth of that of gold.

Response 2.7

First, these electrolytes are strong (i.e. they *completely* dissociate to form ions) so there is a sufficient supply of ions available to carry the charge, thus explaining why the conductivity Λ is high. Secondly, the transport numbers t of the anions and cation in these two salts happen to be just about the same, thereby minimizing the junction potentials (see Chapter 3).

Response 2.8

Figures 6.3 and 6.4 show such concentration profiles.

Chapter 3

Response 3.1

From equation (3.4) and the *emf*, $\Delta G' = -212$ kJ mol^{-1}, from equation (3.5) and the temperature coefficient of voltage, $\Delta S' = 96.5$ J K^{-1}mol^{-1}, while from equation (3.6), $\Delta H' = 75$ kJ mol^{-1}.

Response 3.2

By using equation (3.3), we obtain a value for $E^{\ominus}_{Zn^{2+},Zn}$ of -0.76 V.

Response 3.3

From SAQ 3.2, $E^{\ominus}_{Zn^{2+},Zn}$ has a value of -0.76 V. The number line which is obtained should look something like the following:

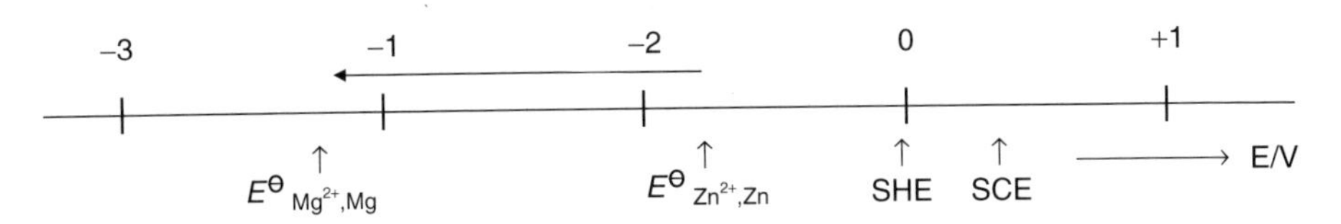

The separation between the two electrode potentials is $(-2.36 \text{ V}) - (-0.76 \text{ V}) = -1.6$ V. The answer is negative because $E^{\ominus}_{Mg^{2+},Mg}$ is positioned to the left of $E^{\ominus}_{Zn^{2+},Zn}$.

Response 3.4

By using equation (3.8), and adopting the same approach as that used in Worked Example 3.4, we obtain $a(Fe^{2+}) = 0.016$.

Response 3.5

By using equation (3.8), $E_{Fe^{2+},Fe} = -0.588$ V. Equation (3.3) then gives an *emf* of 0.830 V.

Response 3.6

$E_{Ag^+,Ag}$ is obtained after having accounted for the second electrode by using equation (3.3). The activity is then obtained from the Nernst equation (equation (3.8)) as $a(Ag^+) = 10^{-5}$.

Response 3.7

$E_{Ag^+,Ag}$ is obtained after having accounted for the second electrode by using equation (3.3). The activity is then obtained from the Nernst equation (equation (3.8)) as $x_{Ag} = 0.68$. The mole fraction and activity are the same for solid mixtures of this type, so the amalgam filling is 68 mol% silver.

Response 3.8

The activity of hydrogen is obtained from the Nernst equation (equation (3.8)); remember that two protons are involved in the balanced redox reaction, so the numerator within the bracket is written as $(H^+)^2$. Pressure and activity are the same for gaseous mixtures of this type, so the pressure of hydrogen gas is 6.5 Pa.

Response 3.9

By adopting the approach described in Worked Example 3.7, or using the data presented in Table 3.1, we find $I = 4 \times c$.

Response 3.10

The *activity*, $a(Zn^{2+})$, is readily calculated from the Nernst equation (equation (3.8)) to be 8.8×10^{-5}. Next, we assume that only the H_3PO_4 contributes to the ionic strength, and so $I = 5 \times c$ (from Table 3.1 for a simple 1:3 electrolyte). Accordingly, by using the extended Debye–Hückel law (equation (3.15)), $\gamma_{\pm} = 0.233$ and, from $c = a/\gamma_{\pm}$ (equation (3.12)), we find $c = 3.78 \times 10^{-4}$ mol dm^{-3}.

Response 3.11

The Nernst equation (equation (3.8)) is required for both of the calculations, with the only difference being that in (i) we say that $a = c$, in which case $E^{\ominus}_{Cu^{2+},Cu} = 0.310$ V because $\gamma_{\pm} = 1$. In (ii), we perform a similar calculation, but say here that $a = c \times \gamma_{\pm}$, and so the activity $= (0.1 \text{ mol dm}^{-3} \times 0.158)/1$ mol dm^{-3}, which gives $E^{\ominus}_{Cu^{2+},Cu} = 0.287$ V when $\gamma_{\pm} = 0.158$. We divided by 1 mol dm^{-3} merely to ensure that the activity was dimensionless.

Response 3.12

Activity a and concentration c are only the same at extremely small concentrations ($\gamma_{\pm} = 1$ if $c < 10^{-4}$ mol dm^{-3}), so in such cases the values of E will be the same whichever variable is inserted into the Nernst equation. Conversely, as the concentration rises, so $\gamma_{\pm}$ decreases. We see then that $c \neq a$, so the effects of $\gamma_{\pm}$ are seen most prominently at higher concentrations.

Response 3.13

By using equation (3.16), we calculate that the cell constant $K = -0.138$ V and the unknown pH is 10.8.

Response 3.14

By using equation (3.17), $K = -88$ mV – this knowledge of K ensures calibration of the ISE. Next, with this value of K and using equation (3.17), we can calculate that the toothpaste solution contains 8.3×10^{-3} mol dm^{-3} of F^-.

Response 3.15

We will use the equation, $emf = K + RT/F \ [\ln a(F^-) + 1/3000 \ln a(Cl^-)]$, which is akin to equation (3.18). From the pH of the hydrochloric acid, $[Cl^-] = 3.2 \times 10^{-6}$ mol dm^{-3}, which gives $K = 1.176$ V. Therefore, knowing K, we can readily calculate that $a(F^-) = 5 \times 10^{-7}$.

Response 3.16

We use a similar procedure to that described in Worked Example 3.14: from the Nernst equation (equation (3.8)), we obtain $a(Fe^{2+})$ and then, recalling that *two* hydroxides are generated per Fe^{2+}, we can say that $K_s = [a(Fe^{2+})]^3$, which gives a value of 1.6×10^{-14}.

Response 3.17

The extreme values of the electrode potential are 0.303 and 0.297 V, so the range of activities (as calculated from the Nernst equation (equation (3.8)), is $0.0352 < a(Cu^{2+}) < 0.0562$.

Response 3.18

By using the variant of the Nernst equation in equation (3.19), and remembering that there are two protons, and thus the activity of the proton should be squared, we calculate that $E_{H^+,H_2} = -0.0013$ V.

Response 3.19

From the Nernst equation (equation (3.8)), $E_{Ag^+,Ag} = 0.471$ V before the electrolysis. Next, 5×10^{-6} mol of electrons is passed oxidatively, so Ag^+ ion is *generated*; thus, from the stoichiometry, 5×10^{-6} mol of electrons generates 5×10^{-6} mol of Ag^+. The new concentration of Ag^+ is therefore 6×10^{-6} mol. We will assume that the activity of the silver metal remains constant. The new electrode potential is also obtained from the Nernst equation, and we find that after electrolysis the potential has risen to 0.494 V.

Response 3.20

The leads to the electrodes can be made from a metal of high electronic conductivity, e.g. copper. The solution resistance will decrease if we maintain a high ionic strength, although we need always to appreciate that additional ions will cause $\gamma_{\pm}$ to alter.

Response 3.21

(i) By using the uncompensated value of $E_{Ag^+,Ag}$, we obtain the activity from the Nernst equation (equation (3.8)) as $a(Ag^+) = 6.61 \times 10^{-3}$.

(ii) Having compensated for the value of $E_{Ag^+,Ag}$ by taking account of E_j, its value becomes 0.648 V, and we then obtain an activity from the Nernst equation of 2.81×10^{-3}. By ignoring E_j, the answer in (i) is 135% too high.

Response 3.22

By following a similar procedure to that described in Worked Example 3.18, we can calculate that E_j has a value of 11 mV.

Chapter 4

Response 4.1

By using the relationship, $c_1V_1 = c_2V_2$, i.e. treating the reaction as a straightforward titration, we obtain a value of 0.05 mol dm^{-3}.

Response 4.2

The electrode potential is obtained from equation (3.3) as 0.550 V.

Response 4.3

The end point in Figure 4.3 occurs at an *emf* of −137 mV.

Response 4.4

The reduced form of any of the couples shown in Appendix 3 which have an $E^{\ominus}$ of about 0.9 V or more, i.e. about 0.2 V higher than the $E^{\ominus}$ of the quinone–hydroquinone couple, would be suitable. Good choices would be permanganate or ceric ion. Bromine will also oxidize hydroquinone, but its chemical reactivity precludes its use in such an application.

Response 4.5

By using equation (4.4), the separation found is 0.35 V.

Response 4.6

Insert $K = 10^6$ and $n = 2$ into equation (4.4) to give the required value.

Response 4.7

We obtain a value for K_s of 7.7×10^{-13} by inserting the value of the electrode potential at equivalence into the Nernst equation (equation (3.8)), and then using the relationship, $K_s = [a(Ag^+)]^2$.

Response 4.8

A similar procedure to that used in SAQ 4.7 is adopted, but here we should note that $Co(OH)_2$ has a 2:1 stoichiometry, and therefore $K_s = [a(Co^{2+})]^3$. The Nernst calculation then gives $a(Co^{2+}) = 4.04 \times 10^{-5}$, and so $K_s = 6.6 \times 10^{-14}$.

Response 4.9

By using the equation, $E = E_{in}{}^{\ominus} + RT/nF \ln (25)$, and taking $n = 1$, the minimum separation is 0.083 V (or 83 mV). Note that the figure of '25' in the logarithmic term comes from (5×5).

Chapter 5

Response 5.1

From Appendix 3, it will be seen that the value of $E^{\ominus}$ for the copper couple is more positive than that of the iron couple, so it is the copper couple which undergoes reduction. Accordingly, the iron couple undergoes oxidation, and the balanced redox reaction is $Cu^{2+} + 2Fe^{2+} \rightarrow Cu^0 + 2Fe^{3+}$.

Response 5.2

The Sn^{2+}, Sn couple involves two electrons, so 0.05 mol of Sn^{2+} requires $2 \times$ 0.05 mol of electrons = 0.1 mol = $0.1F$.

Response 5.3

2734 C represents (2734/96 485) of a faraday = 0.0283 mol of electrons. As the reduction reaction represents a two-electron couple, 0.0283 mol of electrons will effect $1/2 \times 0.0283$ mol of reaction = 0.0142 mol. 1 mol of zinc has a mass of 65 g, and so 0.92 g of metal are therefore formed.

Response 5.4

1 mol of silver has a mass of 108 g, and 1 faraday of charge (96 485 C) will generate this mass. By taking ratios, 0.075 g/108 g $mol^{-1} = 6.94 \times 10^{-4}$ mol of silver, and so 6.94×10^{-4} of a faraday are required, which gives a value of 67 C.

Response 5.5

The charge passed represents $0.833F$. The oxidation of iodide is a two-electron reaction, so $1/2 \times 0.833$ (= 0.417) mol of I_2 would have been formed if all of the charge generated iodine. The efficiency is therefore 0.239 mol/0.417 mol = 57%.

Response 5.6

10^{-4} mol of zinc will require $2 \times 10^{-4}F$ of charge (= 19.3 C), because we have a two-electron couple. 77.1% of the total charge that was passed forms this amount of zinc, so 19.3/0.771 of overall charge is passed, which gives a value of 25.0 C.

Response 5.7

First, we must remember to consider *both* sides of the copper plate, so the area is in fact 6.0 cm^2. The overall capacitance = capacitance per unit area × area, so $C = 6.0 \times 25$ μF cm^{-2} = 150 μF cm^{-2}.

Response 5.8

Overall capacitance = capacitance per unit area × area, so in this case $C = 6.0 \times$ 25 μF cm^{-2}, and therefore charging requires 5 cm^2 × 1.43 C cm^{-2} = 7.15 C. (i) A charge of 1.0 C is less than the charging of the capacitance, so none of the charge can be faradaic, and therefore the efficiency = 0%. In (ii), 7.15/320 C is consumed by charging the capacitance, so 2.3% is non-faradaic, and the efficiency is therefore 97.7%.

Response 5.9

Intrinsic capacitance × area = overall capacitance, so area = overall capacitance divided by intrinsic capacitance, which gives an electrochemical area of 1.22 cm^2.

Response 5.10

The fraction of F passed was (3×10^{-4})/96 485 C mol^{-1}, and therefore 3.11×10^{-9} mol of electrons are passed. Electroanalysis of copper involves a two-electron process, and so 1.55×10^{-9} mol of Cu were in the solution.

From the definition, concentration = amount/volume, and therefore $[Cu^{2+}] =$ 3.34 nmol dm^{-3}.

Response 5.11

If the radius is 5 μm, then the area = 7.9×10^{-5} cm^2 (from πr^2). Accordingly, the maximum current is 0.2 A $cm^{-2}/7.9 \times 10^{-5}$ cm^2, so I_{max} in this case is 16 μA.

Response 5.12

If the solution contains exactly 1 μmol of enzyme, then 1 μF of charge is required for *complete* reduction. A current of 1 μA represents 10^{-11} F of charge passed per second. From $I = Qt$ (equation (2.2)), 10^5 s are therefore needed, i.e. 27 h, 46 min and 40 s.

Chapter 6

Response 6.1

Following the ratio approach used in Worked Example 6.2, we obtain $c_{analyte} =$ 1.19×10^{-4} mol dm^{-3}.

Response 6.2

Draw a Heyrovsky–Ilkovic plot of ln $[(I_d - I)/I]$ against E. $I_d = 25.01$ μA, because the current becomes limiting at higher E. The best straight line should have an intercept of $E_{1/2} = -1.00$ V and, from the gradient, the number of electrons (n) passed is 2.

Response 6.3

The gradient of the graph produced using the data should be 79.8 V^{-1}, while the theoretical gradient is 77.2 V^{-1} (with $n = 2$). We can therefore safely say that the two gradients are the same within experimental error, i.e. $\alpha = 1$.

Response 6.4

By using equation (6.8), the shift in $E_{1/2}$ is -0.036 V, or -36 mV.

Response 6.5

After plotting a graph of $\Delta E_{1/2}$ against $\log c_L$, a value of $K = 10^{16.3}$ is obtained from the intercept, with a stoichiometry of 1:1 being obtained from the gradient.

Response 6.6

The Nernst equation (equation (3.8)) is $E_{O,R} = E^{\ominus}_{O,R} + RT/nF \ln(a_O/a_R)$.

From the definition of activity (equation (3.12)) of $a = c \times \gamma$, and substituting for each activity term, we obtain the following:

$$E_{O,R} = E^{\ominus}_{O,R} + \frac{RT}{nF} \ln\left(\frac{\gamma_O \times c_O}{\gamma_R \times c_R}\right)$$

Separating the ln terms within both numerator and denominator then yields:

$$E_{O,R} = E^{\ominus}_{O,R} + \frac{RT}{nF} \ln\left(\frac{c_O}{c_R}\right) + \frac{RT}{nF} \ln\left(\frac{\gamma_O}{\gamma_R}\right)$$

The electrode potential on the left-hand side can be replaced by using equation (6.10), which gives:

$$E^{0'}_{O,R} + \frac{RT}{nF} \ln\left(\frac{c_O}{c_R}\right) = E^{\ominus}_{O,R} + \frac{RT}{nF} \ln\left(\frac{c_O}{c_R}\right) + \frac{RT}{nF} \ln\left(\frac{\gamma_O}{\gamma_R}\right)$$

The log (concentration ratio) term is common to both sides, and can be cancelled out, thus leaving:

$$E^{0'}_{O,R} = E^{\ominus}_{O,R} + \frac{RT}{nF} \ln\left(\frac{\gamma_O}{\gamma_R}\right)$$

Response 6.7

$E^{0'}$ and $E^{\ominus}$ will be the same when the two γ terms are equal, which itself causes the ln term to be zero.

Response 6.8

When the ionic strength is high, the proportional change in γ is decreased, so $E^{\ominus} \approx E^{0'}$ if excess inert electrolyte is added to the voltammetry solution.

Response 6.9

By using equation (6.12), a value of $\tau = 22$ s is obtained.

Response 6.10

By plotting ΔE_p (as 'y') against $\nu^{1/2}$ (as 'x') for each pair of peaks, the intercepts are 24 mV for the first couple and 31 mV for the second. In both cases, therefore, the intercept is consistent with a value of about 59/n mV, provided that each couple is a two-electron transfer. We expect a two-electron couple, however, because the analyte is V_2O_5, so one electron is transferred per vanadium.

Response 6.11

By using equation (6.13), a value of $D = 8.6 \times 10^{-6}$ cm^2 s^{-1} is obtained.

Response 6.12

From the rules described in Section 6.4.4 and using Table 6.4, the notation for this process is 'ECE'.

Response 6.13

By using equation (6.14), a value of $\tau = 18.8$ s is obtained.

Response 6.14

After plotting a graph of $I_{\mathrm{p(back)}}/I_{\mathrm{p(forward)}}$ against $\log \tau$, a value for the 'offset' from k of 10^{-3} s^{-1} is obtained.

Chapter 7

Response 7.1

From the definition of current density i in equation (1.1), we obtain an area of 0.12 cm^2.

Response 7.2

The same current will be passed through the working and counter electrodes (albeit, one oxidative and one reductive). By inserting the relevant values into equation (1.1), we obtain a value of $i = 0.06$ A cm^{-2}, so therefore the maximum value of i is *not* reached.

Response 7.3

By converting the force given in newtons to kilograms, i.e. by saying 10 N = 1 kg, we see that $\eta_s = 0.282$ g s^{-1}. Accordingly, by using equation (7.2), we obtain a value for υ of 0.239 cm^3 s^{-1}.

Response 7.4

By using equation (7.1), $I_{\mathrm{lim}} = 0.29$ mA.

Response 7.5

By using the same value of k, and following the same procedure as described in Worked Example 7.1, a concentration of 0.56×10^{-3} mol dm^{-3} is obtained.

Response 7.6

Start by writing $I_{\lim(1)} = kc_{(1)}$ and $I_{\lim(2)} = kc_{(2)}$. Dividing one equation by the other then gives $I_{\lim(1)}/I_{\lim(2)} = kc_{(1)}/kc_{(2)}$. The two '$k$ terms' will cancel out, thus leaving the ratios.

Response 7.7

By using equation (7.4), $f = 0.95$ Hz $= 57$ rpm.

Response 7.8

By using equation (7.4), $\omega = 75.4$ rad s^{-1}.

Response 7.9

The Levich plot obtained should start to show a loss of linearity (because of turbulence) at frequencies above about 95 Hz.

Response 7.10

By using equation (7.5), $\delta = 3.34 \times 10^{-3}$ cm.

Response 7.11

By using equation (7.5), $\delta = 334$ Hz, which is sufficiently fast for some turbulence to occur.

Response 7.12

By using equation (7.7), $I_{\lim} = 2.31$ A.

Response 7.13

By using equation (7.7), $I_{\lim} = kV_{\mathrm{f}}^{1/3}$, where k is simply the product of n, c, D and x (assuming all are constant). Taking logarithms, we obtain:

$$\log(I_{\lim}) = \log k + \log(V_{\mathrm{f}}^{1/3})$$

so:

$$\log(I_{\lim}) = \log k + \tfrac{1}{3}\log V_{\mathrm{f}}$$

which is equivalent to:

$$y = c + mx$$

i.e. the equation for a straight line, where $m = +1/3$.

Response 7.14

By using equation (7.6), which is a simplified form of equation (7.7), [dye] = 1.0 mmol dm^{-3}.

Response 7.15

By using equation (7.6), which is a simplified form of equation (7.7), [Co] = 4.55 μg cm^{-3}. Note that the values of a and r need not be known in this simplified approach.

Response 7.16

By using equation (7.10), $N_0 = 0.258$.

Response 7.17

The gradient of the linear portion in Figure 7.11 is 1.3×10^{-7} mol dm^{-3} s^{-1}, and therefore (from equation (7.11)), $k_2 = 7.4 \times 10^4$ (mol $dm^{-3})^{-1}$ s^{-1}.

Response 7.18

By using equation (7.14), $I_{\text{oxidative}} = 10.3$ A cm^{-2}.

Response 7.19

By using equation (7.16), $I_{\text{net}} = 0.348$ A.

Response 7.20

By taking logarithms of the terms in equation (7.16), we obtain:

$$\ln (I_{\text{net}}) = \ln (I_0) + \left(\frac{\alpha n F \eta}{RT}\right) - \left(\frac{-\alpha n F \eta}{RT}\right)$$

The value of η at the intercept is zero, so both potential-dependent terms vanish, leaving $\ln (I_{\text{net}}) = \ln (I_0)$, thus showing that the intercept when $\eta = 0$ is $\ln (I_0)$.

In practice, most electrochemists work with $\log_{10}$. If written in terms of $\log_{10}$, the two large brackets on the right-hand side of this logarithmic version of equation (7.16) should be multiplied by a factor of '2.303'.

Response 7.21

By using equation (7.17), $k_{\text{et}} = 1.22 \times 10^{-5}$ cm s^{-1}.

Chapter 8

Response 8.1

By using equation (8.2), $E = 3.31 \times 10^{-19}$ J and 200 kJ mol^{-1}.

Response 8.2

From a suitable textbook: (a) V^{2+} is lavender and V^{3+} is green; (b) Fe^{2+} is very pale green–yellow and Fe^{3+} is yellow; (c) MV^{2+} is colourless and $MV^{+\bullet}$ is blue.

Response 8.3

'Chrom' means colour and 'electro' implies an electrochemical process, so 'electrochromic' means colour change or generation of a new optical absorption band caused by an electron-transfer reaction.

Response 8.4

From the charge passed, 10^{-3} mol of charge forms 10^{-3} mol of bronze. Then, by taking $c = 10^{-3}$ mol dm^{-3}, $\varepsilon = 5200$ cm^{-1} mol^{-1} dm^{3} is obtained from the Beer–Lambert relationship (equation (8.3)).

Response 8.5

The optical absorbance data give $[An^{-\bullet}] = 4.66 \times 10^{-4}$ mol dm^{-3} (via the Beer–Lambert law, equation (8.3)), while the electrochemical data (via Faraday's laws) give $[An^{-\bullet}] = 5.18 \times 10^{-4}$ mol dm^{-3}. Accordingly, the efficiency = 90%.

Response 8.6

By using equation (8.2), $E = 1.99 \times 10^{-23}$ J.

Response 8.7

Taking logarithms of equation (8.4) yields:

$$\ln\,(V_{\mathrm{f}}) = \ln\,(k) - \frac{2}{3}\ln\left(\frac{S}{I_{\mathrm{lim}}}\right)$$

which is equivalent to:

$$y = c + mx$$

i.e. the equation of a straight line, where $m = -2/3$.

Response 8.8

The lack of scratch marks will decrease the incidence of turbulent flow, as well as ensuring that the fractal area of the electrode is essentially the same as the electrochemical area.

Response 8.9

By using the analogue of Ohm's law (equation (8.5)), $Z = 0.028\ \Omega$.

Response 8.10

The resistance R is found to be $7 \times 10^{-4}\ \Omega$. Because the platinum is a pure resistor, we can use equation (8.7) and thus $Z = R$, which also gives a value for Z of $7 \times 10^{-4}\ \Omega$.

Response 8.11

By using equation (8.8), and converting from linear to angular frequency, we obtain a value for Z'' of $1.59 \times 10^{4}\ \Omega$.

Response 8.12

By using equations (8.7) and (8.9), we obtain a value for Z_{total} of $46\ \Omega$.

Response 8.13

By using equations (8.7) and (8.10), we obtain a value for Z_{total} of $0.95\ \Omega$.

Response 8.14

By using equation (8.11), $\omega = 20\,000$ rad s^{-1}.

Response 8.15

By using equation (8.13), the duration of one cycle (τ) is 0.21 μs.

Response 8.16

By using equation (8.14), $i_0 = 0.206$ A cm^{-2}.

Response 8.17

By using equation (8.15), $D = 5.5 \times 10^{-8}$ cm^{2} s^{-1}.

Response 8.18

By using equation (8.1), we obtain values of ΔAbs for (a) 0.011 and (b) 0.301.

Chapter 9

Response 9.1

The name 'aqua regia' is Latin for 'water of kings', and is so-called because it is sufficiently oxidizing that it can dissolve precious metals such as gold or platinum if contact is prolonged.

Response 9.2

Fine abrasive will remove the 'scratch marks' formed by the coarser particles, i.e. decreasing the *fractal* area of the electrode (see Section 5.1.2).

Response 9.3

Concentrated nitric acid oxidizes HCl in aqueous solution to form elemental chlorine, which is the active oxidant in this system. Accordingly, the reaction is $Au_{(s)} + 2Cl_{2(soln)} \rightarrow [AuCl_4]^{3-}$.

Response 9.4

If the oxide layer is repeatedly formed and removed, the surface of the electrode will become pitted (*fractal*), which itself causes the electrode surface area to increase (see Section 5.1.2).

Response 9.5

Total internal reflection (TIR) of light within the thin ITO layer removes light of frequencies corresponding to the thickness of the film so, if light of $\lambda = 500$ nm is removed, then only red and blue remain, which therefore gives the purple colour.

Response 9.6

Referring back to Chapter 3, the Nernst equation (equation (3.8)) for this system is:

$$E_{\text{AgCl,Ag}} = E^{\ominus}_{\text{AgCl,Ag}} + \frac{RT}{F} \ln \left[\frac{a(\text{AgCl})}{a(\text{Ag})a(\text{Cl}^-)} \right]$$

Response 9.7

The spectral distribution of light emitted from a fluorescent strip light contains more UV component than does a normal tungsten bulb, and thus it will emit more photons of a sufficiently high energy to effect the degradation reaction shown in equation (9.6).

Response 9.8

By taking the Nernst equation used in Response 9.6, and appreciating that both Ag and AgCl are solid phases, and from the laws of logarithms that $\ln [1/a(\mathrm{Cl}^-)] = -\ln [a(\mathrm{Cl}^-)]$, we may rewrite the equation as follows:

$$E_{\mathrm{AgCl,Ag}} = E^{\ominus}_{\mathrm{AgCl,Ag}} - \frac{RT}{F} \ln [a(\mathrm{Cl}^-)]$$

which is equivalent to:

$$y = c + mx$$

i.e. the equation of a straight line, where $m = -RT/F$.

Chapter 10

Response 10.1

$E^{\ominus}$ is found at the midpoint between the forward and reverse peaks, provided that the couple is reversible (see Table 6.3).

Bibliography[†]

Books

General Analytical Texts

Undergraduate texts for courses in analytical chemistry usually contain at least a couple of chapters on electroanalysis, with typically one each covering potentiometric and voltammetric analyses. The following texts contain suitable material.

Higson, S. P. J., *The Elements of Analytical Chemistry*, Oxford University Press, Oxford, 2001. This new book is part of the Oxford Primer Series, so it will be a cheap and affordable addition to an analyst's library since it covers all of the basics.

Christian, G. F., *Analytical Chemistry*, 5th Edn, Wiley, New York, 1994. This book is fairly comprehensive and will cover some of the material covered. Its sections on voltammetric determinations are much poorer than its sections on potentiometric methods.

Fifield, F. W. and Kealey, D., *Principles and Practices of Analytical Chemistry*, International Textbook Company, Edgeware, UK, 1975. A trusted work-horse of a text. It is showing its age a bit now, but is an excellent first look at analytical chemistry. Its 'compare and contrast' style of potentiometric and voltammetric analyses is widely copied, it seems.

General Electrochemistry Texts

While not explicitly written with the requirements of an analyst in mind, most electrochemistry texts do contain rudiments of electroanalysis.

[†] The opinions expressed within this bibliography are not those of the publisher.

Crow, D. R., *Principles and Applications of Electrochemistry*, 4th Edn, Blackie Academic, Glasgow, 1994. This text has quite a long history now (first published in 1974). It is pleasure to read and is therefore recommended highly, both for potentiometric and voltammetric study. Some of its examples and self-assignment questions are ideal practise material.

Koryta, J., Dvorák, J. and Kavan, L., *Principles of Electrochemistry*, 2nd Edn, Wiley, Chichester, 1993. The release of this second edition of a now-classic work was much welcomed. The text is well structured and always worth reading. Some of the examples are particularly relevant, although the voltammetric sections are inferior to those concentrating on potentiometric studies.

Electroanalytical Texts

Bard, A. J. and Faulkner, L. R., *Electrochemical Methods*, Wiley, New York, 1980. Without doubt, the best 'all round' book for the electroanalyst, and already regarded as a modern classic. Any library not owning a copy of this book is not a real library. Its treatment is generally mathematical, so it is probably not suitable for many undergraduates. Slightly dated in many aspects, although the release of a second edition is imminent.

Brett, C. M. A. and Brett, A. M. O., *Electroanalysis*, Oxford University Press, Oxford, 1998. This text is a recent addition to the Oxford Primer Series and, as such, is affordable and good value for money. Although not long, it does provide a clear and concise introduction to electroanalysis. Probably on the difficult side for many undergraduates of analytical chemistry, but nevertheless is still worth a look.

Greef, R., Peat, R., Peter, L. M., Pletcher, D. and Robinson, J., *Instrumental Methods in Electrochemistry*, Ellis Horwood, Chichester, 1990. 'The Southampton Electrochemistry Book' – now out of print, but well worth a look. Its treatment is considerably less mathematical than Bard and Faulkner's text (see above) and, in consequence, is more readable for the average student and analyst. Some of the book is just about suitable for undergraduate work, although most of its content should be thought of as being postgraduate level.

Galus, Z., *Fundamentals of Electrochemical Analysis*, 2nd Edn, Ellis Horwood, Chichester, 1994. Don't be fooled by the title: 'fundamentals' here means 'as derived from first principles', rather than a book for beginners. An unashamedly mathematical read, and not for the faint-hearted. Although out of print now, it must still be considered the ultimate book on the subject. Essentially for postgraduates with great self-confidence.

Vanysek, P. (Ed.), *Modern Techniques in Electroanalysis*, Wiley-Interscience, New York, 1996. Each chapter of this anthology commences with a thorough introduction before surveying recent progress. The chapters are naturally of variable quality, but Dewald's description of stripping analyses in Chapter 4 is well worth reading. Although expensive, this monograph is, nevertheless, good value for money.

Kissenger, P. T. and Heinemann, W. R. (Eds), *Laboratory Techniques in Electroanalytical Chemistry*, Marcel Dekker, New York, 1984. This book is perhaps a little dated, but is nevertheless quite readable and a good introduction to the topic.

Wang, J., *Electroanalytical Techniques in Clinical Chemistry and Laboratory Medicine*, VCH, New York, 1988. A short book from the 'master' himself. Not an easy read, though, as it seems to presuppose quite a lot of background material – but it is, after all, *applied* electroanalysis, as the title suggests. The chapter on ion-selective electrodes is dated in content but full of useful insights.

Riley, T. and Tomlinson, C., *Principles of Electroanalytical Methods*, ACOL Series, Wiley, Chichester, 1987. This is the first of the three Analytical Chemistry by Open Learning (ACOL) texts on electroanalysis (all of which are now out of print), and the lesser of the three in terms of quality, so few of today's readers will bother to search its pages now. It is nevertheless quite a good introduction to the field, although its age is showing quite badly. The lack of an index, together with other period features, detract from an otherwise readable book.

Specific Bibliography

Chapter 1

The Oxford Dictionary for Scientific Writers and Editors, Oxford University Press, Oxford, 1992.

Mills, I., Cvitaš, T., Homann, K., Kallay, N. and Kuchitsu, K., *Quantities, Units and Symbols in Physical Chemistry*, 2nd Edn, IUPAC, Blackwell Science, Oxford, 1995.

Chapter 2

Fisher, A. C., *Electrode Dynamics*, Oxford University Press, Oxford, 1996. As part of the Oxford Primer Series, this book is a readily affordable and very readable introduction to mass transport. Highly recommended.

Chapter 3

Compton, R. G. and Sanders, G. H. W., *Electrode Potentials*, Oxford University Press, Oxford, 1996. This book is another in the Oxford Primer Series, and thus represents good value for money. The treatment of the Nernst equation, in particular, is thorough and straightforward. This book contains copious examples and exercises in the form of self-assessment questions (SAQs). Note, however, that it does not cover sensors.

Evans, A., *Potentiometry and Ion Selective Electrodes*, ACOL Series, Wiley, Chichester, 1987. Another ACOL text which again shows those traits which drew criticism to the first editions of this series. In fact, this is a shame, since as an open learning text this provides a really good introduction to this subject. This book is now out of print.

Buffle, J., *Complexation Reactions in Aquatic Systems: An Analytical Approach*, Ellis Horwood, Chichester, 1990. This gem of a book covers all of the theoretical and practical aspects of speciation. Although out of print now, its wealth of insight and clear-sighted use of examples makes it a 'must'. It is somewhat mathematical in parts, although most of this material can generally be avoided.

Diamond, D. (Ed.), *Principles of Chemical and Biological Sensors*, Wiley, New York, 1998. This is Volume 150 in the Wiley-Interscience Chemical Analysis Series. The first chapter by Diamond is a useful overview of sensors, both their construction and usage. The other chapters vary in terms of usefulness, and are also of variable quality.

Cass, A. E. G. (Ed.), *Biosensors: A Practical Approach*, Oxford University Press, Oxford, 1990. Although now somewhat dated, many practitioners still regard this as the best book available on the subject. Its style is non-mathematical and almost 'chatty' in parts, thus making it both clear and accessible.

Janata, J., *Principles of Chemical Sensors*, Plenum Press, New York, 1989. Some sensor specialists regard this as the definitive work on the subject. While extremely dated, its introductory sections provide a clear, uncluttered introduction to the different modes of sensor operation.

Chapter 4

The text by Buffle (above) also covers potentiometric systems in which electron transfer does occur.

Chapter 5

There are no books explicitly dedicated to coulometry, but many analytical and electrochemistry texts do offer a thorough introduction. In particular, the following text describes coulometry in some depth.

LaCourse, W. R., *Pulsed Electrochemical Detection in High Performance Liquid Chromatography*, Wiley, New York, 1997. The chapters on the various types of stripping with pulsed wave forms are very good.

Chapter 6

Electrode Dynamics (Fisher), mentioned above, contains a readable and affordable introduction to voltammetry and cyclic voltammetry. It is intended more for the chemist wanting mechanistic data rather than for the analyst. In addition, note that it does not explicitly cover polarography, although it does describe a few electrochemical techniques that are not included in this present book, so is still a useful text.

Riley, T. and Watson, A., *Polarography and Other Voltammetric Methods*, ACOL Series, Wiley, Chichester, 1987. This is the best of the three ACOL electroanalysis texts (although again out of print). Its structure and examples are well chosen and well executed. Unfortunately, as with the other ACOL texts, the lack of an index and the irritating format may prove annoying to many readers.

Chapter 7

Levich, V. G., *Physicochemical Hydrodynamics*, Prentice Hall, Englewood Cliffs, NJ, USA 1962. Levich's book is *the* text on hydrodynamic electrodes. Although not at all up-to-date, most of the principles described are timeless. The text is quite mathematical, and somewhat fearsome-looking for the uninitiated, but is nevertheless an excellent and reliable reference source.

Albery, W. J., *Electrode Kinetics*, Clarendon Press, Oxford, 1975. In many respects, this is the best book on the subject of electrode kinetics. It can be quite mathematical, but otherwise it provides a lucid and user-friendly approach to this subject.

Albery, W. J. and Hitchman M. L., *Ring-Disc Electrodes*, Oxford University Press, Oxford, 1971. This now-classic book describes one of the most formidable tools in the arsenal of the electroanalyst, i.e. the rotated ring-disc electrode (RRDE). Its first two chapters are a clear and lucid introduction to the basic rotated disc electrode (RDE) and the multi-faceted problems of mass transport. Well worth a read, if only for the occasional 'dip' into this field.

Brett, C. M. A. and Brett A. M. C. F. O., 'Hydrodynamic Electrodes', in *Comprehensive Chemical Kinetics*, Vol. 27, Bamford, C. H. and Compton R. G. (Eds), Elsevier, Amsterdam, 1986, pp. 355–441. This monograph provides a thorough and useful introduction to the topics of mass transport and convection-based electrodes. It also contains one of the better discussions on flow systems, in part because it can be read quite easily despite the overall treatment being so overtly mathematical.

Trojanowicz, M., 'Electrochemical Detectors in Automated Analytical Systems', in *Modern Techniques in Electroanalysis*, Vanysek P. (Ed.), Wiley-Interscience, New York, 1996, pp. 187–239. This chapter contains a fairly thorough discussion of the possible arrangements of electrodes within flow systems, and for a variety of applications.

Chapter 9

Selley, N. J., *Experimental Approach to Electrochemistry*, Edward Arnold, London, 1977. While desperately dated in parts, much of its introductory material is timeless, particularly in terms of preparing electrode surfaces for routine analyses (e.g. in its Section 8.3). The last chapter contains some projects in electrochemistry which can be tackled as worked examples. It also contains numerous examples for self-assessment. While long out of print, many libraries may still have a copy.

Chapter 10

Electrochemical Methods (Bard and Faulkner), *Instrumental Methods in Electrochemistry* ('The Southampton Electrochemistry Book') (Greef *et al.*) and *Modern Techniques in Electroanalysis* (Vanýsek), all cited above, present suitable simulation programs.

Britz, D., *Digital Simulation in Electrochemistry*, Springer-Verlag, Berlin, 1981 and MacDonald, D., *Transient Techniques in Electrochemistry*, Plenum Press, New York, 1977. Both of these books contain copious details concerning electrochemical simulations. Although these texts are extremely mathematical (as all simulation work has to be), the basic concepts are not too difficult to follow. The application notes to Condecon (see URL on page 301) are also a feast of detail.

Articles and Reviews

The *Current Separations* Series (produced free of charge by Bioanalytical Systems, Inc.)[†] is an excellent source of detailed and reliable information.

Chapter 2

Bott, A. W., 'Mass transport', *Current Separations*, **14**, 104–109 (1995) provides a good introduction to the subject.

Chapter 3

Hitchman, M. L. and Hill, H. A. O., 'Electroanalysis and electrochemical sensors', *Chemistry in Britain*, **22**, 1117–1124 (1986), provides a lively, general introduction to this subject, giving details of sensors based on potentiometry, such as ISEs, together with some historical background.

Chapter 5

Mediation is mentioned in Bott, A. W., 'Redox properties of electron-transfer metalloproteins', *Current Separations*, **18**, 47–54 (1999). In addition, Bott, A. W., 'Electrochemical titrations,' *Current Separations*, **19**, 128–132 (2000), and Bott, A. W., 'Controlled current techniques,' *Current Separations*, **19**, 125–127 (2000), are both worth consulting.

[†] Bioanalytical Systems, Inc., 2701 Kent Avenue, West Lafayette, IN 47906-1382, USA *or* http://www.current separations. com (where some articles can be downloaded as .pdf documents).

Chapter 6

Bott, A.W., 'Practical problems in voltammetry. 3: 'Reference electrodes for voltammetry', *Current Separations*, **14**, 64–68 (1995) is an excellent first stop for the novice, as is Bott, A. W., 'Characterization of chemical reactions coupled to electron-transfer reactions using cyclic voltammetry', *Current Separations*, **18**, 9–16 (1999), which also introduces simulations. In addition, the article by Hitchman and Hill in *Chemistry in Britain* (see above) contains a low-level general introduction to cyclic voltammetry for analyses.

Chapter 7

The article by Hitchman and Hill (above) is again useful here in that it also contains an introduction to the rotated ring-disc and wall-jet electrodes.

Chapter 8

The two reviews, 'Spectroelectrochemistry: Applications' and 'Spectroelectrochemistry: Methods and instrumentation', both by Mortimer, R. J., appear in *The Encyclopedia of Spectroscopy and Spectrometry*, Lindon, J. C., Trantor, G. E. and Holmes J. L. (Eds), Academic Press, London, 2000, pp. 2161–2174 and 2174–2181, respectively, and give excellent coverage of this combined spectroscopic and electrochemical technique.

While somewhat complicated, Bott, A. W., 'Electrochemical impedance spectroscopy using the BAS-Zahner IM6 and IM6e impedance analyzers', *Current Separations*, **17**, 53–59 (1998), is a helpful introduction to the subject, and also mentions computer simulations.

Chapter 9

The following articles in the *Current Separations* Series are a useful start: Bott, A. W., 'Practical problems in voltammetry. 3: Reference electrodes for voltammetry; 4: Preparation of working electrodes', *Current Separations*, **14**, 64–68 (1995); **16**, 79–83 (1997).

Chapter 10

The following articles in the *Current Separations* Series are an excellent introduction to the topic of simulation: Bott, A. W., 'Fitting experimental cyclic voltammetry data with theoretical simulations using DigiSim® 2.1', *Current Separations*, **15**, 67–71 (1996); Bott, A. W. and Jackson, B. P., 'Study of ferricyanide by cyclic voltammetry using the CV-50w,' *Current Separations*, **15**, 25–30 (1996).

Web-Based Resources[†]

The following web site:

http://electrochem.cwru.edu/estir/chap.htm#2000

is an excellent resource for all aspects of electrochemistry and electroanalysis, and is well worth looking at. In addition, the site:

www.nico2000.net

represents an extremely comprehensive introduction to ion-selective electrodes, and contains a large and comprehensive glossary.

[†] As of January 2001. The products displayed are not endorsed by the author or the publisher.

Glossary of Terms

This section contains a glossary of terms, all of which are used in the text. It is not intended to be exhaustive, but to explain briefly those terms which often cause difficulties or may be confusing to the inexperienced reader.

Activity, a This is a form of concentration; it is the concentration *perceived* thermodynamically at an electrode.

Activity coefficient, $\gamma_{\pm}$ The ratio of activity a and concentration c, i.e. $\gamma_{\pm} = a/c$.

Ammeter A device for measuring current.

Amperometry The techniques and methodology of determining a concentration as a function of current; the most popular amperometric measurement technique is voltammetry.

Analyte The material to be quantified, characterized or in any other way investigated.

Anion A negatively charged ion.

Anode The electrode at which oxidation occurs during electrolysis.

Auxiliary redox couples (*see* Mediation)

Cathode The electrode at which reduction occurs during electrolysis.

Cation A positively charged ion.

Channel cell An electrochemical cell in which analyte solution flows at a velocity V_f over flat stationary electrode(s).

Charge, Q The quantity of unbalanced electricity in a body such as an electron or an ion.

Charge Density, q The quotient of charge Q and electrode area A, i.e. $q = Q/A$.

Chemically modified electrode (CME) An electrode covered with a special coating (of a conductor or semiconductor unless very thin), thereby altering its electrochemical properties.

Chronoamperometry The techniques and methodology of studying current as a function of time.

Convection That form of mass transport in which the solution containing electroanalyte is moved: natural convection occurs predominantly by heating of solution, while forced convection occurs by careful and deliberate movement of the solution, e.g. at a rotated disc electrode or by the controlled flow of analyte solution over a channel electrode.

Coulometer A device for measuring the charge Q.

Coulometry The techniques and methodology of measuring the charge Q.

Counter electrode (CE) The third electrode in voltammetry and polarography, where the current is measured between the counter and working electrodes. It is rare to monitor the potential of the counter electrode.

Current, *I* The rate of charge flow or passage, i.e. $I = (\mathrm{d}Q/\mathrm{d}t)$.

Current density, *i* The quotient of current I and electrode area A, i.e. $i = I/A$.

Current maximum suppresser A chemical, usually a surfactant (detergent), added to polarography solutions in order to decrease the incidence of polarographic 'peaks' (see Section 6.8.1).

Cyclic voltammogram (CV) A plot of current (as 'y') against potential (as 'x') obtained during a voltammetric experiment in which the potential is ramped twice, once forward to the switch potential E_λ, and then back again.

Depletion region (depletion layer) That layer around the electrode within which analyte is consumed during current flow (see Section 6.2.1).

Differential pulse voltammetry A form of voltammetry in which a linear potential ramp of $\mathrm{d}E/\mathrm{d}t$ is applied to the working electrode, superimposed on which is a succession of pulses.

Diffusion That form of mass transport in which motion occurs in response to a gradient in concentration or composition, itself caused by a gradient of the chemical potential μ. Diffusion is ultimately an entropy-driven process.

Diffusion coefficient, *D* A measure of the velocity of electroanalyte as its diffuses through solution prior to an electron-transfer reaction.

Diffusion current, I_d The maximum current in polarography (see Section 6.3.2).

Double-layer capacitance The capacitance C owing to the two Helmholtz layers at the electrode | solution interface.

Dropping mercury electrode (DME) Historically, a popular choice of working electrode in polarography.

Dynamic An electroanalytical or electrochemical measurement accompanied by current flow and hence by changes in the concentrations of analyte.

Electroanalyte An analyte investigated by electrochemical means.

Electrochemical area The fractal area of an electrode; the area that an electrode is 'perceived' to have.

Electrochemical cell A pair of electrodes (usually metallic) in contact with a solution containing an electrolyte.

Electrochromism The electrochemical generation of colour in accompaniment with a redox reaction, e.g. as displayed by a redox indicator.

Electrode A conductor employed either to determine an electrode potential (at zero current, i.e. for potentiometric experiments), or to determine current during a dynamic electroanalytical measurement. The electronic conductivity of most electrodes is metallic.

Electrode potential, E The energy, expressed as a voltage, of a redox couple at equilibrium. E is the potential of the electrode when measured relative to a standard (ultimately the SHE). E depends on temperature, activity and solvent. By convention, the half cell must first be written as a reduction, and the potential is then designated as positive if the reaction proceeds spontaneously with respect to the SHE. Otherwise, E is negative.

Electrolyte An ionic salt to be dissolved in a solvent, *or* the solution formed by dissolving an ionic salt in a solvent.

Electromodification The process of altering a molecule, ion, etc. of analyte by the passing of a current. The simplest examples are reduction or oxidation reactions.

Electron mediation (*see* Mediation)

emf The potential that develops between two electrodes in an electrochemical cell when a state of frustrated equilibrium has been reached; the value of the *emf* is always defined as positive (*emf* derives from the somewhat archaic term 'electromotive force').

Equilibrium In an electroanalytical context, equilibrium is best defined as *zero current* during a potentiometric measurement.

Equivalent circuit In impedance analyses, a collection of electrical components used to mimic the frequency behaviour of a cell or electrochemical system.

ex situ Measurement performed away from the site of formation (i.e. afterwards).

Exchange current, I_0 The rate constant of electron transfer (expressed as a current) at zero overpotential.

Exchange current density, i_0 The quotient of exchange current I_0 and electrode area A, i.e. $i_0 = I_0/A$.

Exhaustion (exhaustive electrolysis) The complete removal of analyte, for example, during stripping.

Faradaic charge That component of the total charge Q that follows Faraday's laws; the charge that does not follow these laws is termed 'non-faradaic' (see Section 5.1.1).

Faradaic efficiency The quotient of faradaic current to total current passed.

Flow cell Electrochemical cell in which analyte solution flows at a constant velocity V_f through stationary tubular electrode(s).

Flux, j The amount of electroanalyte that reaches an electrode surface (from the solution bulk), usually expressed in moles per unit time.

Formal electrode potential, $E^{0'}$ An electrode potential measured at STP, in which all reagents and products are present at unit *concentration*.

Fractal area The area, e.g. of an electrode, that is greater than the simple geometric area on account of surface roughness.

Glassy carbon (GC) A form of carbon that is extremely hard and highly conductive to electrons, and is hence a good choice of material for constructing inert electrodes (also known as 'vitreous' carbon).

Gran plot A commonly employed multiple-addition method, used to correct for unknown amounts of contaminant and for dilution errors (see Section 4.3.2).

Half-cell reactions Oxidation or reduction reactions occurring at a single electrode in a cell.

Half-wave potential, $E_{1/2}$ An electrode potential from polarography that is characteristic of the analyte (see Section 6.3.2).

Hanging mercury-drop electrode (HMDE) A commonly employed working electrode in polarography. The HMDE is often preferred to the experimentally simpler dropping mercury electrode (DME) because resultant polarograms do not have a 'sawtoothed' appearance and because accumulation (for 'stripping' purposes) is readily achieved at its static surface.

Heterogeneous A process occurring between phases, as for example, electron transfer between a solid electrode and a solution-phase analyte.

Homogeneous A process occurring in one phase, as for example, electron transfer during a potentiometric titration.

Impedance AC resistance.

in situ Measurement performed in the same place as the site of formation (i.e. performed at the same time).

Indirect coulometry (*see* Mediation)

Indium–tin oxide (ITO) Tin-doped indium oxide, used as a thin solid film on glass when constructing optically transparent electrodes.

Inert electrode An electrode which does not comprise any part of the redox couple under investigation. The best examples are platinum, gold or glassy carbon.

Interface In electrochemistry and electroanalysis, the region between an electrode and the phase containing analyte (usually indicated with a vertical line, |).

Ion A charged species; ions should always be regarded as solvated.

Ion-selective electrode (ISE) In potentiometry, an electrode having a nernstian response to one ion, ideally to the exclusion of others.

***IR* drop** A potential developed when a current I flows through an electrochemical cell. It occurs as a consequence of the cell resistance R, and has a magnitude IR (from Ohm's law). The IR drop is always subtracted from the theoretical cell potential.

International Union of Pure and Applied Chemistry (IUPAC) The body which defines terminology in chemistry and the other sciences.

Junction potential (*see* Liquid junction potential)

Kinematic viscosity, υ The quotient of viscosity η_s and density ρ, i.e. $\upsilon = \eta_s/\rho$.

Limiting current, I_{lim} Current that is directly proportional to the bulk concentration of analyte.

Linear-sweep voltammogram A normal voltammogram, in which the voltage ramp stops at the switch potential (in contrast to a cyclic voltammogram, in which it does not).

Liquid junction potential, E_j A potential developed across a boundary between electrolytes which differ in concentration or chemical composition. E_j is caused by differing rates of ionic migration across the boundary, thereby leading to a charge separation. This potential is usually some tens of millivolts, with the magnitude being variable, but it may be minimized by the use of a salt bridge connection containing, e.g. the electrolytes KCl or NH_4NO_3 (occasionally gelled in agar). With such a bridge, E_j will be about 1–2 mV.

Luggin capillary A device for allowing the reference electrode to be positioned extremely close to the surface of the working electrode, thereby decreasing the *IR* drop (see Section 6.8.2).

Mass transport The means of effecting movement of electroanalyte to an electrode prior to electromodification. The three forms ('modes') of mass transport are convection, migration and diffusion.

Mediation The *electrochemical* generation of an oxidant or reductant which can then effect redox changes *chemically* (see Section 5.4).

Mediator titration A titration in which the titre relates to a coulometric titration with a mediator.

Mercury film electrode (MFE) A thin layer of elemental mercury, usually on graphite, prepared by electrochemical deposition from an aqueous solution of mercury(II).

Microelectrode An electrode of surface area 10^{-3} mm^2 or less, used because its double-layer charging current is negligible.

Migration That form of mass transport that occurs in response to coulombic attraction between charged ions and an electrode bearing a charge of the opposite sign to that of the ions that move.

Nernst layer (Nernst depletion layer) (*see* Depletion region)

Nernstian Obeying the Nernst equation (equation (3.8)) at all times.

Non-faradaic charge That component of the total charge Q that does not follow Faraday's laws; the charge that does follow these laws is termed 'faradaic' (see Section 5.1.1).

Normal hydrogen electrode (*see* Standard hydrogen electrode)

Ohmic Obeying Ohm's law (equation (2.3)).

Ohmic drop *see IR* drop

Optically transparent electrode (OTE) An electrode used for *in situ* spectroelectrochemistry.

Overpotential, η Deviation of the potential at an electrode from the equilibrium potential, with the deviation usually being effected at a working electrode by a potentiostat.

Oxidation The loss of electrons, resulting in the analyte having a higher oxidation number:

$$R \longrightarrow O + ne^-$$

for example:

$$Cu \longrightarrow Cu^{2+} + 2e^-, \text{ or } 2Br^- \longrightarrow Br_2 + 2e^-$$

Peak current, I_p In voltammetry and cyclic voltammetry, the current that is directly proportional to the bulk concentration of analyte.

Polarograph A form of potentiostat.

Polarography The techniques and methodology of a form of voltammetry in which the working electrode is liquid, elemental mercury, usually employed within a dropping mercury electrode (DME) or a hanging mercury-drop electrode (HMDE).

Potential 'window' That range of potentials in which it is safe and convenient to perform electroanalyses. Either the solvent or the electrolyte are likely to undergo electron-transfer reactions outside of this 'window'.

Potentiometer A device for determining a potential (likely to be called a 'voltmeter').

Potentiometry The techniques and methodology of determining an activity as a function of potential (at zero current). Activity and concentration can often be interchanged at low ionic strength.

Potentiostat The instrument employed in dynamic electrochemistry such as voltammetry, allowing three electrodes to be used.

Quiescent (*see* Still solution)

Quiet solution (*see* Still solution)

Ramp A jargon term, meaning to smoothly increase at a constant rate of $d(\text{variable})/dt$. A voltage ramp is therefore dE/dt.

Randles–Sevčik plot A graph derived from the Randles–Sevčik equation, showing a plot of the cyclic voltammetry peak height I_p, (as 'y') as a function of $\nu^{1/2}$ (as 'x'), where ν is the scan rate.

Random array of microelectrodes (RAM) A microelectrode system comprising about 1000 carbon fibres embedded randomly within an inert adhesive such as an epoxy resin.

Redox An abbreviation of 'reduction and oxidation'; a type of reaction in which both reduction and oxidation occur concurrently.

Redox couple Two redox states of the same material, e.g. Cu^{2+} and Cu. A couple sometimes comprises those ions which are necessary to maintain a balanced redox equation, e.g. $AgBr + e^- = Ag + Br^-$.

Redox electrode An electrode, the metal of which comprises one half of a redox couple.

Redox indicator A chemical that changes colour (is 'electrochromic') when undergoing redox change: since the change occurs at a specific potential, the substance is able to function as an indicator.
Redox potential (*see* Electrode potential)
Reduction The gain of electrons, resulting in the analyte having a lower oxidation number:

$$O + ne^- \longrightarrow R$$

for example:

$$Cu^{2+} + 2e^- \longrightarrow Cu, \text{ or } Br_2 + 2e^- \longrightarrow 2Br^-$$

Reduction potential (*see* Electrode potential)
Reference electrode (RE) A constant-potential device (e.g. an SHE or SCE) used as a half cell of known potential.
Reinmuth notation A useful shorthand notation for abbreviating the order of a sequence of electrochemical ('E') and chemical ('C') reactions (see Section 6.4.3). The order of a reaction is read from the left of the Reinmuth code to the right.
Rise time The length of time needed for the double-layer at the electrode|solution interface to charge (c. milliseconds) (see Section 6.2).
Rotated ring-disc electrode (RRDE) Disc electrode within a concentric ring electrode.
Salt bridge A device employed to minimize the liquid junction potential E_j. The bridge connects two half cells, with one end being immersed in each. The bridge is occasionally a piece of paper soaked in KCl or NH_4NO_3, or more commonly a glass U-tube filled with the same electrolytes, gelled with agar or terminated at either end with a frit.
Sampled DC polarography A form of polarography in which a linear potential ramp of dE/dt is applied to the working electrode, but where the current is measured only during the last 15% or so of each drop cycle.
Saturated calomel electrode (SCE) A reference electrode in which the redox couple comprises elemental mercury and mercury(I) chloride, Hg_2Cl_2 (which also has the archaic name of 'calomel').
Scan rate, ν In cyclic voltammetry, the rate at which the potential of the working electrode is varied, i.e. $\nu = (dE/dt)$. The value of ν is always cited as positive.
Sensor A device having a response (ideally) for one particular analyte. Potentiometric sensors are typically ion-selective electrodes, while amperometric sensors rely on Faraday's laws.
Silver paint A conductive material comprising colloidal silver suspended in a polymeric base, which is liquid when applied to form an electrical contact, but then sets rapidly to form a hard and highly conductive contact. The material is commonly applied with a brush, hence the term 'paint'.

Sparging The removal of oxygen dissolved in a solution of analyte, carried out, e.g. by bubbling purified nitrogen gas through the solution for c. 10 min before analysis.

Square-wave voltammetry A form of voltammetry in which the potential waveform applied to the working electrode is a square wave. Pairs of current measurements are made for each wave cycle. The current associated with the forward part of the pulse is termed $I_{forward}$, while the current associated with the reverse part is $I_{reverse}$. A square-wave voltammogram is then just a graph of the difference between these two currents as a function of the applied potential.

Standard electrode potential, $E^{\ominus}$ An electrode potential measured at standard temperature and pressure, where all products and reagents are present at unit *activity*.

Standard hydrogen electrode (SHE) The standard against which redox potentials are measured. The SHE consists of a platinum electrode electroplated with Pt black (to catalyse the electrode reaction), over which hydrogen at a pressure of 1 atm is passed. The electrode is immersed in a solution containing hydrogen ions at unit activity (e.g. 1.228 mol dm^{-3} of aqueous HCl at 20°C). The potential of the SHE half cell is *defined* as 0.000 V at all temperatures.

Still solution Solution of analyte wholly free of convective effects.

Stripping (General), the techniques and methodology of analyte pre-concentration used in voltammetry to improve the sensitivity; (specific), the electrochemical re-removal of analyte accumulated at an electrode.

Supporting electrolyte (*see* Swamping electrolyte)

Swamping electrolyte An ionic salt added to a solution of analyte in order to minimize migration effects and increase the ionic conductivity.

Sweep rate (*see* Scan rate)

Switch potential, E_{λ} In cyclic voltammetry, the potential at which the voltage ramp changes direction.

Tafel relationship The equation which relates the current I and overpotential η, as $\log I \propto \eta$.

Temperature coefficient (*see* Voltage coefficient)

Temperature voltage coefficient (*see* Voltage coefficient)

Theoretical cell potential The algebraic sum of the individual redox potentials of an electrochemical cell at zero current, i.e. $emf = E_{\text{positive electrode}} - E_{\text{negative electrode}}$. In practice, when current flows in a cell, a liquid junction potential is present, and the cell potential is larger than this theoretical value.

Transfer coefficient, α A measure of the symmetry within the transition state 'complex' formed during an electron-transfer reaction between an electrode and a moiety of electroanalyte. A wholly symmetrical complex has $\alpha = 0.5$.

Voltage coefficient In thermodynamics, the quantity $(\mathrm{d}\,emf\,/\mathrm{d}T)$ (see Section 3.2.1).

Voltammetry The techniques and methodology of measuring current as a function of applied potential.

Voltammogram A trace of current (as '*y*') against applied potential (as '*x*').

Voltmeter A device for measuring a potential. A good-quality voltmeter has as high an internal resistance ('impedance') as possible.

Wall-jet electrode Electrochemical cell in which analyte solution is squirted at high pressure on to a flat circular electrode.

Window (*see* Potential window)

Working electrode (WE) The principal electrode during dynamic analyses. The potential of the WE is measured with respect to a reference electrode, while the current passing through the WE is measured with respect to the counter electrode.

SI Units and Physical Constants

SI Units

The SI system of units is generally used throughout this book. It should be noted, however, that according to present practice, there are some exceptions to this, for example, wavenumber (cm^{-1}) and ionization energy (eV). The use of length (in cm), as an exception, is discussed in Chapter 1.

Base SI units and physical quantities

Quantity	Symbol	SI unit	Symbol
length	l	metre	m
mass	m	kilogram	kg
time	t	second	s
electric current	I	ampere	A
thermodynamic temperature	T	kelvin	K
amount of substance	n	mole	mol
luminous intensity	I_v	candela	cd

Prefixes used for SI units

Factor	Prefix	Symbol
10^{21}	zetta	Z
10^{18}	exa	E
10^{15}	peta	P
10^{12}	tera	T
10^{9}	giga	G
10^{6}	mega	M
10^{3}	kilo	k

(continued overleaf)

Prefixes used for SI units *(continued)*

Factor	Prefix	Symbol
10^2	hecto	h
10	deca	da
10^{-1}	deci	d
10^{-2}	centi	c
10^{-3}	milli	m
10^{-6}	micro	μ
10^{-9}	nano	n
10^{-12}	pico	p
10^{-15}	femto	f
10^{-18}	atto	a
10^{-21}	zepto	z

Derived SI units with special names and symbols

Physical quantity	SI unit		Expression in terms of base or derived SI units
	Name	Symbol	
frequency	hertz	Hz	$1\ \text{Hz} = 1\ \text{s}^{-1}$
force	newton	N	$1\ \text{N} = 1\ \text{kg m s}^{-2}$
pressure; stress	pascal	Pa	$1\ \text{Pa} = 1\ \text{N m}^{-2}$
energy; work; quantity of heat	joule	J	$1\ \text{J} = 1\ \text{N m}$
power	watt	W	$1\ \text{W} = 1\ \text{J s}^{-1}$
electric charge; quantity of electricity	coulomb	C	$1\ \text{C} = 1\ \text{A s}$
electric potential; potential difference; electromotive force; tension	volt	V	$1\ \text{V} = 1\ \text{J C}^{-1}$
electric capacitance	farad	F	$1\ \text{F} = 1\ \text{C V}^{-1}$
electric resistance	ohm	Ω	$1\ \Omega = 1\ \text{V A}^{-1}$
electric conductance	siemens	S	$1\ \text{S} = 1\ \Omega^{-1}$
magnetic flux; flux of magnetic induction	weber	Wb	$1\ \text{Wb} = 1\ \text{V s}$
magnetic flux density; magnetic induction	tesla	T	$1\ \text{T} = 1\ \text{Wb m}^{-2}$
inductance	henry	H	$1\ \text{H} = 1\ \text{Wb A}^{-1}$
Celsius temperature	degree Celsius	°C	$1°\text{C} = 1\ \text{K}$
luminous flux	lumen	lm	$1\ \text{lm} = 1\ \text{cd sr}$
illuminance	lux	lx	$1\ \text{lx} = 1\ \text{lm m}^{-2}$

Derived SI units with special names and symbols *(continued)*

Physical quantity	SI unit		Expression in terms of base or derived SI units
	Name	Symbol	
activity (of a radionuclide)	becquerel	Bq	$1\ \mathrm{Bq} = 1\ \mathrm{s}^{-1}$
absorbed dose; specific energy	gray	Gy	$1\ \mathrm{Gy} = 1\ \mathrm{J\ kg}^{-1}$
dose equivalent	sievert	Sv	$1\ \mathrm{Sv} = 1\ \mathrm{J\ kg}^{-1}$
plane angle	radian	rad	1[a]
solid angle	steradian	sr	1[a]

[a]rad and sr may be included or omitted in expressions for the derived units.

Physical Constants

Recommended values of selected physical constants[a]

Constant	Symbol	Value
acceleration of free fall (acceleration due to gravity)	g_n	$9.806\ 65\ \mathrm{m\ s}^{-2}$[b]
atomic mass constant (unified atomic mass unit)	m_u	$1.660\ 540\ 2(10) \times 10^{-27}\ \mathrm{kg}$
Avogadro constant	L, N_A	$6.022\ 136\ 7(36) \times 10^{23}\ \mathrm{mol}^{-1}$
Boltzmann constant	k_B	$1.380\ 658(12) \times 10^{-23}\ \mathrm{J\ K}^{-1}$
electron specific charge (charge-to-mass ratio)	$-e/m_e$	$-1.758\ 819 \times 10^{11}\ \mathrm{C\ kg}^{-1}$
electron charge (elementary charge)	e	$1.602\ 177\ 33(49) \times 10^{-19}\mathrm{C}$
Faraday constant	F	$9.648\ 530\ 9(29) \times 10^{4}\ \mathrm{C\ mol}^{-1}$
ice-point temperature	T_{ice}	$273.15\ \mathrm{K}$[b]
molar gas constant	R	$8.314\ 510(70)\ \mathrm{J\ K}^{-1}\ \mathrm{mol}^{-1}$
molar volume of ideal gas (at 273.15 K and 101 325 Pa)	V_m	$22.414\ 10(19) \times 10^{-3}\ \mathrm{m}^3\ \mathrm{mol}^{-1}$
Planck constant	h	$6.626\ 075\ 5(40) \times 10^{-34}\ \mathrm{J\ s}$
standard atmosphere	atm	$101\ 325\ \mathrm{Pa}$[b]
speed of light in vacuum	c	$2.997\ 924\ 58 \times 10^{8}\ \mathrm{m\ s}^{-1}$[b]

[a] Data are presented in their full precision, although often no more than the first four or five significant digits are used; figures in parentheses represent the standard deviation uncertainty in the least significant digits.

[b] Exactly defined values.

The Periodic Table

1 H	2 He

3 Li	4 Be											5 B	6 C	7 N	8 O	9 F	10 Ne
11 Na	12 Mg											13 Al	14 Si	15 P	16 S	17 Cl	18 Ar
19 K	20 Ca	21 Sc	22 Ti	23 V	24 Cr	25 Mn	26 Fe	27 Co	28 Ni	29 Cu	30 Zn	31 Ga	32 Ge	33 As	34 Se	35 Br	36 Kr
37 Rb	38 Sr	39 Y	40 Zr	41 Nb	42 Mo	43 Tc	44 Ru	45 Rh	46 Pd	47 Ag	48 Cd	49 In	50 Sn	51 Sb	52 Te	53 I	54 Xe
55 Cs	56 Ba	57 La	72 Hf	73 Ta	74 W	75 Re	76 Os	77 Ir	78 Pt	79 Au	80 Hg	81 Tl	82 Pb	83 Bi	84 Po	85 At	86 Rn
87 Fr	88 Ra	89 Ac	104 Rf	105 Db	106 Sg	107 Bh	108 Hs	109 Mt	110 Uun	111 Uuu	112 Uub						

58 Ce	59 Pr	60 Nd	61 Pm	62 Sm	63 Eu	64 Gd	65 Tb	66 Dy	67 Ho	68 Er	69 Tm	70 Yb	71 Lu
90 Th	91 Pa	92 U	93 Np	94 Pu	95 Am	96 Cm	97 Bk	98 Cf	99 Es	100 Fm	101 Md	102 No	103 Lr

Index

General

Chemical Species